T0189957

Intelligent Systems Reference Library

Volume 141

The aim of this series is to publish a Reference Library, including novel advances and developments in all aspects of Intelligent Systems in an easily accessible and well structured form. The series includes reference works, handbooks, compendia, textbooks, well-structured monographs, dictionaries, and encyclopedias. It contains well integrated knowledge and current information in the field of Intelligent Systems. The series covers the theory, applications, and design methods of Intelligent Systems. Virtually all disciplines such as engineering, computer science, avionics, business, e-commerce, environment, healthcare, physics and life science are included.

More information about this series at http://www.springer.com/series/8578

Rajeeb Dey · Goshaidas Ray
Valentina Emilia Balas

Stability and Stabilization of Linear and Fuzzy Time-Delay Systems

A Linear Matrix Inequality Approach

Rajeeb Dey
National Institute of Technology Silchar
Silchar, Assam
India

Valentina Emilia Balas
"Aurel Vlaicu" University of Arad
Arad
Romania

Goshaidas Ray
Department of Electrical Engineering
Indian Institute of Technology
Kharagpur
India

ISSN 1868-4394 ISSN 1868-4408 (electronic)
Intelligent Systems Reference Library
ISBN 978-3-319-88892-7 ISBN 978-3-319-70149-3 (eBook)
https://doi.org/10.1007/978-3-319-70149-3

Printed on acid-free paper

This Springer imprint is published by Springer Nature
The registered company is Springer International Publishing AG
The registered company address is: Gewerbestrasse 11, 6330 Cham, Switzerland

This Book is dedicated to my:
Maa, wife Mou and daughter Raajika
in grateful appreciation

—Rajeeb Dey

Preface

Time-delays inherently occur in physical, industrial and engineering systems as a consequence of limitations in information processing time, data transmission among various parts of the system, arising as feedback delays in the control loops, etc. Time-delays have a considerable influence on the stability of the dynamical systems leading to instability and even degradation in the performances of the systems. So, in order to understand the behaviour of the systems properly, it is important to include time-delays in the mathematical representation of the dynamical systems that can lead to accurate stability analysis and consequently proper controller designs. The mathematical representation of dynamical system with time-delay information embedded in it leads to a model which is referred as delay-differential equations (DDE) or functional differential equations (FDE). In this thesis, we consider the problem of stability analysis and controller synthesis of linear retarded time-delay systems. The motivation behind such study and investigation is the presence of time-delay in wide variety of engineering systems.

The book deals with the problem of stability analysis and controller synthesis of linear and Fuzzy time-delay system. The first four chapters of the book discuss stability and stabilization of linear time-delay system based on Lyapunov–Krasovskii (LK) functional approach in a linear matrix inequality (LMI) framework. The proposed and existing delay-dependent stability as well as robust stability conditions brings out the fact that the conservatism in the analysis and synthesis of such problems lies in the selection of appropriate LK functional approach and subsequently use suitable tighter bounding inequality to yield a quadratic stability condition that can be recast in the LMI framework. The conditions are provided with emphasis on (i) achieving less conservative delay upper bound estimate compared to the existing criteria and (ii) use of lesser free matrix variables in LMIs such that derived conditions obtained are computationally efficient when solved using LMI toolbox of MATLAB. Next, the problem of controller synthesis using state-feedback control law for nominal and uncertain time-delay systems is discussed in both LMI and nonlinear LMI (NLMI) frameworks. As NLMI condition is not a convex problem, it is solved as a linear minimization problem called cone complementarity algorithm, whereas LMI condition is solved along with multi-objective optimization algorithm, with the following objectives in mind (i) to achieve less conservative delay upper bound and (ii) to design controller that stabilizes the system with less control effort for a given delay value. In sequel, an application to load-frequency control (LFC) problem for a two-area interconnected power system with communication delay is presented here. Another variant of controller synthesis problems for nominal as well as uncertain linear time-delay system with actuator saturation is discussed here that highlights that improved proposing a new delay-dependent local (regional) stabilization criterion leads to improved estimation of delay range as well as gives enhanced domain of attraction (DOA). The controller design takes into account the saturation function by its equivalent linear approximations with two different design techniques, namely (i) polytopic and (ii) sector nonlinearities. The last chapter of the book deals about stability analysis and controller synthesis of Fuzzy T-S time-delay system in an LMI framework. The advantage of using Fuzzy T-S modelling approach is that, for any given nonlinear system if operating points are well defined or known by the user, then the given nonlinear system can be universally approximated as a piecewise linear model using Fuzzy T-S models. Once they are converted into linearized models, then the existing analysis and control methods for linear time-delay system can be readily used for solving the system.

This book is intended for the readers who are naive in the field of time-delay control system as it provides extensive review of research in this area and also discusses systematically the development of various integral inequalities and LK functionals that lead to the improvement of delay bound results. This book can be used as a text for teaching time-delay systems and control of postgraduate students in control engineering as well as used by advanced researchers involved in analysis and synthesis of such systems. The contents of the book have been used to teach Ph.D. students of National Institute of Technology, Silchar, India, for the course fractional order and time-delay systems and control. Furthermore, some MATLAB codes are given at the Appendix for the beginners to acquaint them with the solution of LMI conditions.

Silchar, India — Rajeeb Dey
Kharagpur, India — Goshaidas Ray, Professor (Rtd.)
Arad, Romania — Valentina Emilia Balas, Professor
August 2017

Acknowledgements

I would like to acknowledge my sincere thanks to my co-editors and my supervisor Prof. G. Ray, Professor, Electrical Engineering, IIT Kharagpur for consistent encouragement and valuable inputs while writing this book.

Great thanks to Springer Editor Dr. Thomas Ditzinger and Prof. Lakshmi Jain who have allowed us to prepare this book after going through our proposal in this lecture series. Furthermore, I would like to express my great sense of gratitude and thanks to handling Editor Ms. Varsha Prabhakaran of this book for her patience and support while preparing the manuscript.

Last but not least, my warmest and sincere thanks goes to my family and friends for their consistent support, love, encouragement and patience.

Rajeeb Dey

Contents

1 Introduction 1
1.1 Time-Delay Systems and Its Classification 3
1.1.1 Retarded Time-Delay Systems (RTDS) 3
1.1.2 Neutral Time-Delay Systems (NTDS) 6
1.1.3 Stability Definition of Time-Delay Systems 6
1.1.4 Characterization of Stability of Time-Delay Systems . . . 7
1.2 Basic Stability Theorems 10
1.2.1 Lyapunov-Razumikhin Theorem 10
1.2.2 Lyapunov-Krasovskii Theorem 11
1.3 Effect of Time-Delay in the System Dynamics: A Simulation Study 11
1.4 Linear Matrix Inequality (LMI) 13
1.4.1 The LMI Control Toolbox of MATLAB 16
1.4.2 LMI Formulation of Some Standard Problems 16
1.5 Time-Domain Stability Conditions 17
1.5.1 Delay-Independent Stability Condition 18
1.5.2 Delay-Dependent Stability Condition 19
1.6 Review on Delay-Dependent Stability Condition of Time-Delay Systems 21
1.7 Linear Systems with Actuator Saturation 23
1.8 Outline of the Book 27
References 28

2 Stability Analysis of Time-Delay Systems 33
2.1 Introduction 33
2.2 Description of Time-Delay Systems 34
2.2.1 Nominal Time-Delay System for Stability Analysis 35
2.2.2 Uncertain Time-Delay Systems for Robust Stability Analysis 36

2.3 Delay-Dependent Stability Condition 37
2.3.1 Model Transformation Approach (Based on Newton-Leibniz Formula) 37
2.3.2 Bounding Techniques 45
2.3.3 Descriptor System Approach 48
2.3.4 Free Weighting Matrix Approach 54
2.4 Delay-Range-Dependent Stability Condition 65
2.5 Main Results on Stability Analysis of Time-Delay System 72
2.5.1 Stability Analysis of TDS with Single Time Delay 72
2.5.2 Stability Analysis of TDS with Two Additive Time-Varying Delays 82
2.5.3 Stability Analysis of TDS with Interval Time Varying Delay 87
2.6 Robust Stability Analysis of Time-Delay System 94
2.6.1 Characteristic of Structured Uncertainties 94
2.6.2 Delay-Dependent Robust Stability Analysis 96
2.7 Main Results on Delay-Dependent Robust Stability Analysis of TDS 103
2.7.1 Delay-Dependent Robust Stability Analysis of TDS with Single Time Delay 103
2.7.2 Robust Stability Analysis of TDS with Delay Varying in Ranges 109
2.8 Delay-Range-Dependent Stability Analysis of Uncertain TDS by Delay Partitioning Approach 114
2.9 Conclusion 120
References 121

3 Stabilization of Time-Delay Systems 125
3.1 Introduction 125
3.2 Problem Statement 126
3.3 Delay-Dependent Stabilization of Nominal TDS 127
3.4 Main Results on Delay-Dependent Stabilization of Nominal TDS 135
3.5 Delay-Dependent Robust Stabilization of an Uncertain TDS . . . 142
3.6 Main Results on Delay-dependent Robust Stabilization of an Uncertain TDS 145
3.7 State Feedback H_∞ Control of TDS 161
3.7.1 Problem Statement 161
3.8 Stabilization of LFC Problem for Time-Delay Power System Based on H_∞ Approach 162
3.8.1 Load-Frequency Control (LFC) of Power Systems with Communication Delay 164
3.8.2 Existing H_∞ Control Design For LFC Model 166
3.9 Main Results on H_∞ Based LFC of an Interconnected Time-Delay Power System 168

3.9.1 One-Term H_∞ Control ... 168
3.9.2 Two-Term H_∞ Control ... 169
3.9.3 Simulation Results ... 175
3.10 Conclusions ... 179
References ... 181

4 Control of Time-Delay Systems with Actuator Saturation ... 185
4.1 Introduction ... 185
4.2 Characterization of Actuator Saturation ... 186
4.2.1 Polytopic Representation ... 187
4.2.2 Sector Nonlinearities Representation ... 187
4.3 Problem Formulation and Preliminaries ... 191
4.4 Main Result on Stabilization of TDS with Actuator Saturation ... 193
4.4.1 Stabilization Using Sector Nonlinearities Approximation ... 193
4.4.2 Optimization Algorithm ... 199
4.4.3 Simulation Results ... 202
4.4.4 Stabilization Using Polytopic Approximation ... 204
4.5 Main Result on Robust Stabilization of TDS with Actuator Saturation ... 212
4.5.1 Uncertain TDS with Actuator Saturation ... 212
4.5.2 Robust Stabilization Using Sector Nonlinearities Approach ... 213
4.6 Conclusion ... 221
References ... 222

5 Fuzzy Time-Delay System ... 225
5.1 Introduction ... 225
5.2 Problem Formulation and Preliminaries ... 226
5.3 Stability Analysis ... 230
5.4 State Feedback Stabilization ... 234
5.5 Numerical Examples ... 236
5.6 Conclusion ... 241
References ... 241

Appendix ... 243

About the Editors

Rajeeb Dey is presently working with National Institute of Technology, Silchar, Assam, India, as Assistant Professor in the Department of Electrical Engineering. Before joining NIT Silchar, he has served Sikkim Manipal University, Sikkim, for 12 years in various positions (Lecturer, Reader and Associate Professor). His research interests are time-delay systems and control, robust control, control of biomedical systems and application of wireless communication in control. He is presently reviewer of many SCI(E) journals related to control engineering and applied mathematics. He is senior member of IEEE, CSS, Life member of System Society of India, Member Institution of Engineers (India). E-mail: rajeeb.iitkgp@gmail.com.

Goshaidas Ray is presently full Professor in Heritage Institute of Technology, Kolkata. Before joining Heritage Institute of Technology, he has served for over 35 years as a Professor in the Department of Electrical Engineering at Indian Institute of Technology Kharagpur, India. He has published over 100 papers in reputed international and national journals. E-mail: gray@ee.iitkgp.ernet.in.

Valentina Emilia Balas is currently full Professor in the Department of Automatics and Applied Software at the Faculty of Engineering, "Aurel Vlaicu" University of Arad, Romania. She holds a Ph.D. in Applied Electronics and Telecommunications from Polytechnic University of Timisoara. Dr. Balas is author of more than 250 research papers in refereed journals and International Conferences. Her research interests are in Intelligent Systems, Fuzzy Control, Soft Computing, Smart Sensors, Information Fusion, Modeling and Simulation. She is the Editor-in Chief to ***International Journal of Advanced Intelligence Paradigms (IJAIP) and to International Journal of Computational Systems Engineering (IJCSysE)***, member in Editorial Board member of several national and international journals and is the director of Intelligent Systems Research Centre in Aurel Vlaicu University of Arad. She is a member of EUSFLAT, SIAM and a Senior Member IEEE, member in TC – Fuzzy Systems (IEEE CIS), member in TC – Emergent Technologies (IEEE CIS), member in TC – Soft Computing (IEEE SMCS). E-mail: balas@drbalas.ro.

Symbols and Acronyms

Symbols

$\mathcal{R}$	The set of real numbers
$\mathcal{R}^n$	The set of real n vectors
$\mathcal{R}^{m\times n}$	The set of real $m \times n$ matrices
$\lVert\cdot\rVert_2$	Euclidean norm of a vector or a matrix induced 2-norm
$\in$	Belongs to
$< (\leq)$	Less than (Less than equal to)
$> (\geq)$	Greater than (Greater than equal to)
$\neq$	Not equal to
$\forall$	For all
$\rightarrow$	Tends to
$y \in [a, b]$	$a \leq y \leq b; y, a, b \in \mathcal{R}$
[0]	A null matrix with appropriate dimension
I	An identity matrix with appropriate dimension
X^T	Transpose of matrix X
X^{-1}	Inverse of matrix X
$\lambda(X)$	Eigenvalues of matrix X
$\lambda_{\max}(X)$	Maximum eigenvalue of matrix X
$\lambda_{\min}(X)$	Minimum eigenvalue of matrix X
$\det(X)$	Determinant of matrix X
$\mathrm{diag}\{x_1, \ldots, x_n\}$	A diagonal matrix with diagonal elements as $x_1, x_2, \ldots, x_n$
$X > 0$	Positive definite matrix X
$X \geq 0$	Positive semidefinite matrix X
$X < 0$	Negative definite matrix X
$X \leq 0$	Negative semidefinite matrix X
$Sat(u(t))$	Saturation function of vector $u(t)$
$d(t)$	Time-varying state delay
d_u	Maximum delay upper bound

$\|\cdot\|_\infty$	H_∞ Norm
$\mathcal{C}$	Set of continuous function
$\mathcal{C}_-$	left half of s-plane
$\lvert\cdot\rvert$	Absolute value

Acronyms

ARE	Algebraic Riccati Equation
BMI	Bi-Linear Matrix Inequality
DDE	Delay-Differential Equation
DDS	Delay-Dependent Stability
DIS	Delay-Independent Stability
DOA	Domain of Attraction
DRDS	Delay-Range-Dependent Stability
LFC	Load-Frequency Control
LK	Lyapunov–Krasovskii
LMI	Linear Matrix Inequality
LTI	Linear Time-Invariant
NCS	Network-Control System
NLMI	Non-Linear Matrix Inequality
NTDS	Neutral Time-Delay System
p.u	Per-Unit
RTDS	Retarded Time-Delay System
TDS	Time-Delay System

List of Figures

Fig. 1.1 Model of regenerative chatter in a machine tool 4
Fig. 1.2 Block diagram of chatter loop 4
Fig. 1.3 Case-I: Solution does not vanish asymptotically 12
Fig. 1.4 Case-II: Stable solution 12
Fig. 1.5 Case-III: Oscillatory response of the system 13
Fig. 3.1 Numerical implementation of Minimization Problem 138
Fig. 3.2 Variation of α with $\| K \|$ 140
Fig. 3.3 Open-loop simulation of system in Example 3.1 141
Fig. 3.4 Closed-loop simulation of system in Example 3.1 141
Fig. 3.5 Flow-chart for Cone Complementarity Algorithm 150
Fig. 3.6 Closed-loop simulation of system in Example 3.2 157
Fig. 3.7 Control input of system in Example 3.2 157
Fig. 3.8 Time-varying delay considered for Example 3.3 158
Fig. 3.9 Open-loop simulation of system in Example 3.3 158
Fig. 3.10 Closed-loop simulation of system in Example 3.3 159
Fig. 3.11 Two-area LFC system 165
Fig. 3.12 Deviation in frequency for open-loop system 176
Fig. 3.13 Deviation in frequency for the closed-loop system for unit-step load disturbance for feedback delay $(\tau) = 0.7$ Sec 178
Fig. 3.14 Control inputs for unit-step load disturbance for feedback delay $(\tau) = 0.7$ Sec 178
Fig. 3.15 Deviation in frequency for the closed-loop system for time-varying load disturbance of $w(t) = \sin(2\pi t)$ and feedback delay $\tau = 0.7$ Sec 179
Fig. 3.16 Control inputs for time-varying load disturbance time-varying load disturbance of $w(t) = \sin(2\pi t)$ and feedback delay $\tau = 0.7$ Sec 180
Fig. 4.1 Saturation function 187
Fig. 4.2 Sector interpretation for local stability 188
Fig. 4.3 Sector interpretation for global stability 189

Fig. 4.4 Equivalent representation of saturation nonlinearity 189
Fig. 4.5 Saturation nonlinearity 190
Fig. 4.6 Dead zone nonlinearity 190
Fig. 4.7 Estimated DOA inside the ellipsoid 202
Fig. 4.8 Convergence of state trajectories inside the DOA 203
Fig. 4.9 Solution of state $x_1(t)$ of Example 4.1 203
Fig. 4.10 Solution of state $x_2(t)$ of Example 4.1 204
Fig. 4.11 Control input for Example 4.1 204
Fig. 4.12 Solution of state $x_1(t)$ of Example 4.1 by Corollary 1 of [19] 211
Fig. 4.13 Solution of state $x_2(t)$ of Example 4.1 by Corollary 1 of [19] 211
Fig. 4.14 Phase plane trajectory obtained for Example 4.1 by Corollary 1 of [19] 212
Fig. 4.15 Solution of state $x_1(t)$ of Example 4.3 219
Fig. 4.16 Solution of state $x_2(t)$ of Example 4.3 220
Fig. 4.17 Phase plane trajectory converging to DOA 220
Fig. 4.18 Control input of Example 4.3 221
Fig. 5.1 State responses of the system given in Example 2 for $\tau(t) \in [0\ \ 1.7027]$ and $\mu = 0.4$ 238
Fig. 5.2 State responses of the system given in Example 1 for $\tau(t) \in [1\ \ 1.9624]$ and $0 < \mu < 1$ 239
Fig. 5.3 State responses of the system given in Example 3 240
Fig. 5.4 Control signal for stabilization of Example 3 240

Chapter 1
Introduction

In many applications one assumes that the future state of the dynamical system is determined solely by the present state of the system and is independent of the past state information. However, it's apparent that the above principle is only the first approximation to the actual situations and that the more realistic realization of the system dynamics would include some delayed state information. The theory for such system has been extensively developed around 1960s [1, 2]. The class of system which includes past states information along with present states is termed as delay-differential equation (DDE) or more generic name is time-delay systems (TDS) and it can be expressed as,

$$\dot{x}(t) = f(t, x(t), x(t-d))$$

where, $x(t) \in \mathcal{R}^n$ is the state variable vector and 'd' is the delay in the state of the system. It is a well recognized fact that, time-delays are natural component of any physical systems (e.g., chemical processes, process control, population dynamics and aerospace engineering etc.). The presence of delay may be either beneficial or detrimental to the operation of dynamic systems. For an example, judicious introduction of delay may stabilize a dynamical system or otherwise make it unstable, while on the other hand a feedback control system is stable without delay may become unstable for some delays involved in the system. Through simulation of TDS we have attempted to show the effect of time-delay into the system stability in the succeeding section of this chapter.

The literature on stability and stabilization of TDS (or DDE) are exhaustive and can be found in the monograph [1, 3, 4] and also in the survey papers [5, 6]. Broadly there are three major approaches to carry out the stability analysis of TDS, they are as follows:

- The **frequency-domain approach**, which includes the techniques like, frequency sweeping test [3], D-decomposition and τ-decomposition methods, modified

R. Dey et al., *Stability and Stabilization of Linear and Fuzzy Time-Delay Systems*, Intelligent Systems Reference Library 141, https://doi.org/10.1007/978-3-319-70149-3_1

Rouths-Hurwitz method [7, 8], Nyquist stability method [4]. In frequency-domain technique one can compute exact values of the time-delay analytically or graphically but they can deal with only time-invariant delays.

- Another method to analyze the stability of time-delay systems is based on **time-domain approach**, this approach includes generalization of Lyapunov's second method. Following Lyapunov's second method, two main stability theorems have been developed; namely (i) Lyapunov-Krasovskii (LK) theorem and (ii) Lyapunov-Razumikhin theorem. The early work on stability of DDE or time-delay system (TDS) using Lyapunov's second method can be found in [2]. Recent literature on stability analysis of time-delay system are found to be based on Lyapunov-Krasovskii theorem. The stabilization and destabilization effects of delays on stability and control of TDS are important issues in the control literature. The stabilization of TDS based on finite eigenvalues spectrum technique has received a considerable attention and can be found in literature [9–11].

The stability analysis of time-delay systems by Lyapunov-Krasovskii theorem involves finding the sufficient stability condition that guarantees the asymptotic stability of such systems. The sufficient conditions derived are broadly classified into three types, depending upon the size and bound of the delay, they are

(i) **Delay-independent stability (DIS) condition**
(ii) **Delay-dependent stability (DDS) condition**
(iii) **Delay-range-dependent stability (DRDS) condition**

The stability conditions that are independent of the size of the time-delay are called **delay-independent conditions**, and they guarantees the asymptotic stability of the time-delay systems for any arbitrary large delay 'd'. Such conditions are far from practical realization owing to the fact that any physical systems (aerospace, electrical, biological, process control etc.) can sustain stability for a certain finite and small delay value, and hence they are treated as conservative framework of stability analysis of TDS. The delay-independent stability (and/or robust stability) and stabilization (and/or robust stabilization) of TDS can be found in [12–17], and in the monographs [8, 18, 19] and references cited therein.

The stability criteria, which depend on the size of time-delay or in other words that carries the information of the delay value in it, is called **delay-dependent stability condition**. The development of this stability condition has received increasing attention in last few decades due to the fact that, almost all the physical systems exhibit stability up to certain finite value of time-delay rather than delay being unbounded. The delay-dependent stability (and/or robust stability) as well as stabilization (and/or robust stabilization) conditions for time-delay system using Lyapunov-Razumikhin theorem are derived in [20]. The use of Lyapunov-Razumikhin theorem for deriving the delay-dependent stability conditions yields conservative delay upper bound results compared to Lyapunov-Krasovskii theorem [21] and references cited therein.

Remark 1.1 The content of the book deals on deriving delay-dependent as well as delay-range-dependent stability sufficient conditions based on Lyapunov-Krasovskii theorem. In this context, a brief literature review of all the existing work on delay-dependent stability analysis are discussed in the succeeding sections of this chapter, so as to enable readers or researcher working in this field of research for easy understanding. Mathematical definitions and terminologies associated with the analysis are presented in am attempt to ease the explanation presented in the following chapters.

1.1 Time-Delay Systems and Its Classification

The ordinary differential equations (ODE) is mostly used to describe the dynamics of the system and its generic expression is given by,

$$\dot{x}(t) = f(t, x(t))$$

where, $x(t) \in \mathcal{R}^n$ is the state vector. The future evolution of the states are determined from information of the states at current time for which an initial value of $x(t)$ at $t = t_0$ is required.

But in practice there are many dynamical systems whose future value of states $x(t)$ depends on the current information of states as well as past instants of time and such systems are called time-delay systems (TDS). To evaluate the state response of such systems, the initial value of $x(t)$ at $t = t_0$ is not sufficient, rather information of the initial function of $x(t)$ over all the past instants of time is required (i.e., $x(\theta),\ t_0 - d \leq \theta \leq t_0$). The generic representation of the time-delay systems takes the form of,

$$\dot{x}(t) = f(t, x(t), x(t-d), \dot{x}(t-d))$$

1.1.1 Retarded Time-Delay Systems (RTDS)

If the value of the highest derivatives at time 't' depends only on the values of lower derivatives at preceding times $(t + \theta)$, $-\infty < \theta < 0$, then it is called RTDS, the general mathematical representation is expressed as,

$$\dot{x}(t) = f(t, x(t), x(t-d))$$

A practical example of time-delay system is provided below to illustrate how delay arise in practice in a metal cutting process.

Regenerative chatter in a Machine Tool [3, 8, 22]: The metal cutting process of typical machine tool like lathe machine is shown in Fig. 1.1. The cylindrical workpiece rotates with an angular velocity of 'ω rad/sec' and the cutting tool translates along the axis of the workpiece with constant linear velocity of '$\frac{\omega f}{2\pi}$', where $f =$

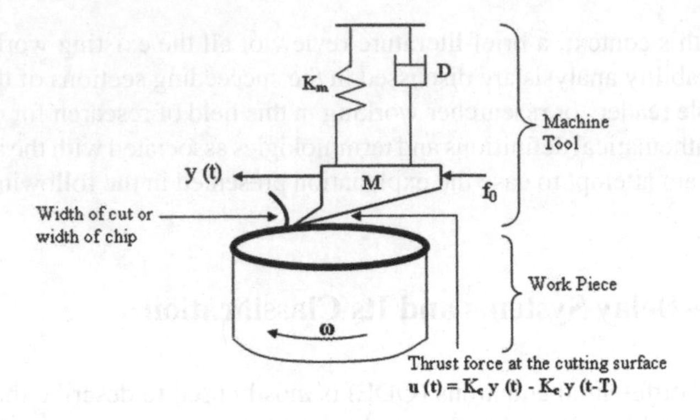

Fig. 1.1 Model of regenerative chatter in a machine tool

constant feed rate in length/revolutions corresponding to normal width of the cut (this parameter is pre-designated). The tool generates a surface as the material is removed and any disturbance (or vibration) in the tool will be reflected on the surface.

In Fig. 1.1, one can observe that in addition to the constant (steady state) force 'f_0' (proportional to the constant feed rate, 'f') to the tool piece some other disturbance forces act at the interface of workpiece and the machine tool (due to chip breakage, non-homogeneity of the workpiece material etc.), such forces lead to change in relative displacement between machine tool and the workpiece that consequently may lead to change in the cutting parameter (which may be mainly the width of the chip or width of the cut, cutting stiffness etc.) and therefore affects the resulting force applied to the cutting process. This interaction between machine tool and the workpiece can be represented by a closed-loop system as shown in Fig. 1.2. If the closed-loop system becomes unstable then it will lead to self-excited vibration between the machine tool and the workpiece which is called **chatter**. This chatter becomes regenerative as the surface generated in previous pass becomes the upper surface of the chip on the subsequent pass which is indicated in Fig. 1.2 by the

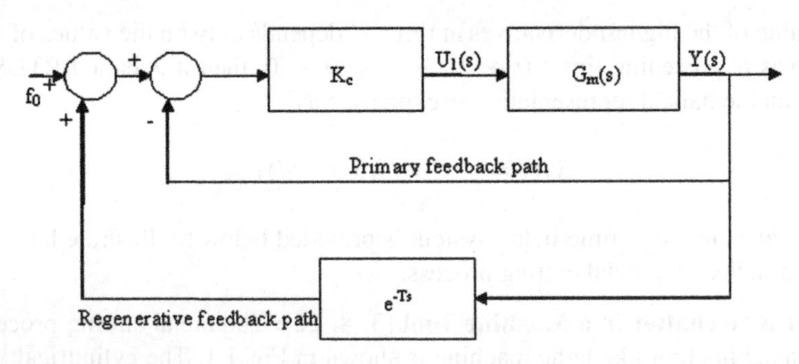

Fig. 1.2 Block diagram of chatter loop

secondary (or regenerative) loop. This phenomenon can lead to precision errors, poor quality of surface finish in the work piece and possible damage to the machine tool system.

Referring to Fig. 1.1 one can model the dynamics of the machine tool and the cutting process at the interface of the machine tool and the workpiece as,

$$\begin{aligned} M\ddot{y}(t) + D\dot{y}(t) + K_m y(t) &= u_1(t) \\ \ddot{y}(t) + \frac{D}{M}\dot{y}(t) + \frac{K_m}{M}y(t) &= \frac{f_0}{M} - \frac{1}{M}u(t) \end{aligned} \tag{1.1}$$

where,

$$\begin{aligned} u_1(t) &= -u(t) + f_0, \ and \\ \frac{1}{M}u(t) &= -\frac{K_c}{M}y(t) + \frac{K_c}{M}y(t-T) \end{aligned}$$

where the notations for the various symbols used in (1.1) are as follows, M, D and K_m indicates the inertial mass, damping and the stiffness characteristics of the machine tool, 'T' is the time-delay or called time for one revolution which is inverse of the spindle speed 'N' of the lathe to which workpiece is connected, 'K_c' is the stiffness characteristic of the cutting process and is liable to change, '$y(t)$' is the displacement of the machine tool along the axis of the workpiece and '$\dot{y}(t)$' is the velocity of the tool piece.

The total thrust force applied for the cutting process is the algebraic sum of the thrust force at the current time instant, thrust applied at all the previous pass (or past time instant) and the constant force 'f_0'. The delayed force term causes the regenerative action in the system.

If we consider '$y(t)$' and '$\dot{y}(t)$' are the two states of the system, then defining the state vector as

$$x(t) = \begin{bmatrix} y(t) \ \dot{y}(t) \end{bmatrix}^T \tag{1.2}$$

In view of (1.2), one can rewrite (1.1) in state space form as,

$$\dot{x}(t) = Ax(t) + A_d x(t-T) + Bf_0 \tag{1.3}$$

where, $A = \begin{bmatrix} 0 & 1 \\ -\frac{K_c+K_m}{M} & -\frac{D}{M} \end{bmatrix}$, $A_d = \begin{bmatrix} 0 & 0 \\ \frac{K_c}{M} & 0 \end{bmatrix}$, $B = \begin{bmatrix} 0 \\ \frac{1}{M} \end{bmatrix}$

With $f_0 = 0$ and denoting $T = d$ in (1.3) one can get an autonomous time-delay systems in state space form as,

$$\dot{x}(t) = Ax(t) + A_d x(t-d) \tag{1.4}$$

where, $x(t) \in \mathcal{R}^n$ is the state vector, $A,\ A_d \in \mathcal{R}^{n\times n}$ are the constant system matrices associated with instantaneous and delayed states respectively.

1.1.2 Neutral Time-Delay Systems (NTDS)

If the value of the highest derivatives at time 't' depends not only on the values of lower derivatives but also on the highest derivatives at preceding times $(t+\theta)$, $-\infty < \theta < 0$, then it is called NTDS whose general mathematically representation is,

$$\dot{x}(t) = f(t, x(t), x(t-d), \dot{x}(t-d))$$

The state-space representation of such systems are expressed as,

$$\dot{x}(t) - F\dot{x}(t-d) = Ax(t) + A_d x(t-d) \tag{1.5}$$

Note: One can find substantial examples on time-delay systems (RTDS and NTDS) in [2, 3, 5, 8, 19] and references therein.

Remark 1.2 The nature of time-delay in both the systems described in (1.4) and (1.5) can be time-varying or time-invariant. In this thesis, stability and stabilization of RTDS is considered. In all succeeding sections and chapters from now onwards we refer RTDS as time-delay systems (TDS) only.

1.1.3 Stability Definition of Time-Delay Systems [3]

A time-delay systems is described as a *functional differential equation* of the form [3, 5],

$$\dot{x}(t) = f(x_t, t), \quad t \in \mathcal{R}^+, \tag{1.6}$$

where $x(t) \in \mathcal{R}^n$ is the state vector; $x_t = x(t+\theta)$, $-d \leq \theta \leq 0$ are the delayed states, $d > 0$ being the time-delay; $f(x_t, t) : \mathcal{C} \times \mathcal{R} \to \mathcal{R}^n$, where $\mathcal{C}$ is the set of continuous functions mapping from $\mathcal{R}^n$ in the time-interval $t - d \leq \theta \leq t$ to $\mathcal{R}^n$. Clearly, if the evolution of $x(t)$ is sought at time instant $t \geq t_0$ then one must first know the history i.e., x_t for $-d \leq \theta \leq 0$, which therefore defines the initial condition and is denoted as $x_{t_0} \in \mathcal{C}$. Defining a state norm as $||x_t||_c = \max_{t-d \leq \theta \leq t} ||x(\theta)||$, the stability definitions for (1.6) in the sense of Lyapunov are as follows.

Definition 1.1 (i) The system (1.6) is *stable* if for scalar $\epsilon > 0$ there exists a $\delta = \delta(t_0, \epsilon) > 0$ such that $||x_{t_0}||_c < \delta$ implies $||x_t||_c < \epsilon$ for all $t \geq t_0$. (ii) It is *uniformly stable* if for scalar $\epsilon > 0$ there exists a $\delta = \delta(\epsilon) > 0$ such that $||x_{t_0}||_c < \delta$ implies $||x_t||_c < \epsilon$ for all $t \geq t_0$. (iii) It is *asymptotically stable* if there exists a $\delta(t_0) > 0$ such that $||x_{t_0}||_c < \delta(t_0)$ implies $\lim_{t\to\infty} x(t) = 0$. (iv) It is *uniformly asymptotically stable* if for every $\epsilon > 0$ there exists a $\delta > 0$ and a $T(\epsilon) > 0$ such that $||x_t||_c < \epsilon$ for all $t \geq t_0 + T(\epsilon)$ whenever $||x_{t_0}||_c < \delta$.

Based on the above definitions, stability of (1.6) can be ascertained using (i) Lyapunov-Krasovskii (LK) theorem and (ii) Lyapunov-Razumikihin theorem (both of them provide only sufficient conditions for stability of TDS).

1.1.4 Characterization of Stability of Time-Delay Systems

Consider a time-delay systems (TDS) (1.4), taking its laplace transform with zero initial condition gives,

$$(sI_n - A - A_d e^{-ds})X(s) = 0$$
$$\Delta(s)X(s) = 0 \tag{1.7}$$

The expression $det(\Delta(s)) = p(s, e^{-ds})$ is called the characteristic quasi-polynomial and can be expressed as,

$$p(s, e^{-ds}) = det(sI_n - A - A_d e^{-ds}) = 0 \tag{1.8}$$

and the relation $det(\Delta(s)) = 0$ provides the poles of the system under consideration. It must be mentioned here that due to the presence of the exponential term in (1.8) it is a transcendental equation, which possess an infinite number of roots in the complex plane $\mathbb{C}$, called the characteristic roots. One can find from (1.8) that the location of roots in the complex s-plane depends upon the size of the delay 'd' and 'A_d' matrix in addition to 'A' matrix. The stability of the time-delay system depends upon the size of the delays can be classified into two categories.

- Delay-independent stability, means that the system (1.4) is stable for any arbitrary large delay.
- Delay-dependent stability, means that the system (1.4) is stable up to some finite delay value.

In this section, we briefly present the qualitative test to characterize the above two stability notions in terms of the eigenvalues of given 'A' and 'A_d' matrices without actually solving the characteristic quasi-polynomial. This qualitative test is based on the pseudo-delay method presented in [3, 8].

The stability of time-delay systems can be characterized bilinear transformation expressed below,

$$z = \frac{1 - sT}{1 + sT}, \; T > 0$$

thus a new polynomial can be formed substituting above transformation into $p(s, e^{-ds})$,

$$c(T, s) = (1 + sT)^q p(s, \frac{1 - sT}{1 + sT})$$

where, q is the order of the multiplicity of the e^{-sd} term and $c(T, s)$ is a polynomial in 's' with the coefficients characterized by the pseudo-delay parameter 'T' and this polynomial turns out to be free from delay terms. Substituting $z = e^{-ds}$ in $p(s, e^{-ds})$ gives $p(s, z)$ which is a bivariate polynomial.
As the parameter 'T' is varied from 0 to ∞ for the polynomial $c(T, s) = p(s, \frac{1-sT}{1+sT})$ the polynomial $p(s, z)$ varies from $p(s, 1)$ at $T = 0$ to $p(s, -1)$ at $T = \infty$. Following steps will explain the method for testing the stability of time-delay systems in (1.4) qualitatively,

1. At $T = 0$, one can get the condition $c(0, s) = p(s, 1) = det(sI - (A + A_d))$. This condition tell us about the eigenvalues of the matrix $(A + A_d)$. As, the necessary condition for the stability of time-delay systems is that the matrix $(A + A_d)$ must be Hurwitz at $d = 0$, so it is assumed to be stable.
2. When the parameter T is increased infinitely, then one can get a condition at $T = \infty$ as, $c(T, s) = p(s, \frac{1-sT}{1+sT}) = p(s, -1) = det(sI - (A - A_d))$. At this stage using Routh-Hurwitz criteria one can find the location of the roots of $p(s, -1)$ in the complex plane, which may have two possibilities. This approach is the basis of stability analysis of (1.4) using (1.8) [3].
   - If the roots of $p(s, -1)$ do not lie in the open right half unit disk of the complex plane, then it implies that the eigenvalues of $(A - A_d)$ matrix is Hurwitz, thus indicating that for all $0 \leq T \leq \infty$ roots of $c(T, s)$ lie on left half of s-plane $\mathcal{C}_-$ and no imaginary roots exists, hence the system in (1.4) is characterized as delay-independently stable.
   - If some of the roots of $p(s, -1)$ are found to lie in the open right half unit disk of the complex plane, then it implies that eigenvalues of $(A - A_d)$ matrix is unstable, thus indicating that for some finite value of $T_i > 0$, there exist purely imaginary roots $s_i = j\omega_i$ for $c(T, s) = p(s, \frac{1-sT}{1+sT})$, the corresponding delay value can be estimated using the bivariate polynomial [18].

$$z_i = \frac{1 - j\omega_i T_i}{1 + j\omega_i T_i} = e^{-j\theta_i}$$

where, $\theta_i = \omega_i d_i$, and one can get,

$$\begin{aligned} \theta_i &= 2\, tan^{-1}\omega_i T_i \\ \omega_i d_i &= 2\, tan^{-1}\omega_i T_i \\ d_1 &= min\, \left(\frac{2}{\omega_i} tan^{-1}\omega_i T_i \,:\, \omega_i > 0 \right) \end{aligned}$$

One can conclude from above that by examining the eigenvalues of $(A + A_d)$ and $(A - A_d)$ matrices the stability characterization of time-delay systems in (1.4) can be qualitatively ensured,

a. If $(A + A_d)$ and $(A - A_d)$ are stable then the TDS is delay-independently stable.
b. If $(A + A_d)$ is stable and $(A - A_d)$ is unstable then the TDS is delay-dependently stable.

We present few numerical examples to characterize their stability using the qualitative test discussed above.

Numerical Example 1.1 [3]. *Consider the system in (1.4) with the following constant matrices*

$$A = \begin{bmatrix} -2 & 0 \\ 0.5 & -2 \end{bmatrix}, A_d = \begin{bmatrix} 0 & 0.5 \\ 0 & 0 \end{bmatrix}$$

Form quasi-characteristic equation (1.8) we have,

$$\begin{aligned} p(s, e^{-ds}) &= det(sI - A - A_d e^{-sd}) = 0 \\ &= s^2 + 4s + 4 - 0.25e^{-sd} = 0 \end{aligned}$$

and the corresponding characteristic equation $c(T, s) = 0$ can be written as,

$$\begin{aligned} c(T, s) = (p(s, \frac{1 - sT}{1 + sT})) &= s^2 + 4s + 4 - 0.25(\frac{1 - sT}{1 + sT}) = 0 \\ &= s^3T + (1 + 4T)s^2 + (4 + 4.25T)s + 3.75 = 0 \end{aligned} \tag{1.9}$$

using the Routh-Hurwitz criteria the stability of $c(T, s) = 0$ is found to be stable for all values of T (i.e., $0 \leq T \leq \infty$). Subsequently both the polynomials $p(s, 1) = 0$ and $p(s, -1) = 0$ are found to be stable, which in turn implies that the eigenvalues of $(A + A_d)$ and $(A - A_d)$ matrices are Hurwitz, thus it is characterized as delay-independently stable system.

Numerical Example 1.2 [23]. *Consider the system in (1.4) with the following constant matrices*

$$A = \begin{bmatrix} -2 & 0 \\ 0 & -0.9 \end{bmatrix}, A_d = \begin{bmatrix} -1 & 0 \\ -1 & -1 \end{bmatrix}$$

Form quasi-characteristic equation (1.8) we have,

$$\begin{aligned} p(s, e^{-ds}) &= det(sI - A - A_d e^{-sd}) = 0 \\ &= s^2 + 2.9s + 1.8 + 2.9e^{-sd} + 2se^{-sd} + e^{-2sd} = 0 \end{aligned}$$

and the corresponding characteristic equation $c(T, s) = 0$ can be written, now Routh-Hurwitz criteria is used to find the stability of $c(T, s) = 0$ and it is found to be stable for certain finite range of T (and not for $0 \leq T \leq \infty$). Next stability of $p(s, 1) = 0$ and $p(s, -1) = 0$ are carried out using Routh criteria and whereas the polynomial $p(s, 1)$ is found to be stable and whereas the polynomial $p(s, -1) = 0$ is found to be unstable. This indicates existence of some finite value of $T_i > 0$ for which there exist imaginary roots of $c(T, s)$, the corresponding delay value for this finite value of T_i can be computed as mentioned above. Thus, when $(A + A_d)$ is Hurwitz but $(A - A_d)$ matrix is unstable then the system is characterized as delay-dependently stable.

Remark 1.3 In general, it may be noted that the characteristic equation $c(T, s) = 0$ is a continuous function of the time-delay 'd' which in turn is a function of 'T'. It follows that the system (1.4) can switch from stability to instability or vice-versa, only when at least one characteristic root crosses to the imaginary axis as 'T' changes.

1.2 Basic Stability Theorems

The stability analysis of time-delay system has been presented in this book using Lyapunov's method, this method requires the use of Lyapunov function (for non delayed system) or Lyapunov-Krasovskii functional (for time-delay system). As the definition of time-delay system (TDS) is already stated in Sect. 1.1.3, we proceed to introduce two basic Lyapunov's based stability theorems for TDS below.

1.2.1 Lyapunov-Razumikhin Theorem [3].

Theorem 1.1 *The system (1.6) is uniformly stable if there exists a continuous differentiable function $V(x(t))$, $V(0) = 0$, such that*

$$u(||x(t)||) \leq V(x(t)) \leq v(||x(t)||) \tag{1.10}$$

and

$$\dot{V}(x(t)) \leq -w(||x(t)||) \quad \textit{whenever} \quad V(x(t+\theta)) \leq V(x(t)), \forall\theta \in [-d, 0], \tag{1.11}$$

where u, v, w are continuous nondecreasing scalar functions with $u(0) = v(0) = w(0) = 0$ and $u(r) > 0$, $v(r) > 0$, $w(r) \geq 0$ for r>0. If $w(r) > 0$ for r>0 then it is uniformly asymptotically stable. If, in addition, $\lim_{r\to\infty} u(r)=\infty$, then it is globally uniformly asymptotically stable.

Moreover, if there exists a continuous nondecreasing function $p(r) > 0$ for $r > 0$ then (1.11) may be replaced by

$$\dot{V}(x(t)) \leq -w(||x(t)||) \quad \textit{whenever} \quad V(x(t+\theta)) \leq p(V(x(t))), \forall\theta \in [-d, 0]. \tag{1.12}$$

Remark 1.4 The above theorem tries to ensure that $\bar{V}(t) = \max_{t-d\leq\theta\leq t} V(x(\theta))$ does not increase with time. Note that, if $\bar{V}(t) \neq V(x(t))$ and $\dot{V}(x(t)) \geq 0$, $\bar{V}(t)$ does not increase. Clearly, $\bar{V}(t)$ increases if and only if $\bar{V}(t) = V(x(t))$ and $\dot{V}(x(t)) \geq 0$ which is denied if the condition (1.11)/(1.12) is satisfied.

1.2.2 Lyapunov-Krasovskii Theorem [3].

Theorem 1.2 *The system (1.6) is uniformly stable if there exists a continuous differentiable function $V(x_t)$, $V(0) = 0$, such that*

$$u(||x(t)||) \leq V(x_t) \leq v(||x_t||_c), \tag{1.13}$$

and

$$\dot{V}(x_t) \leq -w(||x(t)||), \tag{1.14}$$

where u, v, w are continuous nondecreasing scalar functions with $u(0) = v(0) = w(0) = 0$ and $u(r) > 0$, $v(r) > 0$, $w(r) \geq 0$ for $r > 0$. If $w(r) > 0$ for $r > 0$, then it is uniformly asymptotically stable and if, in addition, $\lim_{r \to \infty} u(r) = \infty$, then it is globally uniformly asymptotically stable.

1.3 Effect of Time-Delay in the System Dynamics: A Simulation Study

In this section, it will be demonstrated through simulation studies, how the delayed term in TDS may drastically change the qualitative behavior of TDS [24]. The numerical simulation of the TDS has been carried out using standard routine "dde23" of MATLAB (or fourth order R-K routine may be used too).
Consider a linear scalar TDS,

$$\begin{aligned} \dot{x}(t) &= \lambda x(t) + \mu x(t-1), t \geq 0 \\ x(t) &= -t + 1, t \leq 0 \end{aligned} \tag{1.15}$$

With the real constant coefficients λ and μ. It is known that for $\mu = 0$, the above equation becomes,

$$\begin{aligned} \dot{x}(t) &= \lambda x(t), \quad t \geq 0 \\ x(0) &= 1 \end{aligned} \tag{1.16}$$

the solution of (1.16) will vanish asymptotically for any negative λ, whereas it will grow up for any positive λ. On the other hand, for $\mu \neq 0$ the delayed term $\mu x(t-1)$ in (1.15) acts as forcing term and the above mentioned statement of the solution might

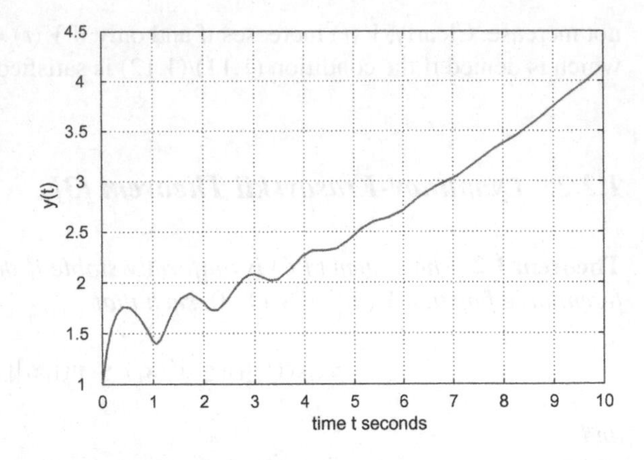

Fig. 1.3 Case-I: Solution does not vanish asymptotically

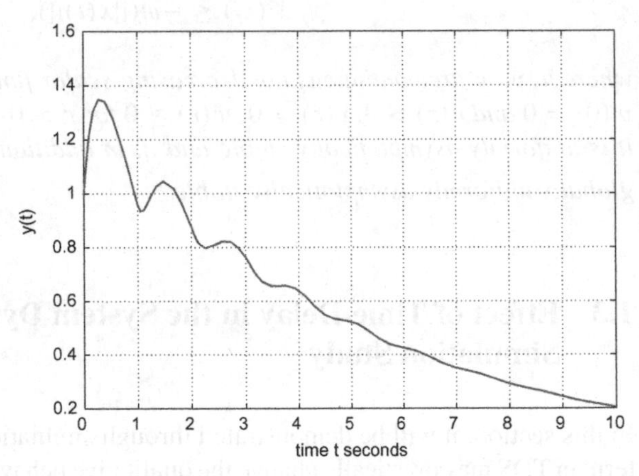

Fig. 1.4 Case-II: Stable solution

not hold. Three different nature of solution of the system with the initial function given in (1.15) for different values of λ and μ are shown in Figs. 1.3, 1.4 and 1.5.

Case I: For the given values of $\lambda = -3.5$ and $\mu = 4$, using the characterization discussed above one can find that $(\lambda + \mu)$ is positive, thus the necessary condition of stability of the TDS in (1.15) is violated hence the solution of the system is unstable, which is shown in Fig. 1.3.

Case II: For the given values of $\lambda = -5$ and $\mu = 4$, using the characterization discussed above one can find that $(\lambda + \mu)$ is Hurwitz, while $(\lambda - \mu)$ is also Hurwitz thus indicating that system (1.15) is delay-independently stable, the solution is shown in Fig. 1.4. For this system it is observed that if the delay term is increased to a very large value the system still remains stable.

Case III: For the values of $\lambda = 0.5$ and $\mu = -1$, using the characterization discussed above one can find that $(\lambda + \mu)$ is Hurwitz, while $(\lambda - \mu)$ is not Hurwitz thus indicating that system (1.15) is delay-dependently stable, the solution is shown in Fig. 1.5. For this system it is observed that increasing the delay after certain value makes the system unstable.

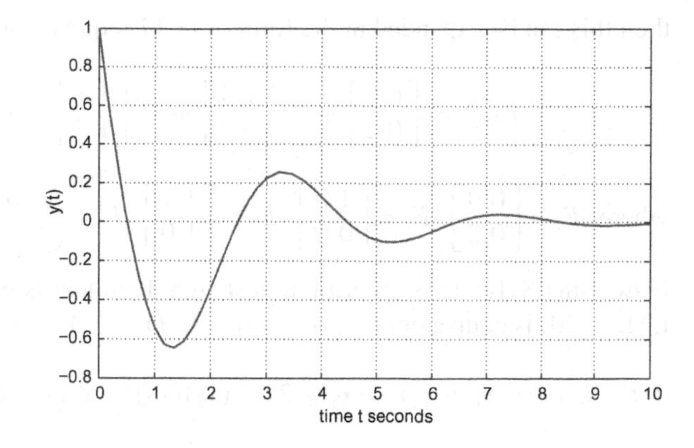

Fig. 1.5 Case-III: Oscillatory response of the system

1.4 Linear Matrix Inequality (LMI) [25, 26]

The stability and stabilization criteria derived using Lyapunov's second method as well as many other problems in systems and control can be formulated as optimization problems involving constraints that can be expressed in LMI frameworks. So, here we proceed to introduce briefly about LMIs that are linear set of matrix variables.

If $F_0, \ F_1, \ldots.., F_n \in \mathcal{R}^{m\times m}$ are symmetric matrices, then the inequality of the form,

$$F(x) = F_0 + \sum_{i=1}^{n} x_i F_i < 0 \tag{1.17}$$

is called the linear matrix inequality (LMI) in the variable of $x \in R^n$ and $F : \mathcal{R}^n \rightarrow \mathcal{R}^{m\times m}$ is an affine mapping function of the variable x, which means that the set,

$$S = \{x : F(x) < 0\} \tag{1.18}$$

is convex. Now, if x_1 and $x_2 \in S$ and $\beta \in (0, 1)$, then

$$F(\beta x_1 + (1-\beta)x_2) = \beta F(x_1) + (1-\beta)F(x_2) < 0 \tag{1.19}$$

The first equality follows from the fact that the function 'F' is assumed to be affine and the last inequality is due to the assumption that $\beta \geq 0$, $(1-\beta) \geq 0$ and $F(x) \leq 0$.

The basic LMI problem is whether or not there exist $x_{n\times 1}$ such that the above equation defined in (1.17) is satisfied. The representation of the LMI in (1.17) is illustrated with an example. Let us consider an $F(x)$ of the form,

$$F(x) = \begin{bmatrix} x_1 + x_2 & x_2 \\ x_2 & x_3 + 2 \end{bmatrix} < 0 \tag{1.20}$$

then this can be expanded in the form of (1.17) equivalently as,

$$F(x) = \begin{bmatrix} 0 & 0 \\ 0 & 2 \end{bmatrix} + x_1 \begin{bmatrix} 1 & 0 \\ 0 & 0 \end{bmatrix} + x_2 \begin{bmatrix} 1 & 1 \\ 1 & 0 \end{bmatrix} + x_3 \begin{bmatrix} 0 & 0 \\ 0 & 2 \end{bmatrix} < 0 \tag{1.21}$$

where, $F_0 = \begin{bmatrix} 0 & 0 \\ 0 & 2 \end{bmatrix}$, $F_1 = \begin{bmatrix} 1 & 0 \\ 0 & 0 \end{bmatrix}$, $F_2 = \begin{bmatrix} 1 & 1 \\ 1 & 0 \end{bmatrix}$, $F_3 = \begin{bmatrix} 0 & 0 \\ 0 & 2 \end{bmatrix}$.

Now using Sylvester's criterion to test sign definiteness of a symmetric matrix, the LMI (1.20) is equivalent to, $(x_1 + x_2) < 0$, $(x_3 + 2) < 0$ and

$$det(F(x)) = (x_1 + x_2) \times (x_3 + 2) - (x_2)(x_2) = x_1 x_3 + 2x_1 + x_2 x_3 - x_2^2 < 0,$$

which is a nonlinear inequality in the variables 'x' expressed as LMI of the form (1.17).

If there are multiple LMIs,

$$F^{(1)}(x), \ldots\ldots, F^{(k)}(x) < 0$$

then one can write equivalent to one single LMI using a block diagonal structure,

$$F(x) = \begin{bmatrix} F_1(x) & 0 & \ldots & 0 \\ 0 & F_2(x) & \ldots & 0 \\ \vdots & \vdots & \ddots & \vdots \\ 0 & 0 & \ldots & F^k(x) \end{bmatrix} < 0 \tag{1.22}$$

In most of the control engineering problems the LMIs do not appear in the form described in (1.17), i.e., the variables in this case are not from the set S but they are matrix variables. The simple example of LMI in control engineering is the stability analysis of input free dynamic system based on Lyapunov stability criterion and it is described as,

$$A^T P + PA < 0 \tag{1.23}$$

which is an LMI in matrix variable P. The matrix P is symmetric but and A is a known matrix. In this case the scalar entries of P can be treated as variables (analogously components of x_i in (1.17)) and thus the LMI in (1.23) can be recast into the form (1.17) as illustrated below,

Let $A = \begin{bmatrix} 1 & 2 \\ -1 & 0 \end{bmatrix}$ and $P = \begin{bmatrix} p_{11} & p_{12} \\ p_{12} & p_{22} \end{bmatrix} = P^T$,

In view of (1.17) we can express (1.23) as,

$$F(P) = A^T P + PA < 0 \tag{1.24}$$

The number of scalar or decision variables will depend upon the structure of the matrix variable P, in our case as the structure is $P = P^T$, so the number of variable is $n = \frac{n(n+1)}{2}$, thus the substituting the value of A and P matrices in (1.23) and then expanding it, one can get the following form,

$$F_0 + p_{11}\begin{bmatrix} 2 & 2 \\ 2 & 0 \end{bmatrix} + p_{12}\begin{bmatrix} -2 & 1 \\ 1 & 4 \end{bmatrix} + p_{22}\begin{bmatrix} 0 & -1 \\ -1 & 0 \end{bmatrix} \tag{1.25}$$

where, $F_0 = \begin{bmatrix} 0 & 0 \\ 0 & 0 \end{bmatrix}$, thus (1.24) is expressed in the form of (1.17). The solution of (1.24) LMI framework means finding the variables $P_{i,j} = P_{j,i}$ $i = 1, 2$ and $j = 1, 2$ that satisfy the LMI or no solution exists.

One can now conclude from the foregoing discussion that, if the objective function of an optimization problem is a convex function in the concerned variable (say 'x') and the constraints are also expressed in LMI (of the same concerned variable x) then the whole problem can be recast into convex optimization problem in an LMI framework. The convex optimization problems are attractive mainly for two reasons: (a) it has unique global minima if it exists and (b) computationally attractive, due to the availability of efficient algorithms (or LMI solvers) for solving the LMI or set of LMIs (e.g., LMI toolbox of MATLAB, LMI-tool, YALMIP etc.).

In fact problems associated with LMI can be classified into three categories:

1. *Feasibility problem:*
   Finding if there exists a solution of an LMI ($F(x) < 0$).
2. *Optimization problem:*
   Minimizing a convex objective $f(x)$ subject to an LMI constraint ($F(x) < 0$).
3. *Generalized eigenvalue problem:*
   Minimizing λ subject to $G(x) - \lambda F(x) < 0$, $F(x) > 0$ and $H(x) < 0$.

Schur Complement:

Often, a class of nonlinear matrix inequalities are confronted in systems and control theory, which can be reformulated as LMIs using *Schur Complement* formula [25]. It states that for matrices $Z_1 = Z_1^T$, $Z_2 = Z_2^T$ and L,

$$\left.\begin{array}{c} Z_2 < 0 \ \text{ and } \ Z_1 - LZ_2^{-1}L^T < 0 \\ \text{is equivalent to} \begin{bmatrix} Z_1 & L \\ L^T & Z_2 \end{bmatrix} < 0, \end{array}\right\}. \tag{1.26}$$

Congruence Transformation:

Sign definiteness of a matrix is invariant under pre- and post-multiplication by a full rank real matrix and its transpose respectively. For example, if $V > 0 \in \mathcal{R}^{n\times n}$ and a real full rank matrix $\Lambda \in \mathcal{R}^{n\times n}$, then following inequalities hold,

$$\Lambda V \Lambda^T > 0$$

This procedure is called congruence transformation. It finds application in removing bilinear terms in matrix inequalities and such cases are often encountered while designing state feedback controller in LMI framework.

1.4.1 The LMI Control Toolbox of MATLAB [27]

The LMI control toolbox provides an LMI Lab to specify and solve user defined LMIs. In this thesis, this LMI Lab has been used for solving LMIs. Some commands of this LMI Lab that are used to obtain the numerical results are presented in the following.
SETLMIS : This initializes the LMI system description.
GETLMIS : It is used when all the LMIs are described and returns the internal description of the defined LMI.
LMIVAR : It is used to declare the LMI variables.
LMITERM : The LMI terms are specified with this command.
FEASP : This is an LMI solver which is used to solve LMI feasibility problems.
MINCX : This LMI solver is used to solve an LMI optimization problem.
GEVP : It is used for solving generalized eigenvalue problem.
A graphical user interface LMIEDIT also exists to define LMIs.

The *FMINSEARCH* command of the Optimization Toolbox of MATLAB [28] has also been used to tune certain parameters that are associated with the derived LMI criteria.

Remark 1.5 The MATLAB command for the stability & stabilization problems formulated in an LMI frame work will be presented in the chapters to follow.

1.4.2 LMI Formulation of Some Standard Problems

Stability analysis and stabilization problems based on Lyapunov approach can be formulated either in linear or nonlinear matrix inequality framework. Many of them can be solved efficiently by formulating them in LMI framework and then solving using available software packages. This section demonstrates the solution of some control problems based on LMI framework.

The Lyapunov stability criterion (1.23) described below

$$PA + A^T P < 0, \tag{1.27}$$

is an LMI on the variable P.
To obtain the state feedback stabilization criterion of a system $\dot{x}(t) = Ax(t) + Bu(t)$ with the state feedback control law $u(t) = Kx(t)$, we replace A by $[A + BK]$ in (1.27) to get

$$PA + A^T P + PBK + K^T B^T P < 0. \tag{1.28}$$

One can see that (1.28) is not in LMI form due to the multiplication of the unknown variables K and P. However, it can be equivalently converted to an LMI by using the matrix property that, if $X > 0$ then $Y^T XY > 0$ provided Y is an nonsingular matrix. In view of this, pre- and post-multiplying (1.28) with $\hat{P} = P^{-1} > 0$ and then defining a new variable as $Y = K\hat{P}$ one may write,

$$A\hat{P} + \hat{P}A^T + BY + Y^T B^T < 0. \tag{1.29}$$

This is an LMI in the variables $\hat{P}$ and Y. Now, one may solve the above (1.28) as an LMI feasibility problem and a stabilizing gain matrix K may be obtained as $K = Y\hat{P}^{-1}$ from any feasible solution of it.

Next, consider the inequality of the form mentioned below, such inequality can be obtained while deriving robust stability criterion for systems with norm bounded parametric uncertainties using Lyapunov's second method

$$PA + A^T P + PBK + K^T B^T P + \epsilon PDD^T P + \epsilon^{-1} E^T E < 0. \tag{1.30}$$

which is a nonlinear matrix inequality not only due to multiplication of P and K but also due to the co-existence of the search variables 'ϵ' and 'ϵ^{-1}'. To express this as an LMI as above, first, we pre- and post-multiply it with $\hat{P} = P^{-1}$ and then defining a new variable as $Y = K\hat{P}$ one obtains

$$A\hat{P} + \hat{P}A^T + BY + Y^T B^T + \epsilon DD^T + \epsilon^{-1}\hat{P}E^T E\hat{P} < 0. \tag{1.31}$$

Finally, to express the above as an LMI one may employ the Schur complement formula (1.26) on (1.31). Defining $Z_1 = A\hat{P} + \hat{P}A^T + BY + Y^T B^T + \epsilon DD^T$, $Z_2 = -\epsilon I$, $L = \hat{P}E^T$ and then employing (1.26) one may equivalently write (1.31) as

$$\begin{bmatrix} A\hat{P} + \hat{P}A^T + BY + Y^T B^T + \epsilon DD^T & \hat{P}E^T \\ * & -\epsilon I \end{bmatrix} < 0, \tag{1.32}$$

where '$*$' represents transpose of corresponding upper diagonal terms. This is an LMI and one may solve this along with $\hat{P} > 0$ as an LMI feasibility problem to obtain $\hat{P}$, Y and ϵ. Then a stabilizing control gain may be computed using $K = Y\hat{P}^{-1}$.

Following the above techniques, other stability and stabilization criteria derived so far may also be formulated as LMI problems.

1.5 Time-Domain Stability Conditions

This section briefly illustrates the mathematical formulation of the stability analysis for the time-delay systems (1.4) using LK theorem in an LMI framework.

1.5.1 Delay-Independent Stability Condition

Assumption 1.1 The necessary condition for obtaining delay-independent stability condition is that the matrix A must be Hurwitz in (1.4).

Choosing Lyapunov-Krasovskii functional candidate for the system (1.4) as [3, 19],

$$V(t) = x^T(t)Px(t) + \int_{t-d}^{t} x^T(s)Rx(s)ds,\ P = P^T > 0,\ R = R^T > 0 \quad (1.33)$$

Note: The system (1.4) with $d = 0$ the LK functional in (1.33) reduces to $V(t) = x^T(t)Px(t)$ and is referred as Lyapunov function.

The functional (1.33) is chosen in such a way that it qualifies as Lyapunov-Krasovskii functional and hence satisfies the condition (see condition (1.13)).

$$\lambda_{min}(P) \parallel x(t) \parallel^2 \leq V(x_t) \leq \lambda_{max}(P) \parallel x(t) \parallel^2 + d\ \lambda_{max}(R) \parallel x_t \parallel_c^2 \quad (1.34)$$

Taking the time-derivative of (1.33) one can get,

$$\dot{V}(t) = 2x^T(t)PAx(t) + 2x^T(t)PA_dx(t-d) + x^T(t)Rx(t) - x^T(t-d)Rx(t-d). \quad (1.35)$$

Now, to separate the cross product term of $x(t)$ and $x(t-d)$ in the second term of (1.35), well-known bounding inequality of [29] stated below can be used,

$$\Upsilon_1^T\Upsilon_2 + \Upsilon_2^T\Upsilon_1 \leq \Upsilon_1^T W^{-1}\Upsilon_1 + \Upsilon_2^T W\Upsilon_2, \quad W = W^T > 0, \quad (1.36)$$

where Υ_1, Υ_2 are matrices or can be vectors and W is a symmetric positive definite matrix of appropriate dimensions. Using (1.36) the cross-product term can be written as,

$$2x^T(t)PA_dx(t-d) \leq x^T(t)PA_dR^{-1}A_d^TPx(t) + x^T(t-d)Rx(t-d). \quad (1.37)$$

Using (1.37) in (1.35) and then to satisfy $\dot{V}(x_t) \leq -w(||x(t)||)$ as per Lyapunov-Krasovskii theorem one obtains the resulting stability criterion as

$$PA + A^TP + R + PA_dR^{-1}A_d^TP < 0. \quad (1.38)$$

Using Schur-complement (1.26), one can rewrite (1.38) into an equivalent LMI condition as,

$$\begin{bmatrix} (1,1) & PA_d \\ \star & -R \end{bmatrix} < 0 \quad (1.39)$$

where, (1,1)=$PA + A^TP + R$.

It is observed that the LMI condition (1.39) does not contain any 'd' term in it and hence it is referred delay-independent stability (DIS) condition.

1.5.2 Delay-Dependent Stability Condition

Assumption 1.2 The necessary condition for delay-dependent stability of time-delay systems (1.4) is that, the matrix $(A + A_d)$ (when $d = 0$) must be Hurwitz.

In case of delay-dependent stability for the time-delay systems, the delay size information is included in the stability criterion. The term $x(t - d)$ in (1.4) is substituted by $x(t - d) = x(t) - \int_{t-d}^{t} \dot{x}(s)ds$ using Newton-Leibniz formula, which is expressed as,

$$\int_{t-d}^{t} \dot{x}(s)ds = x(t) - x(t - d)$$

Thus the time-delay systems (1.4) is transformed to following form,

$$\dot{x}(t) = (A + A_d)x(t) - A_d \int_{t-d}^{t} [Ax(s) + A_d x(s - d)]ds \tag{1.40}$$

Now, it must be noted here that, the transformed system (1.40) and the system (1.4) are different hence the knowledge of the initial function for the system (1.40) is required for $t \in [-d, \ d]$ whereas for system (1.4) the initial function is to be known over the time interval $t \in [-d, \ 0]$. The stability of (1.40) implies the stability of (1.4) [18] and references cited therein.

The selection of the LK functional in this case (delay-dependent stability) is,

$$V(t) = x^T(t)Px(t) + V_2(t) + V_3(t), \ P = P^T > 0 \tag{1.41}$$

where,

$$V_2(t) = \int_{-d}^{0} \int_{t+\theta}^{t} x^T(s)Q_1 x(s)dsd\theta$$

$$V_3(t) = \int_{-2d}^{-d} \int_{t+\theta}^{t} x^T(s)Q_2 x(s)dsd\theta$$

where, $Q_1 = Q_1^T > 0$ and $Q_2 = Q_2^T > 0$. The choice of such double integral terms in (1.41) is to take care of the integral terms arising out of the transformed system while computing the derivative of the LK functional. Finding the time-derivative of (1.41) one can obtain

$$\dot{V}(t) = x^T[(A + A_d)^T P + P(A + A_d) + d\,(Q_1 + Q_2)]x(t) \\ -2\int_{t-d}^{t} x^T(t)PA_dAx(s)ds - 2\int_{t-d}^{t} x^T(t)PA_dA_dx(s-d)ds \\ -\int_{t-d}^{t} x^T(s)Q_1x(s)ds - \int_{t-d}^{t} x^T(s-d)Q_2x(s-d)ds \tag{1.42}$$

Now to approximate the cross term integrals in (1.42) one can use the lemma in (1.36) and rewrite (1.42) as,

$$\dot{V}(t) \le x^T[(A + A_d)^T P + P(A + A_d) + d(Q_1 + Q_2)]x(t) \\ +\int_{t-d}^{t} x^T(t)PA_dX_1^{-1}A_d^TPx(t)ds + \int_{t-d}^{t} x^T(s)A^TX_1Ax(s)ds \\ -\int_{t-d}^{t} x^T(s)Q_1x(s)ds - \int_{t-d}^{t} x^T(s-d)A_d^TX_2A_dx(s-d)ds \\ +\int_{t-d}^{t} x^T(t)PA_dX_2^{-1}A_d^TPx(t)ds + \int_{t-d}^{t} x^T(s-d)Q_2x(s-d)ds \tag{1.43}$$

Let, $Q_1 = A^TX_1A$ and $Q_2 = A_d^TX_2A_d$, thus due to this assumption one can observe that in (1.43) the terms arising out after the application of bounding inequality cancels the last two integral terms in (1.42), thus one can write (1.43) as,

$$\dot{V}(t) \le x^T[(A + A_d)^T P + P(A + A_d) + d(Q_1 + Q_2)]x(t) \\ +\int_{t-d}^{t} x^T(t)PA_dX_1^{-1}A_d^TPx(t)ds + \int_{t-d}^{t} x^T(t)PA_dX_2^{-1}A_d^TPx(t)ds \tag{1.44}$$

$$\dot{V}(t) \le x^T[(A + A_d)^T P + P(A + A_d) + d(Q_1 + Q_2)]x(t) \\ +d(PA_dX_1^{-1}A_d^TP) + d(PA_dX_2^{-1}A_d^TP) \tag{1.45}$$

Using Schur-complement (1.26), one can write (1.45) as,

$$\dot{V}(t) \le x^T(t)\begin{bmatrix} (1,1) & d\,PA_d & d\,PA_d \\ \star & -X_1 & 0 \\ \star & 0 & -X_2 \end{bmatrix} x(t) \tag{1.46}$$

where, $(1, 1) = (A + A_d)P + P(A + A_d)^T + d\,(Q_1 + Q_2)$. To guarantee the asymptotic stability of the system (1.4) or equivalently (1.40), it is required that $\dot{V}(t) < 0$ which is satisfied if the following LMI is satisfied,

$$\begin{bmatrix} (1,1) & d\,PA_d & d\,PA_d \\ \star & -Q_1 & 0 \\ \star & 0 & -Q_2 \end{bmatrix} < 0 \tag{1.47}$$

Note: The information of the delay size 'd' is present in LMI (1.47) and hence it is called delay-dependent stability condition.

1.6 Review on Delay-Dependent Stability Condition of Time-Delay Systems

This section gives a brief review of the existing literature on stability/robust stability and stabilization/robust stabilization of TDS based on LK functional approach in an LMI framework.

The initial work on delay-dependent stability involves transformation of the system (1.4) using Newton-Leibniz formula (i.e.,$(x(t-d)) = x(t) - \int_{t-d}^{t} \dot{x}(\alpha)d\alpha$) into an equivalent system, and then appropriate L-K functional and bounding lemma are chosen to obtain the LMI conditions [30–34] and references cited therein. In [18] and [35], it was reported that the use of such model transformation are not equivalent to the original time-delay system and hence introduce some additional dynamics (eigenvalues), which is the source for conservative results as these additional eigenvalues causes the roots of the original characteristic equation to cross the imaginary axis at a much lower delay value resulting into lesser delay upper bound estimate.

In [36], it is reported that, the conservatism also stems from the use of bounding techniques used to approximate the cross terms arising out of the L-K functional derivatives while formulating quadratic stability conditions. To alleviate this issue, Park's bounding inequality was proposed in [36]. The bounding inequality $-2a^T b \leq a^T X a + b^T X^{-1} b$ in [37] and Park's inequality in [36] are the special cases of generalized bounding inequality proposed in [38]. The structure of bounding inequality in [38] is suitable for control synthesis problems compared to the use of Park's bounding inequality. So, its usage in dealing with stability and related issues can be found in [39–42].

In [32, 35, 43], a new model transformation was proposed to reduce the conservatism that arises due to earlier model transformations. This method transforms the original time-delay systems into descriptor systems. The use of this method requires more number of matrix variables [32, 44–49] and references cited therein.

The use of free weighting matrices along with Newton-Leibniz formula can eliminate the cross terms arising in the LK derivative for formulating delay-dependent results, this technique has effectively reduced the conservatism in existing delay-dependent stability criterion that estimates the delay upper bound [50–59], the demerit of such method lies in an increase in more number of decision variables thus increasing the computational complexity. On the other hand, they prove to be useful in formulating LMI conditions which can support the case of fast time-varying delays involved in the system.

Another important inequality (Jensen's integral inequality) that has been introduced in [3] for delay-dependent stability analysis and further this inequality has been widely used in the literature [23, 60–64] and references cited therein pertaining to stability and stabilization of time-delay systems, with a view to reduce the conservatism in the delay upper bound results. The use of Jensen's integral inequality approximates the quadratic integral term without introducing any free weighting matrices, consequently the condition derived is computationally efficient with the appropriate choice of Lyapunov-Krasovskii functional.

The stability and robust stability conditions formulated using the above techniques can be directly extended to solve state feedback stabilization problems in two simple steps, (i) replace A matrix by $(A + BK)$ in the stability condition and (ii) perform congruence transformation on the LMI condition obtained in (i) and subsequently adopting change of matrix variables that allows one to compute the controller gain K. It is apparent from the literature on state feedback stabilization that, two variants of stabilizing condition can be obtained depending on the type of stability criteria formulated (i) LMI framework and (ii) Nonlinear LMI framework. The stabilizing conditions for nominal time-delay system based on state feedback control law have been discussed in [21, 23, 40, 46, 65], whereas the state feedback stabilization for an uncertain time-delay systems can be found in [13–15, 20, 23, 37, 38, 48, 55, 64, 66, 67].

Another variant of stabilization (and/or robust stabilization) problem that has been studied as well as investigated in the thesis for time-delay systems is state feedback stabilization along with disturbance rejection. This problem is simultaneously solved for state feedback stabilization and H_∞ norm minimization of the transfer function between regulated output and the exogenous input (disturbance input) i.e., T_{wz}, the relevant works on this thesis are reported in [32, 35, 41, 68–71]. Application of the above theories (H_∞ control synthesis) to interconnected power systems with time-delay can be found in very few literature like [70, 72–74].

Very recently another variant of delay-dependent stability condition has been reported in literature that is referred as delay-range dependent stability (DRDS) condition. In this method the delay is assumed to vary in intervals i.e., $d_1 \leq d \leq d_2$, whereas in case of delay-dependent stability (DDS) the delay lower bound 'd_1' is always restricted to 0. In this context one can say that, DRDS condition is a generalization of DDS condition. DRDS conditions are addressed by many authors and some mature methods have been widely used to deal with the stability and synthesis problems in [34, 62, 63] and references cited therein. DRDS conditions are of significance for stability analysis of many practical applications and one such application is networked control systems.

Remark 1.6 The research on delay-dependent stability analysis of time-delay systems with time varying state delays have following objectives, (i) to propose stability conditions that can yield less conservative delay upper bound estimate (ii) to derive conditions with fewer decision variables by not sacrificing the conservativeness and (iii) to propose a new Lyapunov-Krasovskii functional with the use tighter bounding inequalities such that resulting LMI condition with a view to achieve objective in (i). The extensive literature review suggests that, the attempt to narrow down the gap between estimated & exact delay value is still continuing and thus this is still an open field of research at a fundamental or applied level.

1.7 Linear Systems with Actuator Saturation

In control systems and other physical systems the saturation phenomenon is unavoidable as all the actuators or sensors in the system have its own operating ranges. The presence of actuator saturation nonlinearities could cause performance degradation, instability for feedback control system and may also lead to generation of limit cycles.

The purpose of this section is to briefly present the mathematical framework for stability analysis or controller synthesis of a linear system subjected to actuator saturation using Lyapunov's second method.

Consider the closed-loop linear system with actuator saturation,

$$\dot{x}(t) = Ax(t) + BSat(u(t)) \tag{1.48}$$

where, the control law is $u(t) = Kx(t)$, $x(t) \in \mathcal{R}^n$ is the state vector, $u(t) \in \mathcal{R}^m$ is the control input vector, $A \in \mathcal{R}^{n\times n}$ is the constant matrix, $B \in \mathcal{R}^{n\times m}$ is the input matrix and $K \in \mathcal{R}^{m\times n}$ is the feedback gain matrix.

A symmetrical saturation function is defined as,

$$Sat(u_{(i)}) = \begin{cases} u_{0(i)}, & \text{if } u_{(i)} > u_{0(i)}; \\ u_{(i)}, & \text{if } -u_{0(i)} \leq u_{(i)} \leq u_{0(i)}; \\ -u_{0(i)}, & \text{if } u_{(i)} < -u_{0(i)} \end{cases}$$

where $u_{(i)}$ is the ith row of $u(t)_{m\times 1}$. The system (1.48) is locally linear, which means that for some values of $x(t) \in \mathcal{R}^n$ the ith control input $\mid u_{(i)}(t) \mid = |k_i x(t)|$ is linear, after which it enters into the saturation region, this attribute can be defined by a linear region as,

$$\mathcal{L}(K, u_0) = \left\{x(t) \in \mathcal{R}^n : -u_{0(i)} \leq k_i x(t) \leq u_{0(i)},\ i = 1, ...m\right\} \tag{1.49}$$

If the ith component of control input $u_{(i)}$ is in linear region (1.49) then,

$$\dot{x}(t) = (A + BK)x(t) \tag{1.50}$$

For the system (1.48), if $x(0) \in \mathcal{L}(K,\ u_0)$ then $x(t)$ would not necessarily belong to $\mathcal{L}(K,\ u_0)$, $\forall\, t \geq 0$, even if (A+BK) is Hurwitz.

Now, to ensure that $\forall\ x(0) \in \mathcal{L}(K,\ u_0)$, $x(t,\ x(0)) \to 0$ when $t \to \infty$, it is necessary to take into account the nonlinear behaviour of the closed-loop system (1.48). Thus one can infer that the stability or stabilization problem of such system is related to the initial conditions. Finding the set of initial condition i.e., the domain of attraction is the main task for such system while ensuring the local stability.

A set $\mathcal{E}$ containing the origin in its interior is said to be region of asymptotic stability (or the region of attraction) for system (1.48) if $\forall\ x(0) \in \mathcal{E}$ such that the corresponding state trajectory converges to origin, while the region of attraction can be expressed mathematically as,

$$R_a(0) = \{x(0) \in \mathcal{R}^n, x(t, x(0)) \to 0, t \to \infty\}$$

hence, $\mathcal{E} \subset R_a(0)$. Now two conditions of stability for such system can be defined mathematically in terms of region of attraction of origin $R_a(0)$ as,

- If $R_a(0)$ represents the whole n-dimensional real space, then it is ensuring global stability of (1.48)
- If $R_a(0)$ is restricted within certain region in an n-dimensional real space with origin in its interior, then it is ensuring local stability of (1.48).

For global stability or stabilization of the system (1.48), it is required that the pair (A, B) must be stabilizable and also matrix 'A' must be Hurwitz. If the matrix 'A' is not Hurwitz (i.e., open-loop system is unstable) then it is not possible to find the global stability of the system [32, 75] and references cited therein. In such a situation local stability of (1.48) has to be ensured, which in turn requires to compute the set of initial condition and is referred as an estimate of region of asymptotic stability.

The estimate of the region of asymptotic stability (or the domain of stability) is usually found out by Lyapunov's second method by establishing the set invariance condition [76]. A set is said to be invariant if all the trajectories starting the set will remain in it.

The connection between the set invariance condition, stability and estimate of domain of attraction using Lyapunov's second method is briefly presented for solving the stability and stabilization problems of (1.48).

Let us consider the Lyapunov function for the system (1.48) as,

$$V(x) = x(t)^T Px(t), \ P = P^T > 0 \tag{1.51}$$

Taking the time-derivative of (1.51) one can obtain,

$$\dot{V}(x(t)) = 2x(t)^T P[Ax(t) + BSat(Kx(t))] \tag{1.52}$$

One can further write (1.52) as,

$$\dot{V}(x(t)) = \begin{bmatrix} x^T(t) & (Sat(Kx(t))^T \end{bmatrix} \begin{bmatrix} A^T P + PA & PB \\ \star & 0 \end{bmatrix} \begin{bmatrix} x(t) \\ Sat(Kx(t)) \end{bmatrix} \tag{1.53}$$

For the stability of the system (1.48), we need to establish $\dot{V}(x) < 0$ and to satisfy this the LMI given below must have feasible solution.

$$\begin{bmatrix} A^T P + PA & PB \\ \star & 0 \end{bmatrix} < 0 \tag{1.54}$$

The matrix inequality in (1.54) is not solvable directly. Thus, we need to reformulate the mathematical expression of the saturation function $Sat(Kx(t))$ in (1.48), such that a tractable LMI condition can be obtained. This in turn, ultimately facilitates the estimate of the region of asymptotic stability.

Broadly two mathematical approximations of saturation function are available in the literature.

(i) **Polytopic representation:** In this method, the saturation function in (1.48) is placed in the convex hull of the group of linear feedbacks. The formation of the convex hull out of two given vectors of same dimension can be done using the lemma presented in [76].

Lemma 1.1 *Let vectors* $u,\ v\ \in\ \mathcal{R}^m$*, suppose that* $|\ v_i\ |\leq u_{0(i)},\ \forall i \in [1,\ m]$*, then*

$$Sat(u) \in co\{D_i u + D_i^- v : i \in [1,\ 2^m]\}$$

where, D be the set of $m \times m$ *diagonal matrices with 0 or 1 as diagonal entries, i.e.,* $D = \{D_i\ :\ i\ \in\ [1,\ 2^m]\}$ *and* $D_i^- = I - D_i$*. For example, if* $m = 2$*, then*
$$D = \left\{ \begin{bmatrix} 0\,0 \\ 0\,0 \end{bmatrix}, \begin{bmatrix} 0\,0 \\ 0\,1 \end{bmatrix}, \begin{bmatrix} 1\,0 \\ 0\,0 \end{bmatrix}, \begin{bmatrix} 1\,0 \\ 0\,1 \end{bmatrix} \right\}.$$

One can place the saturation function $Sat(Kx)$ into the convex hull of group of linear feedbacks given an auxiliary feedback matrices, say H (the dimension of $K,\ H \in \mathcal{R}^{m \times n}$) with $|\ h_i x\ |\leq u_{0(i)}$, using above lemma,

$$Sat(Kx) \in co\left\{D_i Kx + D_i^- Hx : i \in [1,\ 2^m]\right\},\ \forall x \in \mathcal{R}^n \tag{1.55}$$

thus the closed loop system representation of (1.48) becomes,

$$Ax(t) + BSat(u(t))\ \in\ co\left\{Ax + B(D_i K + D_i^- H)x(t)\ |\ i = [1, ..2^m]\right\} \tag{1.56}$$

that must satisfy $|\ h_i x\ |\leq u_{0(i)},\ i = 1, ...m.$

(ii) **Sector nonlinearity representation:** In this approximation method, the saturation nonlinearity is equivalently replaced with a dead-zone nonlinearity '$\Psi(u(t))$' in combination with a linear element, and it is mathematically expressed as,

$$Sat(u(t)) = u(t) - \Psi(u(t)) \tag{1.57}$$

Now one can express dead-zone nonlinearity equivalently in terms of saturation function as,

$$\Psi(u(t)) = u(t) - Sat(u(t))$$

The ith element of dead-zone nonlinearity function as,

$$\Psi_i(k_i x(t)) = k_i x(t) - Sat(k_i x(t))$$

$$= \begin{cases} k_i x(t) - u_{0(i)} > 0, \text{ if } k_i x(t) > u_{0(i)}; \\ k_i x(t) - u_{(i)} = 0, \text{ if } -u_{0(i)} \leq k_i x(t) \leq u_{0(i)}; \\ k_i x(t) + u_{0(i)} < 0, \text{ if } k_i x(t) < -u_{0(i)}, \; for \; i = 1, 2, ...m \end{cases}$$

As the saturation function is replaced by the dead-zone nonlinearity, one needs to satisfy the modified sector condition as stated in the following lemma.

Lemma 1.2 [75] *Consider a dead-zone nonlinear function $\Psi_i(k_i x(t))$, with $x \in \mathcal{S}$, where $\mathcal{S} = \{x \in \mathcal{R} : |(k_i - g_i)x| \leq u_{0(i)}\}$, then the relation*

$$\Psi_i(k_i x(t)) D_{i,i} \, [\, \Psi(k_i x(t)) - g_i x(t)] \, \leq 0 \tag{1.58}$$

is valid for any scalar $D_{i,i} > 0$. g_i is the ith component of the auxiliary feedback gain matrix.

The lemma (1.2) guarantees the constraint on the control as indicated below,

$$|\, k_i x(t) - g_i x(t) \,| \leq u_{0(i)} \tag{1.59}$$

Note: Whereas the constraint of the control for polytopic representation of the saturation function is given by $|\, h_i x \,| \leq u_{0(i)}$.

Now, in view of (1.57) one can write the closed-loop system (1.48) as,

$$\dot{x}(t) = A_c x(t) + A_d x(t - d(t)) - B\psi(Kx(t)) \tag{1.60}$$

where, $A_c = (A + BK)$.

After replacing the saturation function '$Sat(u(t))$' with either dead-zone nonlinearity or polytopic representation, one has to establish the set invariance condition for the transformed systems (1.56) or (1.60) which is briefly explained by following general steps given below,

(a) Assume a Lyapunov function of the form $V(x) = x^T P x(t)$ with $P = P^T > 0$ for the transformed systems (1.56) or (1.60),
(b) Construct a compact convex (ellipsoidal) set $\mathcal{E}(P, 1)$ defined as,

$$\mathcal{E}(P, 1) = \{x \in \mathcal{R}^n : x^T P x \leq 1\} \tag{1.61}$$

(c) If $\dot{V}(x(t)) < 0$ and the set $\mathcal{E}(P, 1) \subset \mathcal{L}(H)$ which means that $|h_i x| \leq u_{0(i)}$ (for polytopic representation), and $|(k_i - g_i)x| \leq u_{0(i)}$ (for sector nonlinearities representation), $\forall x \in \mathcal{E}(P, 1)$, $i = 1, 2, ...m$, then the set $\mathcal{E}(P, 1)$ is called contractive invariant and inside the $R_a(0)$.
(d) Next, certain measure of the estimate of domain of attraction is obtained using P matrix (associated with the Lyapunov function). This measure is maximized (or equivalently recast the maximization problem as minimization problem) subject to (i) LMI conditions obtained from $\dot{V}(t) < 0$ and (ii) the set invariance condition in (c), the corresponding gain matrix K can be evaluated from the solution of the optimization problem.

1.8 Outline of the Book

The outline of the remaining chapters of the book are presented here.

Chapter 2 deals with delay-dependent stability and robust stability analysis of time-delay system. For assessing the stability of a time-delay systems with a single and two additive time-varying delays in the state, new delay-dependent stability conditions are derived using LK functional approach in an LMI framework. Furthermore, the stability condition of single delay case is extended for robust stability analysis of an uncertain time-delay systems. The stability conditions are derived by exploring new LK functional terms followed by the use of improved and tighter bounding inequalities for approximating the quadratic integral terms arising out of the derivative of LK functional that finally results into convex combination of LMIs. This in turn, yields new an improved stability condition compared to the existing methods. Delay-range-dependent stability (DRDS) condition of time-delay systems has been assessed utilizing delay partitioning and non-partitioning approaches. The effectiveness of the derived LMI conditions are demonstrated through several numerical examples and the results are compared with the existing methods.

Chapter 3 discusses about the delay-dependent state feedback stabilization and robust stabilization in an LMI and NLMI frameworks. New and improved robust stabilization condition for a time-delay system with norm bounded uncertainties in an NLMI framework has been derived with a view to obtain an improved size of delay bound. Subsequently, an LMI based improved state feedback stabilization method is employed with lesser number of matrix variables and in sequel a multi-objective optimization algorithm is adopted for tuning the controller gains in such a way that the control effort required to stabilize the time-delay system is less. Furthermore, simultaneous delay-dependent state feedback stabilization and disturbance rejection problem has been considered by minimizing the H_∞ performance index for load-frequency control of a two-area interconnected power system model with constant communication delays. The derived stabilization condition is delay-dependent in terms of the feedback delay which makes it suitable for load-frequency control problem in presence of load-disturbances and also the formulation is different from other results of LFC problems that are solved based on the delay-dependent H_∞ control design techniques.

Chapter 4 deals with the state feedback stabilization of nominal and uncertain linear time-varying delay systems with actuator saturation. A new delay-dependent local stabilizing condition for achieving asymptotic stability of an unstable open-loop system has been derived using the two different representation of saturation function (namely sector nonlinearities and polytopic approach) via convex combination of LMIs. To the best of the author knowledge, it has not been investigated so far in the literature. The proposed method in comparison to the recent existing results can achieve improved delay upper bound value with marginally sacrificing the estimate of domain of attraction (DOA) and on the other hand for a given delay value less conservative estimate of domain of attraction has also been obtained. The stabilizing results are further extended to derive robust stabilization condition for an uncertain time-delay system with sector nonlinearity approach only. The results are validated by considering several numerical examples.

Chapter 5 presents the concept of extension of stability analysis of linear time-delay system to Fuzzy T-S time-delay system.

References

1. J. Hale, *Theory of Functional Differential Equations* (Academic, New York, 1977)
2. N.N. Krasovskii, *Stability of Motion* (Stanford University Press, Stanford, 1963)
3. K. Gu, V.L. Kharitonov, J. Chen, *Stability of Time-Delay Systems* (Birkhuser, Boston, 2003)
4. J.E. Marshall, *Control of Time-Delay System* (Peter Peregrinus, Cambridge, 1979)
5. J.P. Richard, Time-delay systems: An overview of some recent advances and open problems. Automatica **39**, 1667–1694 (2003)
6. S. Xu, J. Lam, A survey of linear matrix inequalities in stability analysis of delay systems. Int. J. Syst. Sci. **39**, 1095–1113 (2008)
7. N. Olgac, R. Sipahi, An exact method for the stability analysis of time-delayed linear time-invariant systems. IEEE Trans. Autom. Control **47**, 793–797 (2002)
8. R. Sipahi, S. Niculescu, C.T. Abdallah, W. Michiels, K. Gu, Stability and stabilization of systems with time-delay. IEEE Control Syst. Mag. **31**(1), 38–65 (2011)
9. W. Michiels, "Stability and stabilization of time-delay systems." PhD thesis, Belgium, 2002
10. W. Michiels, K. Engelborghs, P. Vansevenant, D. Roose, Continuous pole-placement method of delay equations. Automatica **38**, 747–761 (2002)
11. W. Michiels, T. Vyhlidal, An eigen value based appraoch for the stabilization of linear time-delay systems of neutral type. Automatica **41**, 991–998 (2005)
12. J.S. Luo, A. Johnson, P.P.J. van den Bosch, Delay-independent robust stability of uncertain linear system. Syst. Control Lett. **24**, 33–39 (1995)
13. M.S. Mahmoud, N.F. Al-muthairi, Quadratic stabilization of continuous time system with the state delay and norm bounded uncertainties. IEEE Trans. Autom. Control **39**, 2135–2139 (1994)
14. S. Phoojaruenchanachai, K. Furuta, Memoryless stabilization of uncertain linear system including time-varying sate delay. IEEE Trans. Autom. Control **37**, 1022–1026 (1992)
15. J.C. Shen, B.S. Chen, F.C. Kung, Memoryless stabilization of uncertain dynamic delay system: Ricatti equation approach. IEEE Trans. Autom. Control **36**, 638–640 (1991)
16. E.I. Verriest, M.K.H. Fan, J. Kullstam, Frequency domain robust stability criteria for linear delay systems, in *Proceedings of 32nd IEEE Conference on Decision and Control*, 1993, pp. 3473–3478
17. T. Mori, H. Kokame, Stability of $\dot{x}(t) = \mathrm{A}\,x(t) + \mathrm{B}\,x(t-\tau)$. IEEE Trans. Autom. Control **34**, 460–462 (1989)
18. K. Gu, S.I. Niculescu, Additional dynamics in transformed time-dealy systems. IEEE Trans. Autom. Control **45**, 572–575 (2000)
19. M.S. Mahmoud, *Robust Control and Filtering of Time-Delay Systems* (Marcel Dekker, Cambridge, 2000)
20. M.S. Mahmoud, Delay-dependent robust stability and stabilization of uncertain linear delay system:a linear matrix inequality approach. IEEE Trans. Autom. Control **42**, 1144–1148 (1997)
21. M.S. Mahmoud, "Delay-dependent stability and state feedback stabilization criterion for linear time delay system," in *International Conference on Modeling and Simulation, Coimbator*, Vol. 2, 2007), pp. 963–968
22. S.G. Chen, A.G. Ulsoy, Y. Koren, Computational stability analysis of chatter in turning discrete delay and norm-bounded uncertainty. J. Manuf. Sci. Eng. **119**, 457–460 (1997)
23. M.N.A. Parlakci, Improved robust stability criteria and design of robust stabilizing controller for uncertain linear time-delay system. Int. J. Robust Nonlinear Control **16**, 599–636 (2006)

24. A. Bellen, Z. Marino, *Numerical Methods for Delay-Differential Equations* (Oxford University Press, London, 2005)
25. S. Boyd, L. Ghaoui, E. Feron, V. Balakrishnan, *LMI in Systems and Control Theory* (SIAM, Philadelphia, 1994)
26. C. Scherer, S. Weiland, "Lecture notes for DISC course on linear matrix inequalities in control," http://www.er.ele.tue.nl/SWeiland/lmi99.htm, 1999
27. P. Gahinet, A. Nemirovski, A.J. Laub, M. Chilali, *LMI Control Toolbox Users Guide* (Mathworks, Cambridge, 1995)
28. T. Coleman, M. Branch, A. Grace, *Optimization Toolbox for Use with MATLAB* (Mathworks, Natick, 1999)
29. K. Zhou, J.C. Doyel, K. Glover, Robust and Optimal Control. Prentice Hall, 1995
30. V.B. Kolmanovoskii, On the liapunov-krasovskii functionals for stability analysis for linear time-delay systems. Int. J. Control **72**, 374–384 (1999)
31. J.H. Kim, Delay and its time-derivative robust stability of time delayed linear systems with uncertainty. IEEE Trans. Autom. Control **46**, 789–792 (2001)
32. J.H. Kim, Delay-dependent stability and H_∞ control: Constant and time-varying delays. Int. J. Control **76**, 48–60 (2003)
33. T.J. Su, C. Huang, Robust stability of delay dependence for linear uncertain systems. IEEE Trans. Autom. Control **37**, 1656–1659 (1992)
34. P.L. Liu, T.J. Su, Robust stability of interval time-delay systems with delay-dependence. Syst. Control Lett. **33**, 231–239 (1998)
35. E. Fridamn, U. Shaked, A descriptor system approach to H_∞ control of linear time-delay systems. IEEE Trans. Autom. Control **47**, 253–270 (2002)
36. P. Park, A delay dependent stability criterion for systems with uncertain time-invariant delays. IEEE Trans. Autom. Control **44**, 876–877 (1999)
37. X. Li, C.E. de Souza, Criteria for robust stability and stabilization of uncertain linear systems with state delays. Automatica **33**, 1657–1662 (1997)
38. Y.S. Moon, P. Park, W.H. Kwon, Y.S. Lee, Delay-dependent robust stabilization of uncertain state delayed system. Int. J. Control **74**, 1447–1455 (2001)
39. H. Gao, J. Lam, C. Wang, Y. Wang, Delay-dependent output feedback stabilisation of discrete-time systems with time-varying state delay. IEE Proc. Control Theory Appl. **151**, 691–698 (2004)
40. X.M. Zhang, M. Wu, J.H. She, Y. He, Delay-dependent stabilisation of linear systems with time-varying state and input delays. Automatica **41**, 1405–1412 (2005)
41. V. Suplin, E. Fridman, U. Shaked, H_∞ control of linear uncertain time-delay systems-a projection approach. IEEE Trans. Autom. Control **31**, 680–685 (2006)
42. H.J. Cho, J.H. Park, Novel delay-dependent robust stability criterion of delayed cellular neural networks. Chaos, Solitons Fractals **32**, 1194–1200 (2007)
43. E. Fridman, New-lyapunov-krasovskii functional for stability of linear retarded and neutral type. Syst. Control Lett. **43**, 309–319 (2001)
44. E. Fridman, M. Dambrine, Control under quantization, saturation and delay:an LMI appraoch. Automatica **45**, 2258–2264 (2009)
45. E. Fridman, A. Pila, U. Shaked, Regional stabilization and H_∞ control of time-delay systems with saturating actuators. Int. J. Robust Nonlinear Control **13**, 885–907 (2003)
46. E. Fridman, U. Shaked, An improved stabilization method for linear time-delay system. IEEE Trans. Autom. Control **47**, 1931–1937 (2002)
47. E. Fridman, U. Shaked, Descriptor discretized lyapunov functional method: analysis and design. IEEE Trans. Autom. control **51**, 890–897 (2006)
48. Y. Sun, L. Wang, G. Xie, Delay-dependent robust stability and stabilization for discrete-time switched systems with mode-dependent time-varying delays. Appl. Math. Comput. **180**, 428–435 (2006)
49. E. Fridman, U. Shaked, X. Li, Robust H_∞ filtering of linear systems with time-varying delay. IEEE Trans. Autom. Control **48**, 159–165 (2003)

50. Y. He, M. Wu, J.H. She, G. Liu, Delay-dependent robust stability criteria for uncertain neutral system with mixed delays. Syst. Control Lett. **51**, 57–65 (2004)
51. Y. He, W. Min, J.H. She, G.P. Liu, Parameter dependent lyapunov functional for stability of time delay systems with polytopic uncertainties. Automatica **492**, 828–832 (2004)
52. Y. He, Q.-G. Wang, L. Xie, C. Lin, Further improvements of free weighting matrices technique for systems with time-varying delay. IEEE Trans. Autom. Control **52**, 293–299 (2007)
53. Y. He, Q.G. Wang, C. Lin, M. Wu, Delay-range-dependent stability for systems with time-varying delay. Automatica **43**, 371–376 (2007)
54. R. Dey, G. Ray, S. Ghosh, A. Rakshit, Stability analysis for contious system with additive time-varying delays: a less conservative result. Appl. Math. Comput. **215**, 3740–3745 (2010)
55. R. Dey, S. Ghosh, G. Ray, A. Rakshit, State feedback stabilization of uncertain linear time-delay systems: a nonlinear matrix inequality appraoch. Numer. Linear Algebra Appl. **18**(3), 351–361 (2011)
56. J. Lam, H. Gao, C. Wang, Stability analysis for continuous system with two additive time-varying delay components. Syst. Control Lett. **56**, 16–24 (2007)
57. S. Xu, J. Lam, D.W.C. Ho, Y. Zou, Delay-dependent exponential satbility for a class of neural networks with time-delay. J. Comput. Appl. Math. **183**, 16–28 (2005)
58. S. Xu, J. Lam, X. Mao, Y. Zou, A new LMI condition for delay-dependent robust stability of stochastic time-delay systems. Asian J. Control **7**, 419–423 (2005)
59. L. Zhang, E.-K. Boukas, A. Haider, Delay-range-dependent control synthesis for time-delay systems with actuator saturation. Automatica **44**, 2691–2695 (2008)
60. J.H. Park, O. Kwon, On new stability criterion for delay-differential systems of neutral type. Appl. Math. Comput. **162**, 627–637 (2005)
61. O.M. Kwon, J.H. Park, Guaranteed cost control for uncertain large-sacle systems with time-delays via delayed feedback. Chaos, Solitons Fractals **27**, 800–812 (2006)
62. H. Shao, Improved delay-dependent stability criteria for systems with a delay varying in range. Automatica **44**, 3215–3218 (2008)
63. H. Shao, New delay-dependent stability criteria for systems with interval delay. Automatica **45**, 744–749 (2009)
64. W.H. Chen, W.X. Zheng, Delay-dependent robust stabilisation for uncertain neutral systems with distributed delays. Automatica **43**, 95–104 (2007)
65. M. Wu, Y. He, J.H. She, New delay-dependent stability criteria and stabilizing method for neutral system. IEEE Trans. Autom. Control **49**, 2266–2271 (2004)
66. S. Niculescu, A.T. Neto, J.M. Dion, L. Dugard, "Roust stability and stabilization of uncertain linear systems with state delay: multiple delay case (I)," in *IFAC Symposium on Robust Control Design*, 1994
67. Y.S. Lee, Y.S. Moon, W.H. Kwoon, K.H. Lee, in "Delay-Dependent Robust H_∞ Control of Uncertain System with Time-Varying State Delay," 2001, pp. 3208–3213
68. G.J. Hua, P.M. Frank, C.F. Lin, Robust H_∞ state feedback control for linear systems with state delay and parameter uncertainty. Automatica **32**, 1183–1185 (1996)
69. M.S. Mahmoud, M. Zribi, H_∞ controllers for linearized time-delay systems using LMI. J. Optim. Theory Appl. **100**, 89–112 (1999)
70. M. Zribi, M.S. Mahmoud, M. Karkoub, T.T. Lie, H_∞ controller for linearized time-delay power system. IEE Proc. Gener. Transm. Distrib. **147**, 401–403 (2000)
71. X. Li, E. Fridman, U. Shaked, Robust H_∞ control of distributed delay systems with application to combustion control. IEEE Trans. Autom. Control **46**, 1930–1933 (2001)
72. X. Yu, K. Tomsovic, Application of linear matrix inequalities for load frequency control with communication delay. IEEE Trans. Power Syst. **19**, 1508–1515 (2004)
73. Y. Zhao, Y. Ou, L. Zhang, H. Gao, H_∞ control of uncertain seat suspension systems subject to input delay and actuator saturation, in *IEEE* 48th *Conference on Decision and Control*, vol. 14, 2009, pp. 5164–5169
74. G.J. Li, T.T. Lie, C.B. Soh, G.H. Yang, Decentralized H_∞ control for power system stability enhancement. Electr. Power Energy Syst. **20**, 453–464 (1998)

75. J.M.G.D. Silva, A. Seuret, E. Fridman, J.P. Richard, Stabilization of neutral systems with saturating control inputs. Int. J. Syst. Sci. (2010). https://doi.org/10.1080/00207720903353575
76. J.M.G.D. Silva, A. Seuret, E. Fridman, J.P. Richard, *Control Systems with Actuator Saturation: Analysis and Design* (Birkhauser, Boston, 2001)

Chapter 2
Stability Analysis of Time-Delay Systems

This chapter deals with the stability analysis of linear time-delay systems without and with parametric uncertainties. The stability analysis for both constant and time-varying delay in the states is considered. The focus of this chapter is to review the existing methods on delay-dependent stability analysis in an LMI framework based on Lyapunov-Krasovskii approach and consequently the improved results on delay-dependent stability analysis are presented. The results of the proposed techniques are validated by considering numerical examples and compared with existing results.

2.1 Introduction

Time-delays are often observed in many areas of engineering systems such as networked control systems, chemical processes, neural networks, milling process, nuclear reactors and long transmission lines in power systems and their presence can have effect on system stability and performance [1], so ignoring them can lead to design flaws and incorrect stability analysis. In particular the effect of delays become more pronounced in interconnected and distributed system where multiple sensors, actuators and controller introduce multiple delays. Thus stability analysis becomes the prime objective in a control system design. The stability analysis of time-delay systems using Lyapunov's second method are broadly classified into three major categories

- Delay-independent stability analysis
- Delay-dependent stability analysis
- Delay-range-dependent stability analysis

Delay-independent stability analysis considers the size of the delay to be arbitrarily large (delay value$\rightarrow \infty$) and hence the obtained stability conditions are independent of the delay value. Delay-independent stability results are, in general,

R. Dey et al., *Stability and Stabilization of Linear and Fuzzy Time-Delay Systems*,
Intelligent Systems Reference Library 141, https://doi.org/10.1007/978-3-319-70149-3_2

more conservative for many important applications and especially for engineering (or physical) systems [2]. So, in early 1990s increasing attention has been devoted to delay-dependent stability analysis, which considers finite (bounded) delay value, thus in this case derived stability conditions depends on the size of the delay. For both the delay-independent and delay-dependent analysis the lower bound of the delay is assumed to be zero, whereas for the first case the delay upper bound is unbounded and for the second case, it is bounded to some finite value. Very recently [3–5], another variant of delay-dependent stability analysis has been proposed where the information of the delay ranges are available i.e., the lower bound of the delay is not assumed to be explicitly zero but can possess some finite value and the delay upper value is bounded as in the case of delay-dependent stability analysis, such stability analysis is referred as delay-range-dependent stability analysis. In this chapter, we discuss delay-dependent as well as delay-range-dependent stability analysis as they are of physical significance.

The stability analysis has been carried out for (i) nominal time-delay systems i.e., systems without parametric uncertainties and (ii) uncertain time-delay systems i.e., systems possessing uncertainties in the system matrices, the stability analysis of such time-delay systems is referred as robust stability analysis. The structure of the uncertainty is assumed to be of norm-bounded type. The stability analysis of time-delay systems can be carried out using either

- Lyapunov-Razumikhin theorem
- Lyapunov-Krasovskii (LK) theorem

All the recent literature on stability analysis of time-delay systems adopts latter method as the former method yields conservative estimate of delay upper bound compared to LK theorem [6–9] because of the following reasons:

1. The use of the condition $V(t+\theta, x(t+\theta)) \leq pV(t, x(t)), \forall\theta \in [-d, 0]$ in Lyapunov-Razumikhin theorem.
2. More number of bounding inequalities are used while deriving delay-dependent stability conditions, as in this case Lyapunov function is assumed to be very simple one i.e., $V(t) = x^T(t)Px(t)$.
3. Lyapunov-Razumikhin theorem is incapable of handling slow time-varying delay (i.e., bounded differentiable time-varying delay which implies delay-derivative <1), it can treat fast time-varying delays (i.e., non-differentiable time-varying delay that implies delay derivative ≥ 1) and constant delays.

In this thesis the attention is focused on the delay-dependent stability analysis of time-delay systems using Lyapunov-Krasovskii theorem.

2.2 Description of Time-Delay Systems

In this section, description of time-delay systems for carrying out stability as well as robust stability analysis is presented.

2.2.1 Nominal Time-Delay System for Stability Analysis

1. **System with single, constant delay**

$$\Sigma_1 : \dot{x}(t) = Ax(t) + A_d x(t-d) \tag{2.1}$$
$$x(t) = \phi(t), \forall t \in [-d_u, 0] \tag{2.2}$$

2. **System with single, time-varying delay**

$$\Sigma_2 : \dot{x}(t) = Ax(t) + A_d x(t-d(t)) \tag{2.3}$$
$$x(t) = \phi(t), \forall t \in [-d_u, 0] \tag{2.4}$$

where, $d(t)$ is time-varying delay satisfying following conditions

$$0 \leq d(t) \leq d_u \tag{2.5}$$
$$0 \leq d_l \leq d(t) \leq d_u \tag{2.6}$$

Note: The condition (2.5) refers to delay-dependent stability (DDS) notion and (2.6) refers to delay-range-dependent stability (DRDS) notion.
the delay derivative satisfies the condition

$$\dot{d}(t) \leq \mu < 1 \tag{2.7}$$
$$1 \leq \dot{d}(t) \leq \mu \leq \infty \tag{2.8}$$

Note: The condition (2.7) refers to slowly varying time-delay and (2.8) refers to fast varying time-delay [9] and [10].

Notations: $x(t) \in \mathcal{R}^n$ is the state vector, $\phi(t)$ is the initial function in the banach (norm linear space) space, $d(t)$ is the time-varying delay, d_u is the delay upper bound and d_l is the delay lower bound, $A \in \mathcal{R}^{n\times n}$ and $A_d \in \mathcal{R}^{n\times n}$ are known constant matrices.

3. **System with multiple time-varying delays**

$$\Sigma_3 : \dot{x}(t) = Ax(t) + \sum_{i=1}^{n} A_{di} x(t - d_i(t)) \tag{2.9}$$
$$x(t) = \phi(t), \forall t \in \max[-d_{ui}, 0], i = 1, 2 \ldots n \tag{2.10}$$

Note: Here, $d_i(t)$ indicates the time-varying delay in the states and A_{di} is the associated delayed system matrices. The stability analysis of the system Σ_2 can be extended to multiple time-delay case of Σ_3 in a straight forward manner [11] and [8].

4. **System with two additive time-varying delays**

$$\Sigma_4 : \dot{x}(t) = Ax(t) + A_d x(t - d_1(t) - d_2(t)) \tag{2.11}$$
$$x(t) = \phi(t), \forall t \in \max[-d_u, 0], \tag{2.12}$$

where, $d_1(t)$ and $d_2(t)$ are the two additive delay component in the state, they satisfy following conditions

$$\begin{aligned} 0 \leq d_1(t) &\leq d_{1u} < \infty \\ 0 \leq d_2(t) &\leq d_{2u} < \infty \\ d_u &= d_{1u} + d_{2u} \end{aligned} \tag{2.13}$$

and delay derivatives satisfies the following conditions,

$$\begin{aligned} \dot{d}_1(t) &\leq \mu_1 < \infty \\ \dot{d}_2(t) &\leq \mu_2 < \infty \end{aligned} \tag{2.14}$$

2.2.2 *Uncertain Time-Delay Systems for Robust Stability Analysis*

The uncertain time-delay systems with norm-bounded parametric uncertainties for robust stability analysis is described as

$$\dot{x}(t) = A(t)x(t) + A_d(t)x(t - d(t)) \tag{2.15}$$
$$x(t) = \phi(t), \forall t \in [-d_u, 0] \tag{2.16}$$

The matrices $A(t)$ and $A_d(t)$ are uncertain system matrices and are assumed to be of the form:

$$A(t) = A + \Delta A(t) \tag{2.17}$$
$$A_d(t) = A_d + \Delta A_d(t) \tag{2.18}$$

where, A and A_d are nominal system matrices and $\Delta A(t)$ and $\Delta A_d(t)$ are time-varying matrices, which models the parametric uncertainties present in the system that are Lebesgue measurable and are norm bounded [7, 12–15]. Further these may possibly be decomposed by exploiting their structural description as

$$\Delta A(t) = D_a F_a(t) E_a \tag{2.19}$$
$$\Delta A_d(t) = D_d F_d(t) E_d \tag{2.20}$$

where the time-varying uncertain matrices $F_a(t) \in \mathcal{R}^{n_a \times n_a}$ and $F_d(t) \in \mathcal{R}^{n_d \times n_d}$ are norm bounded and satisfies $\forall\, t$ that,

$$F_a^T F_a(t) \leq I \tag{2.21}$$

$$F_d^T F_d(t) \leq I \tag{2.22}$$

The matrices D_a, E_a, D_d and E_d in (2.19) and (2.20) are constant known matrices and possibly characterizes how the matrices $F_a(t)$ and $F_d(t)$ influence the system dynamics.

2.3 Delay-Dependent Stability Condition

This section brings out the review of some significant existing LMI techniques in deriving delay-dependent stability conditions based on Lyapunov-Krasovskii functional approach. This review is useful to understand the evolution and development of improved techniques while attempting to achieve less conservative estimate of delay upper bound.

Assumption 2.1 The necessary condition for delay-dependent stability of time-delay systems in Sect. 2.2.1 is that, the matrix $[A + A_d]$ (when $d = 0$, $d(t) = 0$) must be Hurwitz.

2.3.1 Model Transformation Approach (Based on Newton-Leibniz Formula)

In this section, we review the delay-dependent stability conditions that are derived using fixed model transformation. The systems Σ_1 and Σ_2 with time-delays are transformed into systems with distributed delays using Newton-Leibniz formula for the analysis. The Newton-Leibniz formula is expressed as,

$$\int_{t-d}^{t} \dot{x}(s)ds = x(t) - x(t-d)$$

$$x(t-d) = x(t) - \int_{t-d}^{t} \dot{x}(s)ds \tag{2.23}$$

A. First Model Transformation

Using (2.23) in Σ_1 one can write

$$\dot{x}(t) = (A + A_d)x(t) - A_d \int_{t-d}^{t} [Ax(s) + A_d x(s-d)]ds \tag{2.24}$$

The transformed system obtained in (2.24) is called first model transformation, the asymptotic stability of (2.24) also guarantees the stability of the system Σ_1 [8]. Based on the transformed systems, a lot of delay-dependent stability results have been obtained.

Theorem 2.1 (Theorem 7 [8], Cor. 1 [16]) *The transformed system (2.24) is asymptotically stable for any delay satisfying* $0 \leq d \leq d_u$ *if there exist matrices* $P > 0$, $Q_1 > 0$ *and* $Q_2 > 0$ *such that,*

$$0 > \begin{bmatrix} \Delta & d_u\ PA_d & d_u\ PA_d \\ \star & -Q_1 & 0 \\ \star & 0 & -Q_2 \end{bmatrix} \tag{2.25}$$

where, $\Delta = (A + A_d)P + P(A + A_d)^T + d_u\ (A^T Q_1 A + d\ A_d^T Q_2 A_d)$.

Proof Lyapunov-Krasovskii functional chosen is given by

$$V(t) = x^T(t)Px(t) + \int_{-d_u}^{0}\int_{t+\theta}^{t} x^T(\beta)A^T Q_1 Ax(\beta)d\beta d\theta$$
$$+ \int_{-d_u}^{0}\int_{t-d+\theta}^{t} x^T(\beta)A_d^T Q_2 A_d x(\beta)d\beta d\theta \tag{2.26}$$

Finding the time-derivative of (2.26) one can obtain

$$\begin{aligned}\dot{V}(t) &= x^T[(A + A_d)^T P + P(A + A_d) + d_u\ (A_d^T Q_2 A_d + A^T Q_1 A)]x(t) \\ &-2\int_{t-d_u}^{t} x^T(t)PA_d Ax(\beta)d\beta - 2\int_{t-d_u}^{t} x^T(t)PA_d A_d x(\beta - d)d\beta \\ &- \int_{t-d_u}^{t} x^T(\beta)A^T Q_1 Ax(\beta)d\beta - \int_{t-d_u}^{t} x^T(\beta - d)A_d^T Q_2 A_d x(\beta - d)d\beta\end{aligned} \tag{2.27}$$

In (2.27) two cross terms $-2\int_{t-d_u}^{t} x^T(t)PA_d Ax(\beta)d\beta$ and $-2\int_{t-d_u}^{t} x^T(t)$ $PA_d Ax(\beta - d)d\beta$ appears that are approximated using the bounding Lemma stated below,

Lemma 2.1 ([2, 8, 16]) *For any* $z, y \in \mathcal{R}^n$ *and any positive definite matrix* $X \in \mathcal{R}^{n\times n}$

$$-2z^T y \leq z^T X^{-1} z + y^T Xy \tag{2.28}$$

Now, using Lemma 2.1 in (2.27) one can obtain

$$
\begin{aligned}
\dot{V}(t) \leq\ & x^T[(A+A_d)^T P + P(A+A_d) + d_u\,(A_d^T Q_2 A_d + A^T Q_1 A)]x(t) \\
&+ \int_{t-d_u}^{t} x^T(t) P A_d Q_1^{-1} A_d^T P x(t) d\beta + \int_{t-d_u}^{t} x^T(\beta) A^T Q_1 A x(\beta) d\beta \\
&- \int_{t-d_u}^{t} x^T(\beta) A^T Q_1 A x(\beta) d\beta - \int_{t-d_u}^{t} x^T(\beta-d) A_d^T Q_2 A_d x(\beta-d) d\beta \\
&+ \int_{t-d_u}^{t} x^T(t) P A_d Q_2^{-1} A_d^T P x(t) d\beta + \int_{t-d_u}^{t} x^T(\beta-d) A_d^T Q_2 A_d x(\beta-d) d\beta \quad (2.29)
\end{aligned}
$$

One can observe in (2.29) that, the terms arising out after using bounding lemma will compensate for the last two integral terms in (2.27), thus yielding quadratic Lyapunov inequality in the form of LMI given in (2.25).

Remark 2.1 It is obvious from the above derivation that, the choice of LK functional for this method leads to two cross bounding terms which are approximated using bounding Lemma 2.1. If this theorem has to be extended for multiple delay case (say m delays) then the number of times bounding lemma have to be used will be '2 m'. Hence, more will be the cross bounding terms present in the LK functional derivative, the use of bounding lemma for its approximation will be more, which is a major source of conservativism in the estimate of delay bound results.

Remark 2.2 One can find the similar choice of LK functional for delay-dependent stability analysis using first model transformation for the systems Σ_1 and Σ_2 in [17] and [18] respectively.

The choice of LK functional in [17] is found to be,

$$
V(t) = x^T(t) P x(t) + V_2(t) + V_3(t) \quad (2.30)
$$

where,

$$
V_2(t) = \int_{-d_u}^{0} \int_{t+\theta}^{t} x^T(s) M_1 x(s) ds d\theta
$$

$$
V_3(t) = \int_{-2d_u}^{-d} \int_{t+\theta}^{t} x^T(s) M_2 x(s) ds d\theta,\ M_1 > 0,\ M_2 > 0
$$

the time-derivative of (2.30) is found to be

$$
\dot{V}(t) = x^T(t)[(A+A_d)^T P + P(A+A_d)]x(t) + \dot{V}_2(t) + \dot{V}_3(t) + \xi_1(t) + \xi_2(t) \quad (2.31)
$$

where,

$$
\xi_1(t) \triangleq -2 \int_{t-d_u}^{t} x^T(t) P A_d A x(\alpha) d\alpha \quad (2.32)
$$

and

$$\xi_2(t) \triangleq -2 \int_{t-d_u}^{t} x^T(t) P A_d A_d x(\alpha - d) d\alpha \tag{2.33}$$

To approximate the cross terms $\xi_1(t)$ and $\xi_2(t)$ bounding Lemma 2.1 is used, remaining integral terms arising out of $\dot{V_2}(t)$ and $V_3(t)$ is canceled by the integral terms that appears after the use of Lemma 2.1 with the assumption that, $M_1 = A^T X_1 A$ and $M_2 = A_d^T X_2 A_d$, where X_1 and X_2 are positive definite matrices thus giving a quadratic LMI formulation.

In case of the system Σ_2 (time-varying delay) satisfying the conditions (2.5), the stability condition in [18] is obtained using first model transformation with the similar choice of LK functional given as

$$V(t) = x^T(t) P x(t) + d_u \int_{-d_u}^{0} \int_{t+\theta}^{t} x^T(s) M_1 x(s) ds d\theta$$
$$+ \frac{d_u}{(1-\mu)^2} \int_{-d(t)-d_u}^{-d(t)} \int_{t+\theta}^{t} x^T(s) M_2 x(s) ds d\theta, M_1 > 0, M_2 > 0 \tag{2.34}$$

The factor $\frac{d_u}{(1-\mu)^2}$ associated in the second term of (2.34) is used to compensate for the derivative of the delay term (i.e., $d(t)$) which arises upon differentiation of $V(t)$ due to presence of time-varying delay term in the limit of integration in (2.34).

Remark 2.3 While deriving stability condition using first model transformation, the number of cross terms to be approximated using bounding lemma is twice the number of delays present in the system Σ_1 or Σ_2, secondly, it is proved in [9, 19, 20] that the first model transformation introduces some additional eigenvalues in the transformed system, hence the characteristics of the transformed system is not equivalent to the original one (i.e., Σ_1 or Σ_2), thus the stability condition derived using this transformation yields conservative result of delay upper bound. In other words, the drawbacks associated with this approach is that all of the transformed system is not equivalent to (2.1) or (2.3).

B. Second model transformation [9, 17, 21]

The rearrangement of first model transformation in (2.24) yields the second model transformation (or neutral type transformation) and it is expressed as

$$\frac{d}{dt}[x(t) + A_d \int_{t-d}^{t} x(s) ds] = (A + A_d) x(t) \tag{2.35}$$

Using second model transformation the delay-dependent stability condition obtained in [21] for the system Σ_1 is presented in the form of following theorem.

Theorem 2.2 ([21]) *The system Σ_1 is asymptotically stable for any delay d_u, if the operator $\mathcal{D}(x_t)$ is stable and there exists symmetric and positive-definite matrices P and Q such that following LMI holds:*

$$\begin{bmatrix} (A+A_d)^T P + P(A+A_d) + d_u Q & d_u(A+A_d)^T P A_d \\ \star & -d_u Q \end{bmatrix} < 0 \quad (2.36)$$

The choice of LK functional candidate for this transformed model (2.35) is of the form (as in [21]),

$$V(t) = \mathcal{D}^T(x_t) P \mathcal{D}(x_t) + \int_{-d_u}^{0} \int_{t+\theta}^{t} x^T(s) Q x(s) ds d\theta \quad (2.37)$$

where, $P = P^T > 0$, $Q = Q^T > 0$ and $\mathcal{D}(x_t) = x(t) + A_d \int_{t-d_u}^{t} x(s)ds$.
Finding the time-derivative of (2.37) one can obtain,

$$\begin{aligned} \dot{V}(t) &= \dot{\mathcal{D}}^T(x_t) P \mathcal{D}(x_t) + \mathcal{D}^T(x_t) P \dot{\mathcal{D}}(x_t) \\ &\quad + d_u x^T(t) Q x(t) - \int_{t-d_u}^{t} x^T(s) Q x(s) ds \end{aligned} \quad (2.38)$$

where,

$$\dot{\mathcal{D}}^T(x_t) = \dot{x}(t) + A_d \int_{t-d_u}^{t} \dot{x}(s) ds$$

Substituting the value of $\dot{\mathcal{D}}^T(x_t)$ and $\mathcal{D}^T(x_t)$ in (2.38) and carrying out algebraic manipulations one can get,

$$\begin{aligned} \dot{V}(t) &= x^T(t)(A+A_d)^T P x(t) + x^T(t) P (A+A_d) x(t) + d_u x^T(t) Q x(t) \\ &\quad + 2\int_{t-d_u}^{t} x^T(t)(A+A_d)^T P A_d x(s) - \int_{t-d_u}^{t} x^T(s) Q s(s) ds \end{aligned} \quad (2.39)$$

Applying bounding Lemma 2.1 on the cross term of (2.39) one can get,

$$\begin{aligned} \dot{V}(t) &\le x^T(t)(A+A_d)^T P x(t) + x^T(t) P (A+A_d) x(t) + d_u x^T(t) Q x(t) \\ &\quad + \int_{t-d_u}^{t} x^T(s) Q s(s) ds - \int_{t-d_u}^{t} x^T(s) Q s(s) ds \\ &\quad + \int_{t-d_u}^{t} x^T(t)(A+A_d)^T P A_d Q^{-1} P A_d (A+A_d) x(t) ds \end{aligned} \quad (2.40)$$

After algebraic simplification and using Schur-complement [22] on (2.40) one can get,

$$\dot{V}(t) \le x^T(t) \begin{bmatrix} (A+A_d)^T P + P(A+A_d) + d_u Q & d_u(A+A_d)^T P A_d \\ \star & -d_u Q \end{bmatrix} x(t) \quad (2.41)$$

The negativity of $\dot{V}(t)$ in (2.41) is not sufficient to guarantee the stability of the transformed system, further it is required to assure the stability of the $\mathcal{D}(x_t)$ also. The stability of $\mathcal{D}(x_t)$ is carried out using frequency domain analysis (refer Remark 14 in [21]), thus this stability analysis yields one more additional sufficient condition on $\mathcal{D}(x_t)$ which is given as,

$$d_u \parallel A_d \parallel < 1$$

Remark 2.4 Applying second transformation on Σ_1 and choosing the LK function in (2.37) for the transformed system, it is found that the derivative of this functional yields only one cross term for system Σ_1 thus it has an advantage of approximating half the number of cross terms using bounding Lemma 2.1 compared to the first model transformation. Whereas the additional constraint introduced to guarantee the stability of $\mathcal{D}(x_t)$ results into conservative estimate of the delay bound. Furthermore, this transformation is not suitable for time-varying delay, as frequency domain stability analysis is adopted for $\mathcal{D}(x_t)$ which is a complicated task for systems with differentiable time-varying delays.

C. Third model transformation

Replacing the value of $x(t-d)$ for Σ_1 and $x(t-d(t))$ for Σ_2 using Newton-Liebniz formula (2.23) into (2.1) and (2.3) respectively, one can get,

$$\dot{x}(t) = (A + A_d)x(t) - A_d \int_{t-d}^{t} \dot{x}(s)ds \tag{2.42}$$

The model expressed in (2.42) is called third model transformation. To derive the sufficient delay-dependent stability condition using (2.42) the choice of Lyapunov-Krasovskii functional is of the following form [10, 23],

$$\begin{aligned} V(t) &= x^T(t)Px(t) + \int_{t-d}^{t} x^T(s)Nx(s)ds \\ &\quad + \int_{-d_u}^{0} \int_{t+\theta}^{t} \dot{x}^T(s)A_d^T M A_d \dot{x}(s)dsd\theta \end{aligned} \tag{2.43}$$

The present author have considered stability analysis for system Σ_2 using third model transformation [6] and bounding Lemma 2.1 by selecting LK functional of the type (2.43) (assumed in [23]) satisfying the condition (2.5) to investigate the conservatism of the different model transformations. Similar results are also available in the literature and the results of delay upper bound for different model transformations are presented in Tables 2.1 and 2.2. The stability condition derived in [6] is presented in the form of following theorem.

Theorem 2.3 ([6]) *If there exist $P = P^T > 0$, $Q_1 = Q_1^T > 0$ and $Q_2 = Q_2^T > 0$, such that the following LMI holds,*

$$\phi = \begin{bmatrix} \phi_{11} & 0 & d_u A^T Q_2 & d_u P A_d \\ 0 & -(1-\mu)Q_1 & d_u A_d^T Q_2 & 0 \\ \star & \star & -d_u Q_2 & 0 \\ \star & 0 & 0 & -d_u Q_2 \end{bmatrix} < 0 \quad (2.44)$$

where, $\phi_{11} = P(A + A_d) + (A + A_d)^T P + Q_1$, then the system Σ_2 is asymptotically stable.

Proof We choose LK functional candidate as

$$V(t) = V_1(t) + V_2(t) \quad (2.45)$$

where,

$$V_1(t) = x^T(t)Px(t)$$
$$V_2(t) = \int_{t-d(t)}^{t} x^T(s)Q_1x(s)ds + \int_{-d_u}^{0}\int_{t+\alpha}^{t} \dot{x}(s)^T Q_2 \dot{x}(s)dsd\alpha$$

Taking time-derivative of (2.45), substituting $\dot{x}(t)$ from (2.42) in $\dot{V}_1(t)$ and approximating the quadratic integral term as,

$$-\int_{t-d_u}^{t} \dot{x}^T(s)Q_2\dot{x}(s)ds \le -\int_{t-d(t)}^{t} \dot{x}^T(s)Q_2\dot{x}(s)ds$$

one can obtain $\dot{V}(t)$ as,

$$\begin{aligned}\dot{V}(t) \le{}& 2x^T(t)P(A + A_d)x(t) - 2x^T(t)PA_d \int_{t-d(t)}^{t} \dot{x}(s)ds \\ &+ x^T(t)Q_1x(t) - (1-\mu)x^T(t-d(t))Q_1x(t-d(t)) + d_u\dot{x}^T(t)Q_2\dot{x}(t) \\ &- \int_{t-d(t)}^{t} \dot{x}^T(s)Q_2\dot{x}(s)ds \end{aligned} \quad (2.46)$$

Applying Lemma 2.1 in (2.46) the cross terms are approximated, and $\dot{x}(t)$ term in $\dot{V}_2(t)$ is substituted with (2.3). Further, algebraic manipulations and use of Schur-complement [24] will lead to,

$$\dot{V}(t) \le \xi^T(t)\phi\xi(t) \quad (2.47)$$

where, $\xi(t)$ is an augmented state vector, i.e., $\xi(t) = \left[\, x^T(t)\; x^T(t-d(t)) \,\right]^T$

Now, to guarantee the asymptotic stability of the system Σ_2, the matrix $\phi < 0$ and ϕ is an LMI defined in (2.44).

Remark 2.5 The third model transformation introduced in [10] and [23] transforms the original system Σ_1 or Σ_2 into system with distributed delay which is equivalent to the original system owing to the fact that, the integration of state dynamics is retained unlike in first model transformation [9]. This transformation is suitable to treat the systems Σ_1 and Σ_2. The additional conservatism arises due to quadratic stability condition formed by considering the state space vector $x(t)$ and $x(t-d)$ independently in the augmented state vector $\zeta(t)$. The substitution of $\dot{x}(t)$ term in (2.46) is carried out using (2.3) and not by (2.42) which is also the source of conservativeness in the stability analysis.

Numerical Example 2.1 ([19]) *Consider the system Σ_1 or Σ_2 with the following constant matrices*

$$A = \begin{bmatrix} -2 & 0 \\ 0 & -0.9 \end{bmatrix}, A_d = \begin{bmatrix} -1 & 0 \\ -1 & -1 \end{bmatrix}$$

The eigenvalues of the matrix $[A + A_d]$ are Hurwitz and the eigenvalues of the matrix $[A - A_d]$ are unstable, thus the given system is delay-dependently stable (i.e., the system is asymptotically stable for certain finite delay value, refer Sect. 1.1.4). The analytical delay upper bound for this system is $d_u = 6.1726$ for $\mu = 0$ [14, 21].

Numerical Example 2.2 ([19]) *Consider the system Σ_2 with the following constant matrices*

$$A = \begin{bmatrix} -6 & 0 \\ 0.2 & -5.8 \end{bmatrix}, A_d = \begin{bmatrix} 0 & 4 \\ -8 & -8 \end{bmatrix}$$

The eigenvalues of the matrix $[A + A_d]$ and $[A - A_d]$ are Hurwitz, the given system is delay-independently stable (i.e., the system is asymptotically stable for arbitrary large delay value, refer Sect. 1.1.4).

Remark 2.6 In [19], it has been shown that the system Σ_1 after being transformed using first model transformation has the characteristic equation of the form $\Delta_t(s) = \Delta_{add}(s)\Delta_{or}(s)$, where $\Delta_{add}(s) = det(I - \frac{1-e^{-ds}}{s} A_d)$, thus indicating that the transformed system contains additional eigenvalues which depends on the delayed matrix and delay size. The presence of these additional eigenvalues makes the stability of the transformed system different from original system when the eigenvalues of the A_d matrix are (i) complex conjugate and (ii) positive real, as in both the cases for a small positive delay the additional eigenvalues will reach the imaginary axis before the original system, but this is not the case for the eigenvalues of A_d matrix being negative real. The degree of conservatism could be better understood by observing the result for Numerical Example 2.2 presented in Table 2.2, as the original system is delay-independently stable but the transformed system is found to be stable up to certain finite value only.

The delay value results obtained using different model transformations and bounding Lemma 2.1 depicts that the third model transformation has the advantage for obtaining better estimate of delay value for different delay derivatives ($0 < \mu < 1$) over the other two transformations.

Table 2.1 Delay upper bound (d_u) results of Example 2.1

Stability methods	$\mu = 0$	$\mu = 0.5$	$\mu = 0.9$	Model Transformation
[16]	0.9999	–	–	First model transformation
[17]	0.9999	–	–	First model transformation
[18]	0.9999	0.6551	0.1743	First model transformation
[21]	0.9999	NA	NA	Second model transformation
[6]	0.9999	0.8210	0.4677	Third model transformation

Table 2.2 Delay upper bound (d_u) results of Example 2.2

Stability methods	$\mu = 0$	$\mu = 0.5$	$\mu = 0.9$	Model Transformation
[16]	0.1514	–	–	First model transformation
[18]	0.1514	0.1016	0.0280	First model transformation
[21]	0.1639	NA	NA	Second model transformation
[6]	0.3891	0.3548	0.2712	Third model transformation

It can be concluded from the above discussion and results (presented in Tables 2.1 and 2.2) that, all the transformed systems (first, second and third) discussed above are not equivalent to the original system Σ_1 and/or Σ_2 as all the transformations possesses additional eigenvalues due to the distributed delayed term and this becomes the main reason for the conservatism in estimating delay bound when the eigenvalues of A_d matrix are present on imaginary or positive real axis. Next, the conservatism in the estimate of delay value due to the adopted bounding Lemma is discussed.

2.3.2 Bounding Techniques

The main purpose of the delay-dependent stability studies of time-delay systems is to find sufficient LMI conditions that can estimate less conservative delay upper bound compared to the existing stability methods using bounding techniques [8]. The stability methods discussed so far utilized bounding Lemma 2.1 for approximating the cross terms arising out of the LK functional derivative. It is validated in [23] that the use of better tighter bounding inequality to represent the cross term arising out of LK functional derivative in the stability analysis can play a key role in reducing conservatism. An improved bounding inequality lemma proposed in [23] is presented below.

Lemma 2.2 (Park's Bounding Lemma [23]) *Assume that* $a(\alpha) \in \mathcal{R}^{n_a}$, *and* $b(\alpha) \in \mathcal{R}^{n_b}$, *are given for* $\alpha \in \Omega$. *Then, for any positive definite matrix* $X \in \mathcal{R}^{n_a \times n_a}$ *and any matrix* $M \in \mathcal{R}^{n_b \times n_b}$, *the following inequality holds*

$$-2\int_{\Omega} b^T(\alpha)a(\alpha)d\alpha \leq \int_{\Omega} \begin{bmatrix} a(\alpha) \\ b(\alpha) \end{bmatrix}^T \begin{bmatrix} X & XM \\ \star & (2,2) \end{bmatrix} \begin{bmatrix} a(\alpha) \\ b(\alpha) \end{bmatrix} \tag{2.48}$$

where, (2, 2)= $(M^T X + I)X^{-1}(XM + I)$.
Then by using this inequality, an improved delay-dependent stability has been reported in [23]. The derived sufficient condition in [23] is restated as:

Theorem 2.4 ([23]) *If there exist $P > 0$, $Q > 0$, $V > 0$ and $W > 0$ then the system defined in Σ_1, satisfying the condition (2.2) is asymptotically stable if the following LMI holds,*

$$\begin{bmatrix} (1,1) & -W^T A_d & A^T A_d^T V & (1,4) \\ \star & Q & A_d^T A_d^T V & 0 \\ \star & \star & -V & 0 \\ \star & 0 & 0 & -V \end{bmatrix} < 0 \tag{2.49}$$

where, (1, 1)=$(A + A_d)^T P + P(A + A_d)$+$W^T A_d + A_d^T W + Q$
(1, 4)= $d_u(W^T + P)$, $W = XMP$, and $V = d_u X$

The stability Theorem 2.4 has been derived using third model transformation. The selection of LK functional is same as (2.45) except that the positive definite matrix corresponding to delay-dependent LK functional term is taken as $A_d^T X A_d$ instead of $Q_2 = Q_2^T > 0$. The cross terms that evolve from LK functional derivative is approximated using bounding Lemma 2.2.

Remark 2.7 The use of bounding Lemma 2.2 in deriving the delay-dependent stability condition in Theorem 2.4 resulted into significant increase in the delay upper bound estimate in an LMI framework. The result obtained for the system considered in Example 2.1 using Theorem 2.4 is $d_u = 4.3588$ for $\mu = 0$ which is a significant improvement in comparison to all the previous delay upper bound results of $d_u = 0.9999$ for $\mu = 0$ and also much closer to actual delay upper bound value of $d_u = 6.172$ for the considered system. This validates the fact that, approximation of cross terms by the bounding technique is one of the major source of conservatism in the delay-dependent stability analysis of TDS.

Another significance of delay-dependent stability Theorem 2.4 is that, for the system considered in Numerical Example 2.2 it could establish that the system is delay-independently stable as the delay upper bound estimate for this system turns out to be arbitrarily large. Thus one can be concluded that, the use of bounding Lemma 2.2 not only enhanced the delay upper bound estimate of delay-dependently stable system, but the delay-dependent stability condition (2.49) derived using this bounding lemma could even establish the delay-independent stability of the system in Numerical Example 2.2, this was not possible using third model transformation and bounding Lemma 2.1 as clear from the result presented in Table 2.2.

The generalization of bounding inequality Lemma 2.1 and Lemma 2.2 was proposed in [25] with an idea to provide a simple LMI structure of delay-dependent stability condition such that it can be easily extended to synthesis problems. The generalized bounding Lemma and delay-dependent stability condition obtained for Σ_1 are discussed below.

Lemma 2.3 (Moon's Bounding Lemma, [25]) *Assume that* $a(\alpha) \in \mathcal{R}^{n_a}$, *and* $b(\alpha) \in \mathcal{R}^{n_b}$, *and* $\mathcal{N}(.) \in \mathcal{R}^{n_a \times n_b}$ *are given for* $\alpha \in \Omega$. *Then, for any positive definite matrix* $X \in \mathcal{R}^{n_a \times n_a}$, $Y \in \mathcal{R}^{n_a \times n_b}$ *and any matrix* $Z \in \mathcal{R}^{n_b \times n_b}$, *the following inequality holds*

$$-2\int_{\Omega} a^T(\alpha)\mathcal{N}b(\alpha)d\alpha \le \int_{\Omega} \begin{bmatrix} a(\alpha) \\ b(\alpha) \end{bmatrix}^T \begin{bmatrix} X & Y-\mathcal{N} \\ \star & Z \end{bmatrix} \begin{bmatrix} a(\alpha) \\ b(\alpha) \end{bmatrix} \quad (2.50)$$

where,

$$\begin{bmatrix} X & Y \\ \star & Z \end{bmatrix} \ge 0$$

Remark 2.8 The bounding Lemma 2.3 is more generalized bounding Lemma and one can obtain bounding Lemma 2.1 and bounding Lemma 2.2 from it with proper selection of matrices Y, Z and $\mathcal{N}$,

Case I Selecting $\mathcal{N} = I$, $Y = I$, and $Z = X^{-1}$ in (2.50) one can get bounding Lemma 2.1.

Case II Selecting $\mathcal{N} = I$, $Y = I + XM$, and $Z = (M^T X + I)X^{-1}(XM + I)$ in (2.50) one can get bounding Lemma 2.2.

The delay-dependent stability condition for the system Σ_1 is derived in [25] using the third model transformation selecting the LK functional as:

$$V(t) = x^T(t)Px(t) + \int_{t-d}^{t} x^T(s)Qx(s)ds + \int_{-d_u}^{0}\int_{t+\alpha}^{t} \dot{x}^T(s)Z\dot{x}(s)dsd\alpha \quad (2.51)$$

Taking the derivative of (2.51) along the trajectory of the transformed system and using the bounding Lemma 2.3, one can get the stability condition in an LMI framework [25]. The delay-dependent stability theorem is restated below.

Theorem 2.5 ([25]) *The system* Σ_1 *is asymptotically stable if there exist symmetric matrices* $P > 0$, $Q > 0$, *matrices* $X > 0$, $Y > 0$ *and any matrix* Z *for time-delay* $d \in [-d_u, 0]$ *such that following LMIs hold,*

$$\begin{bmatrix} A^T P + PA + d_u X + Y + Y^T + Q & -Y + PA_d & d_u A^T Z \\ \star & -Q & d_u A_d^T Z \\ \star & \star & -dZ \end{bmatrix} < 0 \quad (2.52)$$

$$\begin{bmatrix} X & Y \\ \star & Z \end{bmatrix} \ge 0 \quad (2.53)$$

Remark 2.9 The estimate of the delay upper bound result using Theorem 2.5 is found to be $d_u = 4.3588$ for the system considered in Numerical Example 2.1, which is same as the result obtained through Theorem 2.4. The advantage of using bounding Lemma 2.3 in comparison to the use bounding Lemma 2.2 in deriving stability condition is that, the former bounding lemma results into simple LMI structure. This can easily be extended for solution of stabilization and robust stabilization problems.

From the above discussions, it is now clear that the conservatism in the delay-dependent stability analysis are due to (i) the presence of distributed delay term in the model transformation which in turn introduces additional dynamics into the transformed system and (ii) the use of bounding inequalities to approximate the cross terms. In an attempt to reduce the conservatism arising out of the model transformations discussed in Sect. 2.3.1 a new model transformation called descriptor system approach was introduced in [26].

2.3.3 Descriptor System Approach

This section discusses briefly the development and further modification of this method which are available in literature [9, 27–29]. In [26] and [9] the delay-dependent study was done on system Σ_1 (constant time-delay), whereas the method was extended to system Σ_2 (time-varying delays) in [28] and [9].

The model transformation of the original system into descriptor system with distributed delay for the system Σ_1 in (2.1) is discussed briefly below:

$$\begin{aligned} \dot{x}(t) &= y(t) \\ y(t) &= Ax(t) + A_d x(t-d) \end{aligned} \tag{2.54}$$

using Newton-Leibniz formula (2.23), the above equation (2.54) can be rewritten as

$$\begin{aligned} \dot{x}(t) &= y(t) \\ y(t) &= (A + A_d)x(t) - A_d \int_{t-d}^{t} y(s)ds \\ 0 &= -y(t) + (A + A_d)x(t) - A_d \int_{t-d}^{t} y(s)ds \end{aligned}$$

$$E\dot{\xi}(t) = \tilde{A}\xi(t) + \tilde{A}_d \int_{t-d}^{t} y(s)ds \tag{2.55}$$

where, $\xi(t) = \begin{bmatrix} x(t) \\ y(t) \end{bmatrix}$, $E = \begin{bmatrix} I & 0 \\ 0 & 0 \end{bmatrix}$ $\tilde{A} = \begin{bmatrix} 0 & I \\ A + A_d & -I \end{bmatrix}$ and $\tilde{A}_d = \begin{bmatrix} 0 \\ -A_d \end{bmatrix}$

The delay-dependent stability conditions for the transformed system (2.55) in [26] has been derived selecting the LK functional candidate as

$$V(t) = \xi^T(t)EP\xi(t) + \int_{-d_u}^{0}\int_{t+\theta}^{t} y^T(s)Ry(s)dsd\theta \tag{2.56}$$

where, $P = \begin{bmatrix} P_1 & 0 \\ P_2 & P_3 \end{bmatrix} \geq 0$ with $P_1 = P_1^T > 0$.

One can find from (2.56) that the LK functional corresponding to the delay-independent term (single integral term) is not present. Finding the time-derivative of the (2.56) and applying the bounding Lemma 2.1 to approximate the cross terms arising out of the LK functional derivative, one can obtain following LMI stability condition for the descriptor system (2.55) as

$$\begin{bmatrix} P_2^T(A+A_d)+(A+A_d)^T P_2 & P_1 - P_2^T + (A+A_d)^T P_3 & d_u P_2^T A_d \\ \star & -P_3 - P_3^T + d_u R & d_u P_3^T A_d \\ \star & \star & -d_u R \end{bmatrix} < 0 \tag{2.57}$$

In [9] the stability condition of the descriptor system (2.55) for time-varying delay satisfying conditions (2.5) and (2.7) was derived. The selection of the LK functional in this case is considered as,

$$V(t) = \xi^T(t)EP\xi(t) + \int_{t-d(t)}^{t} x^T(s)Sx(s)ds + \int_{-d_u}^{0}\int_{t+\theta}^{t} y^T(s)A_d^T RA_d y(s)ds \tag{2.58}$$

where, $\xi(t)$ and P are defined earlier.

The time-derivative of the (2.58) results into cross terms which is approximated using bounding Lemma 2.2. The introduction of functional corresponding to delay-independent term (single integral term) is for the application of bounding lemma. The resulting LMI stability condition is stated below in the form of theorem.

Theorem 2.6 ([9]) *The time-delay system Σ_2 satisfying (2.7) is asymptotically stable for any delay $d(t) \in [-d_u, 0]$ if there exist $P_1 = P_1^T > 0$, P_2, P_3, $R = R^T > 0$, $S = S^T > 0$, W_1 and W_2, such that the following LMI holds,*

$$\begin{bmatrix} (1,1) & (1,2) & d_u(W_1^T + P_2^T) & W_1^T A_d \\ \star & (2,2) & d_u(W_2^T + P_3^T) & W_2^T A_d \\ \star & \star & -d_u R & 0 \\ \star & \star & 0 & -S(1-\mu) \end{bmatrix} < 0 \tag{2.59}$$

where, $(1,1)$=$P_2^T(A+A_d)+(A+A_d)^T P_2 + A_d^T W_1 + W_1 A_d + S$,
$(1,2)$=$P_1 - P_2^T + (A+A_d)^T P_3 + W_2^T A_d$, $(2,2)$=$-P_3 - P_3^T + d_u A_d^T R A_d$,
W_1=RMP_2, and W_2=RMP_3

A modified version of Theorem 2.6 can be found in [28] by introducing the following modifications, (i) cross bounding Lemma 2.3 was used instead of Lemma 2.2, as this lemma leads to simple structure of LMI condition and (ii) the positive definite matrix associated with delay-dependent term in (2.58) is replaced by a symmetric positive-definite matrix R such that it is suitable for the use of bounding Lemma 2.3. The stability conditions derived is restated in the form of theorem.

Theorem 2.7 ([28]) *The time-delay system Σ_2 satisfying (2.7) is asymptotically stable for any delay $d(t) \in [-d_u, 0]$ if there exist $P_1 > 0$, P_2, P_3, $R = R^T > 0$, $S = S^T > 0$, Y_1, Y_2, Z_1, Z_2 and Z_3 such that the following LMIs hold:*

$$\begin{bmatrix} (1,1) & (1,2) & P_2^T A_d - Y_1^T \\ \star & -P_3 - P_3^T + d_u Z_3 + d_u R & P_3^T A_d - Y_2^T \\ \star & \star & -S(1-\mu) \end{bmatrix} < 0 \tag{2.60}$$

$$\begin{bmatrix} R & Y_1 & Y_2 \\ \star & Z_1 & Z_2 \\ \star & \star & Z_3 \end{bmatrix} \geq 0 \tag{2.61}$$

where, $(1,1) = P_2^T A + A^T P_2 + Y_1 + Y_1^T + S + d_u Z_1$, $(1,2) = P_1 - P_2^T + A^T P_3 + Y_2 + d_u Z_2$

Remark 2.10 Comparing the stability conditions in (2.59) and (2.60)–(2.61) it can be observed that (i) the dimension of main LMI in (2.60) is less compared to that in (2.59) and (ii) the structure of LMI is simpler in terms of product of the Lyapunov matrix variables with the system matrices. Due to these reasons, the extension of stability conditions in (2.60) could be extended easily for synthesis problem in [28].

Remark 2.11 The stability condition in (2.57) (derived using descriptor system and bounding Lemma 2.1) when tested on Example 2.1 gives delay upper bound of $d_u = 0.9999$ which is same as the result obtained using the condition derived in [6] (using third model transformation and bounding Lemma 2.1). But the stability conditions in Theorem 2.6 and Theorem 2.7 when tested on Example 2.1 yielded delay upper bound estimate of $d_u = 4.47$ which is less conservative result compared to the delay value of $d_u = 4.3588$ obtained in [23].

Thus from above discussions and results it can be concluded that, a delay-dependent stability conditions using LK functional requires (i) appropriate choice of LK functional [26] followed by appropriate bounding technique to bind the cross terms arising out of the LK functional derivative, in order to reduce the conservatism in the delay bound results.

The advantage of descriptor method is that, it can be easily extended for state feedback synthesis of controller as in [28] and [27], but the observation reveals that the dimension and structure of the LMI conditions are larger and complicated respectively due to the use of descriptor system instead of original system.

Recently in [29] a new generalized delay-dependent stability condition has been proposed for neutral time-delay system using Finsler's Lemma and the bounding

Lemma 2.3. First we present Finsler's Lemma and then the stability condition for time-delay systems Σ_2 using the method proposed in [29] is presented.

Lemma 2.4 (Finsler's Lemma [24, 29]) *The following statements hold* $x^T Qx + f(x) < 0, \forall \bar{B}x = 0, x \neq 0$, *where* $Q = Q^T$, $\bar{B} \in \mathcal{R}^{m\times n}$ *(such that rank* $(\bar{B}) = m < n$ *and* $f(x)$ *is a scalar function, if there exists matrix* $X \in \mathcal{R}^{n\times m}$, *such that*

$$x^T[Q + X\bar{B} + \bar{B}^T X^T]x + f(x) < 0, \forall x \neq 0$$

Theorem 2.8 ([29]) *System* Σ_2 *satisfying (2.7) is asymptotically stable for the delay* $d(t) \in [-d_u, 0]$ *if there exist* $P_1 = P_1^T > 0$, $S > 0$, $P_i, i = 2, 3, 4, Y_1, Y_2, Z_1, Z_2, Z_3$ *and* $R > 0$ *such that following LMIs hold:*

$$\begin{bmatrix} (1,1) & (1,2) & (1,3) \\ \star & (2,2) & (2,3) \\ \star & \star & (3,3) \end{bmatrix} < 0 \tag{2.62}$$

$$\begin{bmatrix} R & Y_1 & Y_2 \\ \star & Z_1 & Z_2 \\ \star & \star & Z_3 \end{bmatrix} \geq 0 \tag{2.63}$$

where,(1, 1)=$P_2^T A + A^T P_2 + S + Y_1 + Y_1^T + d_u Z_1$,

(1, 2)=$P_1 - P_2^T + A^T P_3 + Y_2 + d_u Z_2$

(1, 3)=$A^T P_4 - Y_1^T + P_2^T A_d$, (2, 2)=$-P_3 - P_3^T + d_u R + d_u Z_3$

(2, 3)=$-P_4 - Y_2^T + P_3^T A_d$, and (3, 3)=$-(1-\mu)S + A_d^T P_4 + P_4^T A_d$

Proof The Lyapunov-Krasovskii functional candidate chosen is

$$V(t) = V_1(t) + V_2(t) + V_3(t) \tag{2.64}$$

where, $V_1(t) = x^T(t)P_1x(t)$, $V_2(t) = \int_{-d_u}^{0} \int_{t+\theta}^{t} \dot{x}^T(s)R\dot{x}(s)dsd\theta$, $V_3(t) = \int_{t-d(t)}^{t} x^T(s)Sx(s)ds$

Time-derivative of the (2.64) is

$$\dot{V}(t) = \dot{V}_1(t) + \dot{V}_2(t) + \dot{V}_3(t) \tag{2.65}$$

$$\dot{V}_1(t) = \begin{bmatrix} x^T(t) & \dot{x}^T(t) \end{bmatrix} \begin{bmatrix} 0 & P_1 \\ P_1 & 0 \end{bmatrix} \begin{bmatrix} x(t) \\ \dot{x}(t) \end{bmatrix} \tag{2.66}$$

$$\dot{V}_2(t) \leq d_u \dot{x}^T(t) R \dot{x}(t) - \int_{t-d_u}^{t} \dot{x}^T(s) R \dot{x}(s) ds \tag{2.67}$$

$$\dot{V}_3(t) \leq x^T(t) S x(t) - (1-\mu) x^T(t-d(t)) S x(t-d(t)) \tag{2.68}$$

Defining, augmented state space vector as

$$\xi(t) \triangleq \begin{bmatrix} x(t) \\ \dot{x}(t) \\ x(t-d(t)) \end{bmatrix}$$

In terms of $\xi(t)$ we can express (2.65) as

$$\dot{V}(t) = \xi^T(t) \begin{bmatrix} 0 & P_1 & 0 \\ P_1 & 0 & 0 \\ 0 & 0 & 0 \end{bmatrix} \xi(t) + \dot{V}_2(t) + \dot{V}_3(t) < 0 \tag{2.69}$$

for $\forall\, \xi(t) \in \mathcal{R}^{3n}$ s.t. $[A, -I, A_d]\, \xi(t) = 0$, this is obtained from (2.3).

Applying Lemma 2.4 in (2.69) yields

$$\begin{aligned} 0 > \xi^T(t) \Bigg\{ & \begin{bmatrix} 0 & P_1 & 0 \\ P_1 & 0 & 0 \\ 0 & 0 & 0 \end{bmatrix} + \begin{bmatrix} P_2 & P_3 & P_4 \end{bmatrix}^T \begin{bmatrix} A & -I & A_d \end{bmatrix} \\ & \begin{bmatrix} A & -I & A_d \end{bmatrix}^T \begin{bmatrix} P_2 & P_3 & P_4 \end{bmatrix} \Bigg\} \xi(t) + \dot{V}_2(t) + \dot{V}_3(t) \end{aligned} \tag{2.70}$$

Substituting the values of $\dot{V}_2(t)$ and $\dot{V}_3(t)$ from (2.67) and (2.68) respectively into (2.70) one can write

$$\begin{aligned} 0 > \xi^T(t) \Bigg\{ & \begin{bmatrix} P_2^T A + A^T P_2 + S & P_1 - P_2^T + A^T P_3 & P_2^T A_d + A^T P_4 \\ \star & -P_3^T - P_3 + d_u R & P_3^T A_d - P_4 \\ \star & \star & -(1-\mu)S + P_4^T A_d + A_d^T P_4 \end{bmatrix} \\ & + \begin{bmatrix} 0 & 0 & P_2^T A_d \\ 0 & 0 & P_3^T A_d \\ \star & \star & 0 \end{bmatrix} \Bigg\} \xi(t) - \int_{t-d_u}^{t} \dot{x}^T(s) R \dot{x}(s) ds \end{aligned} \tag{2.71}$$

Define the last two terms of (2.71) as,

$$\mu(t) \triangleq \xi^T(t) \begin{bmatrix} 0 & 0 & P_2^T A_d \\ 0 & 0 & P_3^T A_d \\ \star & \star & 0 \end{bmatrix} \xi(t) - \int_{t-d_u}^{t} \dot{x}^T(s) R \dot{x}(s) ds$$

and carrying out the algebraic manipulations with Lemma 2.3,

$$\mu(t) = \xi^T(t)\left[P_2\ P_3\ 0\right]^T\left[0\ 0\ A_d\right]\left[x^T(t)\ \dot{x}^T(t)\ x^T(t-d(t))\right]^T + \left[x^T(t)\ \dot{x}^T(t)\ x^T(t-d(t))\right]\left[0\ 0\ A_d\right]^T\left[P_2\ P_3\ 0\right]\xi(t) - \int_{t-d_u}^{t}\dot{x}^T(s)R\dot{x}(s)ds$$

$$\mu(t) = \xi^T(t)\left[P_2\ P_3\ 0\right]^T A_d x(t-d(t)) + x^T(t-d(t))A_d^T\left[P_2\ P_3\ 0\right]\xi(t) - \int_{t-d_u}^{t}\dot{x}^T(s)R\dot{x}(s)ds \tag{2.72}$$

Using Newton-Leibniz formula defined in (2.23) on (2.72), one can obtain

$$\mu(t) = 2\xi^T(t)\left[P_2\ P_3\ 0\right]^T A_d x(t) - 2\int_{t-d(t)}^{t}\xi^T(t)\left[P_2\ P_3\ 0\right]^T A_d\dot{x}(s)ds - \int_{t-d_u}^{t}\dot{x}^T(s)R\dot{x}(s)ds \tag{2.73}$$

Applying bounding Lemma 2.3 on (2.73) one can get,

$$\mu(t) \le 2x^T(t)Y\begin{bmatrix} x(t) \\ \dot{x}(t) \end{bmatrix} + d_u\left[x^T(t)\ \dot{x}^T(t)\right]Z\begin{bmatrix} x(t) \\ \dot{x}(t) \end{bmatrix} - 2x^T(t-d(t))Y\begin{bmatrix} x(t) \\ \dot{x}(t) \end{bmatrix} + 2x^T(t-d(t))A_d^T\left[P_2\ P_3\right]\begin{bmatrix} x(t) \\ \dot{x}(t) \end{bmatrix} \tag{2.74}$$

where, $Y = \left[Y_1\ Y_2\right]$ and $Z = \begin{bmatrix} Z_1 & Z_2 \\ \star & Z_3 \end{bmatrix}$.
Substituting $\mu(t)$ from (2.74) into (2.71) and carrying out further algebraic manipulations, one can easily obtain

$$\dot{V}(t) \le \xi^T(t)\Xi\xi(t) \tag{2.75}$$

where matrix Ξ is defined in (2.62). For asymptotic stability of the system Σ_2, the matrix $\Xi < 0$, an additional LMI (2.63) appears in the stability formulation due to adoption of bounding Lemma 2.3.

Remark 2.12 One can observe that the LMI structure in (2.62) and (2.60) are similar except that in the former condition few additional terms consisting of a free matrix P_4 are involved. Now, if one chooses $P_4 = 0$ in (2.62) then one can obtain the condition (2.60) and thus the condition (2.62) is a generalized one. This generalization of LMI was possible due to the use of Finsler's lemma, but in turn introduces additional free matrix variable (in this case P_4).

When stability conditions in (2.62)–(2.63) are tested on Numerical Example 2.2 the delay upper bound obtained is $d_u = 4.4721$ which is same as the result obtained with Theorem 2.7, thus indicating that the use of additional matrix variable P_4 is redundant here.

2.3.4 Free Weighting Matrix Approach

Recently it was pointed out in [13] and [30] that, the derivative of the LK functional that contained $x(t - d(t))$ term was replaced with $x(t) - \int_{t-d_u}^{t} \dot{x}(s)ds$ (due to third model transformation in [23] and [25]) for obtaining the quadratic stability condition whereas the term $d_u \dot{x}^T(t)Z\dot{x}(t)$ in the LK derivative was not replaced by Newton-Leibniz formula rather $\dot{x}(t)$ was substituted with (2.3) (as in [25]), thus the replacements are not done uniformly everywhere in the formulation. In free weighting matrix method, the term $\dot{x}(t)$ is treated as one of the state in augmented state space vector and the relationship among the terms $x(t)$, $x(t - d(t))$ and $\dot{x}(t)$ are expressed using Newton-Leibniz formula by introducing some free matrices. Free weighting matrix method proposed in [30] plays an important role in deriving delay-dependent stability conditions that is restated below.

Theorem 2.9 ([30]) *The system Σ_2 satisfying the conditions (2.5) and (2.7) is asymptotically stable for any delay $d(t) \in [-d_u, 0]$, if there exist $P = P^T > 0$, $Q = Q^T > 0$ and $Z = Z^T > 0$ along with appropriately dimensioned matrices N_i and T_i for $(i = 1, 2, 3)$ such that following LMIs hold:*

$$\begin{bmatrix} (1,1) & (1,2) & (1,3) & d_u N_1 \\ \star & (2,2) & (2,3) & d_u N_2 \\ \star & \star & (3,3) & d_u N_3 \\ \star & \star & \star & -d_u Z \end{bmatrix} < 0 \tag{2.76}$$

where, $(1,1) = -T_1 A - A^T T_1^T + N_1 + N_1^T + Q$, $(1,2) = -N_1 + N_2^T - A^T T_2^T - T_1 A_d$

$(1,3) = P + N_3^T + T_1 - A^T T_3^T$, $(2,2) = -N_2 - N_2^T - (1-\mu)Q - T_2 A_d - A_d^T T_2^T$

$(2,3) = T_2 - N_3^T - A_d^T T_3^T$, and $(3,3) = d_u Z + T_3 + T_3^T$

Proof The LK functional candidate chosen is

$$V(t) = x^T(t)Px(t) + \int_{t-d(t)}^{t} x^T(s)Qx(s)ds + \int_{-d_u}^{0}\int_{t+\theta}^{t} \dot{x}^T(s)Z\dot{x}(s)dsd\theta \tag{2.77}$$

One can write the time-derivative of (2.77) as

$$\dot{V}(t) \leq 2x^T(t)P\dot{x}(t) + x^T(t)Qx(t) - (1-\mu)x^T(t-d(t))Qx(t-d(t))$$
$$+d_u\dot{x}^T(t)Z\dot{x}(t) - \int_{t-d(t)}^{t} \dot{x}^T(s)Z\dot{x}(s)ds \tag{2.78}$$

The inequality sign in (2.78) is due to fact that, $\dot{d}(t)$ has been approximated as μ and the integral $-\int_{t-d_u}^{t}(\dot{x}^T(s)Z\dot{x}(s)ds)$ arising out of the derivative of double integral term in (2.77) is approximated as $-\int_{t-d_u}^{t}(\dot{x}^T(s)Z\dot{x}(s)ds) \leq -\int_{t-d(t)}^{t}(\dot{x}^T(s)$ $Z\dot{x}(s)ds)$.

Earlier methods that are based on model transformations, replace $\dot{x}(t)$ in LK derivative by (2.3), whereas in this method an appropriately dimensioned free weighting matrices N_i for $i = 1, 2, 3$ have been introduced to express the relationship between the terms $x(t)$, $x(t-d(t))$, and $\dot{x}(t)$ using Newton-Leibniz formula as shown below,

$$0 = 2[x^T(t)N_1 + x^T(t-d(t))N_2 + \dot{x}^T(t)N_3]$$
$$\times\left[x(t) - x(t-d(t)) - \int_{t-d(t)}^{t}\dot{x}(s)ds\right] \tag{2.79}$$

Another set of free weighting matrices T_i for $i = 1, 2, 3$ are introduced using the following relation,

$$0 = 2[x^T(t)T_1 + x^T(t-d(t))T_2 + \dot{x}^T(t)T_3]$$
$$\times[\dot{x}(t) - Ax(t) - A_dx(t-d(t))] \tag{2.80}$$

A semi-positive definite matrix $X = \begin{bmatrix} X_{11} & X_{12} & X_{13} \\ \star & X_{22} & X_{23} \\ \star & \star & X_{33} \end{bmatrix} \geq 0$ is introduced and the following holds

$$d_u\xi^T(t)X\xi(t) - \int_{t-d(t)}^{t}\xi^T(t)X\xi(t)ds \geq 0 \tag{2.81}$$

where, $\xi(t) = \left[x^T(t)\ x^T(t-d(t)\ \dot{x}^T(t)\right]^T$.

Adding the terms (2.79)–(2.81) into $\dot{V}(t)$, one can express $\dot{V}(t)$ as

$$\dot{V}(t) \leq \xi^T(t)\Upsilon\xi(t) - \int_{t-d(t)}^{t}\eta^T(t,s)\Theta\eta(t,s)ds \tag{2.82}$$

where, $\eta(t,s)=\left[\xi^T(t)\ \dot{x}^T(s)\right]^T$

$$\Upsilon = \begin{bmatrix} (1,1) + d_uX_{11} & (1,2) + d_uX_{12} & (1,3) + d_uX_{13} \\ \star & (2,2) + d_uX_{22} & (2,3) + d_uX_{23} \\ \star & \star & (3,3) + d_uX_{33} \end{bmatrix}$$

$$\Theta = \begin{bmatrix} X_{11} & X_{12} & X_{13} & N_1 \\ \star & X_{22} & X_{23} & N_2 \\ \star & \star & X_{33} & N_3 \\ \star & \star & \star & Z \end{bmatrix}$$

Selection of $Z > 0$ and $X = \begin{bmatrix} N_1 \\ N_2 \\ N_3 \end{bmatrix} Z^{-1} \begin{bmatrix} N_1 \\ N_2 \\ N_3 \end{bmatrix}^T$ ensures that $X \geq 0$ and $\Theta \geq 0$.

Thus, one can write $\Upsilon + d_u \begin{bmatrix} N_1 \\ N_2 \\ N_3 \end{bmatrix} Z^{-1} \begin{bmatrix} N_1 \\ N_2 \\ N_3 \end{bmatrix}^T$ using Schur-complement equivalent to LMI (2.76).

The asymptotic stability of the system Σ_2 is guaranteed if the LMI in (2.76) is negative definite.

Remark 2.13 In the Theorem 2.9 as $\dot{x}(t)$ is retained in the formulation of stability condition so (2.80) is introduced such that system matrices A and A_d appear in the stability condition. In [13] similar kind of stability condition as in [30] is proposed, $\dot{x}(t)$ which appears in the derivative of LK functional is now replaced by the RHS of (2.3). The stability condition in [13] is presented next in the form of following theorem.

Theorem 2.10 ([13]) *The system Σ_2 is asymptotically stable for any delay $d(t) \in [-d_u, 0]$, if there exist $P = P^T > 0$, $Q = Q^T > 0$ and $Z = Z^T > 0$, a symmetric semi-positive-definite matrix $X = \begin{bmatrix} X_{11} & X_{12} \\ \star & X_{22} \end{bmatrix} \geq 0$, and appropriately dimensioned matrices Y and T such that following LMIs hold:*

$$\begin{bmatrix} (1,1) & (1,2) & d_u A^T Z \\ \star & (2,2) & d_u A_d^T Z \\ \star & \star & -d_u Z \end{bmatrix} < 0 \tag{2.83}$$

$$\begin{bmatrix} X_{11} & X_{12} & Y \\ \star & X_{22} & T \\ \star & \star & Z \end{bmatrix} \geq 0 \tag{2.84}$$

where, $(1,1) = PA + A^T P + Y + Y^T + Q + d_u X_{11}$, $(1,2) = PA_d - Y + Y^T + d_u X_{12}$

$(2,2) = -T - T^T - (1-\mu)Q + d_u X_{22}$

This theorem can be proved in a similar manner as in Theorem 2.9, except that (2.80) need not to be considered now as $\dot{x}(t)$ is substituted by RHS of (2.3).

The present author has recently investigated a stability condition for system Σ_2 using free weighting matrix approach and introduced the following modifications

(i) without using inequality (2.81) (i.e., avoiding the use of semi-positive definite matrix in the formulation) and (ii) without retaining the $\dot{x}(t)$ term and hence not using condition (2.80) over Theorems 2.9 and 2.10 in [31]. The theorem is stated below.

Theorem 2.11 ([31]) *The system Σ_2 is asymptotically stable for any delay $d(t) \in [-d_u, 0]$, satisfying the conditions (2.7) if there exist $P = P^T > 0$, $Q_1 = Q_1^T > 0$ and $Q_2 = Q_2^T > 0$, with appropriately dimensioned free matrices $T_i (i = 1, 2)$ such that following LMIs holds:*

$$\Omega = \begin{bmatrix} (1,1) & (1,2) & T_1 \\ \star & (2,2) & T_2 \\ \star & \star & -d_u^{-1} Q_2 \end{bmatrix} < 0 \tag{2.85}$$

where, $(1,1)=d_u A^T Q_2 A + A^T P + PA + T_1 + T_1^T + Q_1$, $(1,2)=PA_d + d_u A^T Q_2 A_d - T_1 + T_2^T$
$(2,2)=d_u A_d^T Q_2 A_d - T_2 - T_2^T - (1-\mu)Q_1$

Proof Consider LK functional candidate chosen as

$$V(t) = x^T(t)Px(t) + \int_{t-d(t)}^{t} x^T(s)Q_1 x(s)ds + \int_{-d_u}^{0}\int_{t+\theta}^{t} \dot{x}^T(s)Q_2\dot{x}(s)dsd\theta \tag{2.86}$$

One can write the time-derivative of (2.86) as

$$\begin{aligned} \dot{V}(t) \le\; & 2x^T(t)P\dot{x}(t) + x^T(t)Q_1x(t) - (1-\mu)x^T(t-d(t))Q_1x(t-d(t)) \\ & + d_u\dot{x}^T(t)Q_2\dot{x}(t) - \int_{t-d(t)}^{t} \dot{x}^T(s)Q_2\dot{x}(s)ds \end{aligned} \tag{2.87}$$

For delay-dependent condition, one can use the following expression based on Newton-Leibniz formula in the derivative of LK functional.

$$\begin{aligned} 0 = & 2\begin{bmatrix} x^T(t) & x^T(t-d(t)) \end{bmatrix}\begin{bmatrix} T_1 \\ T_2 \end{bmatrix} \\ & \times \begin{bmatrix} x(t) - x(t-d(t)) - \int_{t-d(t)}^{t} \dot{x}(s)ds \end{bmatrix} \end{aligned} \tag{2.88}$$

where, T_1 and T_2 are free matrices. Expanding (2.88), one can get

$$\xi^T(t)\begin{bmatrix} T_1 + T_1^T & -T_1 + T_2^T \\ \star & -T_2 - T_2^T \end{bmatrix}\xi(t) - \int_{t-d(t)}^{t} 2\xi^T(t)\begin{bmatrix} T_1 \\ T_2 \end{bmatrix}\dot{x}(s)ds = 0 \tag{2.89}$$

where, $\xi(t) = \begin{bmatrix} x^T(t) & x^T(t-d(t)) \end{bmatrix}^T$. Applying bounding Lemma 2.1 in the last term of (2.89) one can obtain

$$-\int_{t-d(t)}^{t} \dot{x}^T(s) Q_2 \dot{x}(s) ds \leq \xi^T(t) \begin{bmatrix} T_1 + T_1^T & -T_1 + T_2^T \\ \star & -T_2 - T_2^T \end{bmatrix} \xi(t)$$
$$+\xi^T(t) d_u \begin{bmatrix} T_1 \\ T_2 \end{bmatrix} Q_2^{-1} \begin{bmatrix} T_1 \\ T_2 \end{bmatrix}^T \xi(t) \qquad (2.90)$$

Substituting the value of $\dot{x}(t) = Ax(t) + A_d x(t - d(t))$ and RHS of (2.90) in (2.87) one can obtain

$$\dot{V}(t) \leq \xi^T(t) \Omega \xi(t) \qquad (2.91)$$

where, matrix Ω is defined in (2.85), if $\Omega < 0$ then it ensures the asymptotic stability of the system under consideration.

This stability analysis can be extended for systems with delay-derivative ($\mu > 1$) i.e., fast time-varying delay satisfying the condition (2.8), which is stated in the following corollary.

Corollary 2.1 *For* $\mu > 1$, *the system* Σ_2 *is asymptotically stable if there exist matrices* $P = P^T > 0$, $Q_2 = Q_2^T > 0$, *any free matrices* T_1 *and* T_2 *of appropriate dimensions, such that the following LMI holds:*

$$\begin{bmatrix} (1,1) & (1,2) & T_1 \\ \star & (2,2) & T_2 \\ \star & \star & -d_u^{-1} Q_2 \end{bmatrix} < 0 \qquad (2.92)$$

where,$(1,1)$=$d_u A^T Q_2 A + A^T P + PA + T_1 + T_1^T$, $(1,2)$=$PA_d + d_u A^T Q_2 A_d - T_1 + T_2^T$
$(1,3)$=$d_u A_d^T Q_2 A_d - T_2 - T_2^T$

Proof The proof of this corollary is straight forward following the proof of Theorem 2.11, the stability result is obtained by considering $Q_1 = 0$ in (2.86), this assumption makes the Lyapunov functional candidate corresponding to delay-independent term zero.

Remark 2.14 Theorem 2.11 provides a generalized framework for stability analysis as it can treat systems Σ_1 (constant delay case), Σ_2 (time-varying delay case) for both the types of time-varying delays-slow ($\mu < 1$) and fast ($\mu > 1$). Advantage of the stability condition obtained in Theorem 2.11 compared to Theorem 2.10 (condition (2.83)–(2.84)) are (i) it consists of lesser matrix variables and (ii) lesser number of LMIs need to be solved, whereas compared to Theorem 2.9 the LMI dimension in (2.85) is smaller. Furthermore, Theorem 2.11 has also been extended for system Σ_3 (system with multiple state delays) in [11].

A stability condition has been proposed recently in [14] for the system Σ_2 satisfying condition (2.7). The stability condition is derived by using (i) augmented

Lyapunov-Krasovskii functional candidate, (ii) Jensen's integral inequality (for eliminating the integral terms arising out of the derivative of LK functional) and (iii) free weighting matrices by utilizing Newton-Leibniz formula.

The delay-dependent stability condition of [14] is presented after stating the Jensens integral inequality Lemma [7] as it is significant in establishing this stability condition.

Lemma 2.5 (Jensens inequality [19]) *For any symmetric positive definite matrix* $M > 0$*, scalar* $\gamma > 0$ *and vector function* $\omega : [0, \gamma] \rightarrow R^n$ *such that the integrations concerned are well defined, the following inequality holds:*

$$\left(\int_0^{\gamma} \omega(s)ds\right)^T M \left(\int_0^{\gamma} \omega(s)ds\right) \leq \gamma \left(\int_0^{\gamma} \omega(s)^T M\omega(s)ds\right) \tag{2.93}$$

Theorem 2.12 ([14]) *The system* Σ_2 *is asymptotically stable for any time-delay* $d(t) \in [-d_u, 0]$ *satisfying (2.7), if there exist symmetric positive definite matrices,* P, Q, R, T *and any matrices* $S_i (i = 1, \ldots 4)$ *with the appropriate dimensions satisfying following LMIs:*

$$P = \begin{bmatrix} P_{11} & P_{12} \\ \star & P_{22} \end{bmatrix} \geq 0 \quad with\ P_{11} > 0 \tag{2.94}$$

$$Q = \begin{bmatrix} Q_{11} & Q_{12} \\ \star & Q_{22} \end{bmatrix} \geq 0 \tag{2.95}$$

$$\Gamma = \begin{bmatrix} \Gamma_{11} & \Gamma_{12} & \Gamma_{13} & \Gamma_{14} & \mu P_{12} \\ \star & \Gamma_{22} & \Gamma_{23} & \Gamma_{24} & 0 \\ \star & \star & -Q_{11} & \Gamma_{34} & \mu P_{22} \\ \star & \star & \star & \Gamma_{44} & 0 \\ \star & \star & \star & \star & -\mu T \end{bmatrix} < 0 \tag{2.96}$$

where, $\Gamma_{11}= A^T P_{11} + P_{11}A + d_u^2(Q_{11} + A^T Q_{12}^T + Q_{12}A + A^T Q_{22}A) + R + S_1 + S_1^T$
$\Gamma_{12}=P_{11}A_d - S_1^T + S_2 + d_u^2(Q_{12}A_d + A^T Q_{22}A_d)$, $\Gamma_{13}=A^T P_{12} + S_3$
$\Gamma_{14}=P_{12} - S_1^T + S_4$, $\Gamma_{22}=-(1-\mu)R + \mu T + d_u^2 A_d^T Q_{22}A_d - S_2^T - S_2$
$\Gamma_{23}=A_d^T P_{12} - S_3$, $\Gamma_{24}=-S_2^T - S_4$, $\Gamma_{34}=P_{22} - Q_{12} - S_3^T$ and
$\Gamma_{44}=-Q_{22} - S_4^T - S_4$

Proof The augmented LK functional candidate chosen here is

$$V(t) = V_1(t) + V_2(t) + V_3(t) \tag{2.97}$$

where, $V_1(t)=\eta^T(t)P\eta(t)$, $V_2(t)=d_u\int_{-d_u}^{0}\int_{t+\theta}^{t}\xi^T(s)Q\xi(s)dsd\theta$, $V_3(t)=\int_{t-d(t)}^{t}x^T(s)Rx(s)ds$
$\eta(t)=\begin{bmatrix} x^T(t) & \left(\int_{t-d(t)}^{t}x(s)ds\right)^T \end{bmatrix}^T$, $\xi(s)=\begin{bmatrix} x^T(s) & \dot{x}^T(s) \end{bmatrix}^T$

Finding time-derivative of (2.97), one can get the following

$$\dot{V}_1(t) = 2\eta^T(t)P\dot{\eta}(t)$$

$$\dot{V}_1(t) = 2\eta^T(t)P[\eta_1(t) + \dot{d}(t)\eta_2(t)] \tag{2.98}$$

where, $\dot{\eta}(t) = \begin{bmatrix} \dot{x}^T(t) & \left(\int_{t-d(t)}^{t}\dot{x}(s)ds\right)^T \end{bmatrix}^T$, $\eta_1(t)=\begin{bmatrix} Ax(t) + A_d x(t-d(t)) \\ \int_{t-d(t)}^{t}\dot{x}(s)ds \end{bmatrix}$ and
$\eta_2(t)=\begin{bmatrix} 0 \\ I \end{bmatrix} x(t-d(t))$

Defining the augmented vector as

$$\tau(t) = \begin{bmatrix} x^T(t) & x^T(t-d(t)) & \left(\int_{t-d(t)}^{t}x(s)ds\right)^T & \left(\int_{t-d(t)}^{t}\dot{x}(s)ds\right)^T \end{bmatrix}^T$$

Now, following vectors can be expressed in terms of $\tau(t)$ as given below
$\eta(t)=\Theta_1\tau(t)$, $\eta_1(t)=\Theta_2\tau(t)$, $[0,\ I]P\eta(t)=\Theta_3\tau(t)$ and $x(t-d(t))=\Theta_4\tau(t)$

where, $\Theta_1=\begin{bmatrix} I & 0 & 0 & 0 \\ 0 & 0 & I & 0 \end{bmatrix}$, $\Theta_2=\begin{bmatrix} A & A_d & 0 & 0 \\ 0 & 0 & 0 & I \end{bmatrix}$, $\Theta_3=\begin{bmatrix} P_{12}^T & 0 & P_{22} & 0 \end{bmatrix}$ and
$\Theta_4=\begin{bmatrix} 0 & I & 0 & 0 \end{bmatrix}$

The term $2\eta^T(t)P\dot{d}(t)\eta_2(t)$ in (2.98) can be rewritten as,

$$2\eta^T(t)P\dot{d}(t)\eta_2(t) = 2\dot{d}(t)\eta^T(t)P\begin{bmatrix} 0 \\ I \end{bmatrix} x(t-d(t))$$

thus in view of above notations, one can equivalently write the above equation as,

$$\dot{d}(t)2\eta^T(t)P\begin{bmatrix} 0 \\ I \end{bmatrix} x(t-d(t)) = \dot{d}(t)2\tau^T(t)\Theta_3^T\Theta_4^T\tau(t)$$

Using the bounding inequality Lemma 2.1 one can write,

$$\dot{d}(t)2\tau^T(t)\Theta_3^T\Theta_4^T\tau(t) \le \mu\tau^T(t)\Theta_3^T T^{-1}\Theta_3\eta(t) + \mu\tau^T(t)\Theta_4^T T\Theta_4\eta(t),\ T = T^T > 0 \tag{2.99}$$

Substituting (2.99) into (2.98) one can obtain

$$\dot{V}_1(t) \leq \tau^T(t)(\Gamma_1 + \mu\Theta_3^T T^{-1}\Theta_3)\tau(t) \quad (2.100)$$

where, $\Gamma_1 = \Theta_1^T P\Theta_2 + \Theta_2^T P\Theta_1 + \mu\Theta_4^T T\Theta_4$

Now, the time-derivative of $V_2(t)$ can be written as,

$$\dot{V}_2(t) = d_u^2\xi^T(t)Q\xi(t) - d_u\int_{t-d_u}^{t} \xi^T(s)Q\xi(s)ds$$

using Jensens integral inequality Lemma 2.5 one can write,

$$\dot{V}_2(t) \leq \tau^T(t)\Gamma_2\tau(t) \quad (2.101)$$

where,$\Gamma_2 = d_u^2\Theta_5^T Q\Theta_5 - \Theta_6^T Q\Theta_6$, $\Theta_5=\begin{bmatrix} I & 0 & 0 & 0 \\ A & A_d & 0 & 0 \end{bmatrix}$, and $\Theta_6=\begin{bmatrix} 0 & 0 & I & 0 \\ 0 & 0 & 0 & I \end{bmatrix}$

Finally $\dot{V}_3(t)$ can be written as

$$\dot{V}_3(t) \leq \tau^T(t)\Gamma_3\tau(t) \quad (2.102)$$

where,$\Gamma_3 = \Theta_7^T R\Theta_7 - (1-\mu)\Theta_4^T R\Theta_4$, $\Theta_7=\begin{bmatrix} I & 0 & 0 & 0 \end{bmatrix}$

Using Newton-Leibniz formula relationship among various states of an augmented state vector $\tau(t)$ is expressed by introducing free matrices S such that following equality is satisfied,

$$2\tau^T(t)S^T\Theta_8\tau(t) = 0 \quad (2.103)$$

where, $\Theta_8 = \begin{bmatrix} I & -I & 0 & -I \end{bmatrix}$ and $S = \begin{bmatrix} S_1 & S_2 & S_3 & S_4 \end{bmatrix}$

One can write (2.103) in terms of $\eta(t)$ as

$$\tau^T(t)S^T\Gamma_4\tau(t) = 0 \quad (2.104)$$

where, $\Gamma_4 = S^T\Theta_8 + \Theta_8^T S$
Adding (2.100), (2.101), (2.102) and (2.104), which yields the expression of $\dot{V}(t)$

$$\dot{V}(t) \leq \tau^T(\Gamma_0 + \mu\Theta_3^T T^{-1}\Theta_3)\tau(t) \quad (2.105)$$

where, $\Gamma_0 = \Sigma_{i=1}^3\Gamma_i$

Taking Schur-complement of (2.105), one can obtain the stability condition in (2.95).

Remark 2.15 The LK functional in [13] can be obtained by setting $P_{11} = P_{22} = 0$ and $Q_{11} = Q_{12} = 0$ in functional (2.97), thus this LK functional is more generalized than in [13]. The construction of new LK functional was attempted such that there is proper distribution of delay information in the obtained LMI structure leading to less conservative estimate of delay upper bound.

An improved method of delay-dependent stability analysis has recently appeared in [32] and [3] where it is pointed out that the approximation of the integral term $-\int_{t-d_u}^{t} f(.)d(.) \leq -\int_{t-d(t)}^{t} f(.)d(.)$ is conservative, in sequel exact replacement of the above mentioned integral was proposed which is expressed as,

$$-\int_{t-d_u}^{t} f(.)d(.) = -\int_{t-d(t)}^{t} f(.)d(.) - \int_{t-d(t)}^{t-d_u} f(.)d(.)$$

In all earlier delay-dependent stability methods the integral term $-\int_{t-d(t)}^{t-d_u} f(.)d(.)$ was ignored leading to conservative estimate of the delay upper bound. In [32] and [3] an improved delay-dependent stability condition was derived considering new LK functional so as to accommodate this integral term. The stability theorem of [32] is presented below:

Theorem 2.13 ([32]) *The system Σ_2 is asymptotically stable for any time-delay $d(t) \in [-d_u, 0]$ satisfying (2.7), if there exist symmetric positive definite matrices, P, Q, R, Z_i $(i = 1, 2)$ and free matrices N_i, M_i and S_i $(i = 1, 2, 3)$ with appropriate dimensions such that following LMI holds:*

$$\begin{bmatrix} \Phi & d_u N & d_u S & d_u M & d_u A_{c1}^T (Z_1 + Z_2) \\ \star & -d_u Z_1 & 0 & 0 & 0 \\ \star & \star & -d_u Z_1 & 0 & 0 \\ \star & \star & \star & -d_u Z_2 & 0 \\ \star & \star & \star & \star & -d_u (Z_1 + Z_2) \end{bmatrix} < 0 \tag{2.106}$$

where, $\Phi = \Phi_1 + \Phi_2 + \Phi_2^T$,

$$\Phi_1 = \begin{bmatrix} PA + A^T P + Q + R & PA_d & 0 \\ \star & -(1-\mu)Q & 0 \\ \star & \star & -R \end{bmatrix},$$

$\Phi_2 = \begin{bmatrix} N + M & -N + S & -M - S \end{bmatrix}$, $A_{c1} = \begin{bmatrix} A & A_d & 0 \end{bmatrix}$,

$N = \begin{bmatrix} N_1^T & N_2^T & N_3^T \end{bmatrix}^T$, $S = \begin{bmatrix} S_1^T & S_2^T & S_3^T \end{bmatrix}^T$, and $M = \begin{bmatrix} M_1^T & M_2^T & M_3^T \end{bmatrix}^T$

Proof The LK functional chosen here is

$$V(t) = x^T(t)Px(t) + \int_{-d_u}^{0} \int_{t+\theta}^{t} \dot{x}^T(s)(Z_1 + Z_2)\dot{x}(s)dsd\theta + \int_{t-d(t)}^{t} x^T(s)Qx(s)ds + \int_{t-d_u}^{t} x^T(s)Rx(s)ds \tag{2.107}$$

The time-derivative of (2.107) along with the exact substitution of quadratic integral term arising out of the LK functional derivative of the double integral (associated with Z_1 matrix) as $-\int_{t-d_u}^{t} f(.)d(.) = -\int_{t-d(t)}^{t} f(.)d(.) - \int_{t-d(t)}^{t-d_u} f(.)d(.)$ one can write,

$$\begin{aligned}\dot{V}(t) \leq\ & 2x^T(t)P\dot{x}(t) + x^T(Q+R)x(t) - (1-\mu)x^T(t-d(t))Qx(t-d(t)) \\ & -x^T(t-d_u)Rx(t-d_u) + d_u\dot{x}^T(t)(Z_1+Z_2)\dot{x}(t) - \int_{t-d(t)}^{t} \dot{x}^T(s)Z_1\dot{x}(s)ds \\ & - \int_{t-d_u}^{t-d(t)} \dot{x}^T(s)Z_1\dot{x}(s)ds - \int_{t-d_u}^{t} \dot{x}^T(s)Z_2\dot{x}(s)ds \end{aligned} \tag{2.108}$$

For delay-dependent condition Newton-Leibniz formula is used, which satisfies following equations involving free matrices

$$2\xi^T(t)N\left[x(t) - x(t-d(t)) - \int_{t-d(t)}^{t} \dot{x}(s)ds\right] = 0$$

$$2\xi^T(t)S\left[x(t-d(t)) - x(t-d_u) - \int_{t-d_u}^{t-d(t)} \dot{x}(s)ds\right] = 0$$

$$2\xi^T(t)M\left[x(t) - x(t-d_u) - \int_{t-d_u}^{t} \dot{x}(s)ds\right] = 0 \tag{2.109}$$

where, $N=\begin{bmatrix} N_1 \\ N_2 \\ N_3 \end{bmatrix}$, $S=\begin{bmatrix} S_1 \\ S_2 \\ S_3 \end{bmatrix}$, $M=\begin{bmatrix} M_1 \\ M_2 \\ M_3 \end{bmatrix}$ and
$\xi(t)=\begin{bmatrix} x^T(t) & x^T(t-d(t)) & x^T(t-d_u) \end{bmatrix}^T$

Adding all the terms of (2.109) to (2.108), and with further rearrangement of terms one can get,

$$\begin{aligned}\dot{V}(t) \leq\ & \xi^T(t)\{\Phi + d_u A_{c1}^T(Z_1+Z_2)A_{c1} + d_u N Z_1^{-1} N^T \\ & + d_u S Z_1^{-1} S^T + d_u M Z_2^{-1} M^T\}\xi(t) - \int_{t-d(t)}^{t} [\xi^T(t)N + \dot{x}^T(s)Z_1] \\ & \times Z_1^{-1}[N^T\xi(t) + Z_1\dot{x}(s)]ds - \int_{t-d_u}^{t-d(t)} [\xi^T(t)S + \dot{x}^T(s)Z_1] \\ & \times Z_1^{-1}[S^T\xi(t) + Z_1\dot{x}(s)]ds - \int_{t-d_u}^{t} [\xi^T(t)M + \dot{x}^T(s)Z_2] \\ & \times Z_2^{-1}[M^T\xi(t) + Z_2\dot{x}(s)]ds \end{aligned} \tag{2.110}$$

It may be noted that, the last three integral terms of (2.110) are all less than zero, so if $\xi^T(t)[\Phi+d_u A_{c1}^T(Z_1+Z_2)A_{c1}+d_u N Z_1^{-1} N^T+d_u S Z_1^{-1} S^T+d_u M Z_2^{-1} M^T]\xi(t) < 0$ than by using Schur complement one can obtain the stability condition in LMI form as in (2.106).

The free weighting matrix method has been widely used in the stability analysis of continuous systems with two additive time-varying delay in the states. The closed-loop operation of a networked controlled systems is an example of systems with two additive time-varying delays [33]. The stability condition of system Σ_4 while considering as a single time delay term yields conservative results of delay upper bound compared to the case when the two additive time-varying delays are treated separately in the formulation, because the delays may have different properties as they occur in different places of the network [33].

The delay-dependent stability condition derived in [33] for system Σ_4 based on free-weighting matrix method is presented below.

Theorem 2.14 ([33]) *System Σ_4 in (2.11) with delays $d_1(t)$ and $d_2(t)$ satisfying (2.13) is asymptotically stable if there exist matrices $P > 0$, $Q_1 \geq Q_2 > 0$, $Q_3 \geq Q_4 > 0$, $M_1 \geq M_2 > 0$, $M_3 \geq M_4 > 0$, $N_i, i = 1, \ldots 8$, such that following LMI holds,*

$$\begin{bmatrix} \Pi_{11} & \Pi_{12} & \Pi_{13} & PA_d & A^T\Pi_{55} & N_1 & 0 & N_5 & 0 \\ \star & \Pi_{22} & 0 & \Pi_{24} & 0 & N_2 & N_3 & 0 & 0 \\ \star & \star & \Pi_{33} & \Pi_{34} & 0 & 0 & 0 & N_6 & N_7 \\ \star & \star & \star & \Pi_{44} & A_d^T\Pi_{55} & 0 & N_4 & 0 & N_8 \\ \star & \star & \star & \star & -\Pi_{55} & 0 & 0 & 0 & 0 \\ \star & \star & \star & \star & \star & -d_{1u}^{-1}M_1 & 0 & 0 & 0 \\ \star & \star & \star & \star & \star & \star & -d_{2u}^{-1}M_2 & 0 & 0 \\ \star & \star & \star & \star & \star & \star & \star & -d_{2u}^{-1}M_3 & 0 \\ \star & \star & \star & \star & \star & \star & \star & \star & -d_{1u}^{-1}M_4 \end{bmatrix} < 0 \quad (2.111)$$

where, $\Pi_{11}=A^TP + PA + Q_1 + Q_3 + N_1 + N_1^T + N_5 + N_5^T$,
$\Pi_{12}=-N_1 + N_2^T$, $\Pi_{13}=-N_5 + N_6^T$
$\Pi_{22}=-(1-\mu_1)(Q_1 - Q_2) - N_2 - N_2^T + N_3 + N_3^T$, $\Pi_{24}=-N_3 + N_4^T$,
$\Pi_{33}=-(1-\mu_2)(Q_3 - Q_4) - N_6 - N_6^T + N_7 + N_7^T$, $\Pi_{34}=-N_7 + N_8^T$,
$\Pi_{44}=-(1-\mu_1-\mu_2)(Q_2 + Q_4) - N_4 - N_4^T - N_8 - N_8^T$
$\Pi_{55}=(d_{1u}M_1 + d_{2u}M_2 + d_{2u}M_3 + d_{1u}M_4)$

Remark 2.16 The LK functional candidate selected in [33] for Theorem 2.14) is,

$$V(t) = V_1 + V_2 + V_3 + V_4 + V_5$$

$$V_1(t) = x^T(t)Px(t)$$

$$V_2(t) = \int_{t-d_1(t)}^{t} x^T(s)Q_1x(s)ds + \int_{t-d_1(t)-d_2(t)}^{t-d_1(t)} x^T(s)Q_2x(s)ds$$

$$V_3(t) = \int_{-d_{1u}}^{0}\int_{\beta}^{0} \dot{x}^T(t+\alpha)M_1\dot{x}(t+\alpha)d\alpha d\beta + \int_{-d_{1u}-d_{2u}}^{-d_{1u}}\int_{\beta}^{0} \dot{x}^T(t+\alpha)M_2\dot{x}(t+\alpha)d\alpha d\beta$$

$$V_4(t) = \int_{t-d_2(t)}^{t} x^T(s)Q_3x(s)ds + \int_{t-d_1(t)-d_2(t)}^{t-d_2(t)} x^T(s)Q_4x(s)ds$$

$$\begin{aligned}V_5(t) = &\int_{-d_{2u}}^{0}\int_{\beta}^{0} \dot{x}^T(t+\alpha)M_3\dot{x}(t+\alpha)d\alpha d\beta\\ &+ \int_{-d_{1u}-d_{2u}}^{-d_{2u}}\int_{\beta}^{0} \dot{x}^T(t+\alpha)M_4\dot{x}(t+\alpha)d\alpha d\beta\end{aligned}$$

The selection of the LK functional considered above contains repetitive delay information in some region that can lead to a conservative estimate of the delay upper bound. Moreover, the dimension of the LMI obtained by this method is more due to the introduction of free weighting matrices for approximating quadratic integral terms using Newton-Leibniz formula.

Also, introduction of semi-positive definite matrices to satisfy the inequalities (24)–(27) in [33] and consequently replacing it with inequalities (32) in [33] are not equivalent. This is turn, leads to conservative delay upper bound estimate.

2.4 Delay-Range-Dependent Stability Condition

It was pointed out in [3] that, in practice the delay lower bound cannot necessarily be always restricted to 0 as in many engineering (or physical) systems, delay may vary in a ranges (or intervals) unlike for the system considered in Σ_1, Σ_2, Σ_3 and Σ_4 satisfying (2.5). The stability conditions derived by restricting the lower delay bound to 0 are referred in literature as delay-dependent stability conditions.

The stability condition in an LMI framework for systems with time delay varying in ranges have been reported in [3–5, 34, 35]. In [3] stability condition has been proposed for the system Σ_2 satisfying (2.6) (i.e., delay lower bound is not restricted to 0) by proposing a new LK functional suitable for the condition (2.6), such stability condition are referred as delay-range-dependent stability condition in the literature. The stability condition derived in [3] is presented below.

Theorem 2.15 ([3]) *The system Σ_2 is asymptotically stable for any time-delay $d(t) \in [-d_u, 0]$ satisfying (2.6), (2.7) and (2.8), if there exist symmetric positive* *definite matrices, P, T, Q, R, Z_i $(j = 1, 2)$ such that following LMI holds:*

$$\Phi < 0 \tag{2.112}$$

$$\text{where, } \Phi = \begin{bmatrix} \Phi_{11} & \Phi_{12} & M_1 & -S_1 & d_u N_1 & d_{lu} S_1 & d_{lu} M_1 & A^T U \\ \star & \Phi_{22} & M_2 & -S_2 & d_u N_2 & d_{lu} S_2 & d_{lu} M_2 & A_d^T U \\ \star & \star & -T & 0 & 0 & 0 & 0 & 0 \\ \star & \star & \star & -R & 0 & 0 & 0 & 0 \\ \star & \star & \star & \star & -d_u Z_1 & 0 & 0 & 0 \\ \star & \star & \star & \star & \star & -d_{lu}(Z_1 + Z_2) & 0 & 0 \\ \star & \star & \star & \star & \star & \star & -d_{lu} Z_2 & 0 \\ \star & \star & \star & \star & \star & \star & \star & -U \end{bmatrix} < 0$$

$\Phi_{11} = PA + A^T P + Q + T + R + N_1 + N_1^T$, $\Phi_{12} = PA_d + N_2^T - N_1 + S_1 - M_1$
$\Phi_{22} = -(1-\mu)Q + S_2 + S_2^T - N_2 - N_2^T - M_2 - M_2^T$, $U = d_u Z_1 + d_{lu} Z_2$ and
$d_{lu} = d_u - d_l$

Proof The LK functional is chosen here as

$$\begin{aligned} V(t) = x^T(t)Px(t) &+ \int_{-d_u}^{0}\int_{t+\theta}^{t} \dot{x}^T(s)Z_1\dot{x}(s)dsd\theta \\ &+ \int_{-d_u}^{-d_l}\int_{t+\theta}^{t} \dot{x}^T(s)Z_2\dot{x}(s)dsd\theta + \int_{t-d(t)}^{t} x^T(s)Qx(s)ds \\ &+ \int_{t-d_u}^{t} x^T(s)Rx(s)ds + \int_{t-d_l}^{t} x^T(s)Tx(s)ds \end{aligned} \tag{2.113}$$

The time-derivative of (2.113) is given by

$$\begin{aligned} \dot{V}(t) \le\, & 2x^T(t)P\dot{x}(t) + x^T(t)(Q + R + T)x(t) - (1-\mu)x^T(t - d(t))Qx(t - d(t)) \\ & - x^T(t - d_u)Rx(t - d_u) - x^T(t - d_l)Tx(t - d_l) + \dot{x}^T(t)(d_u Z_1 + d_{lu} Z_2)\dot{x}(t) \\ & - \int_{t-d_u}^{t} \dot{x}^T(s)Z_1\dot{x}(s)ds - \int_{t-d_u}^{t-d_l} \dot{x}^T(s)Z_2\dot{x}(s)ds \end{aligned} \tag{2.114}$$

As stated in [32] that the conservative estimate of the delay bound is obtained as the term $-\int_{t-du}^{t-d(t)}(.)$ was ignored while approximating the term $-\int_{t-du}^{t}(.)$. Hence, the exact expression for the last two integral terms in (2.114) is considered as,

$$\begin{aligned} -\int_{t-d_u}^{t} \dot{x}^T(s)Z_1\dot{x}(s)ds = & -\int_{t-d(t)}^{t} \dot{x}^T(s)Z_1\dot{x}(s)ds \\ & -\int_{t-d_u}^{t-d(t)} \dot{x}^T(s)Z_1\dot{x}(s)ds \end{aligned} \tag{2.115}$$

$$-\int_{t-d_u}^{t-d_l} \dot{x}^T(s)Z_2\dot{x}(s)ds = -\int_{t-d_u}^{t-d(t)} \dot{x}^T(s)Z_2\dot{x}(s)ds - \int_{t-d(t)}^{t-d_l} \dot{x}^T(s)Z_2\dot{x}(s)ds \quad (2.116)$$

Thus in view of (2.115) and (2.116), one can write (2.114) as

$$\begin{aligned}\dot{V}(t) &\le 2x^T(t)P\dot{x}(t) + x^T(Q+R+T)x(t) - (1-\mu)x^T(t-d(t))Qx(t-d(t)) \\ &-x^T(t-d_u)Rx(t-d_u) - x^T(t-d_l)Tx(t-d_l) + \dot{x}^T(t)(d_uZ_1 + d_{lu}Z_2)\dot{x}(t) \\ &-\int_{t-d_u}^{t} \dot{x}^T(s)Z_1\dot{x}(s)ds - \int_{t-d_u}^{t-d(t)} \dot{x}^T(s)(Z_1+Z_2)\dot{x}(s)ds \\ &-\int_{t-d(t)}^{t-d_l} \dot{x}^T(s)Z_2\dot{x}(s)ds \end{aligned} \quad (2.117)$$

For obtaining delay-dependent condition one can use Newton-Leibniz formula such that it satisfies following equations involving free matrices

$$\begin{aligned} 2\xi^T(t)N\left[x(t) - x(t-d(t)) - \int_{t-d(t)}^{t} \dot{x}(s)ds\right] &= 0 \\ 2\xi^T(t)S\left[x(t-d(t)) - x(t-d_u) - \int_{t-d_u}^{t-d(t)} \dot{x}(s)ds\right] &= 0 \\ 2\xi^T(t)M\left[x(t-d_l) - x(t-d(t)) - \int_{t-d(t)}^{t-d_l} \dot{x}(s)ds\right] &= 0 \end{aligned} \quad (2.118)$$

where, $N=\begin{bmatrix} N_1 \\ N_2 \end{bmatrix}$, $S=\begin{bmatrix} S_1 \\ S_2 \end{bmatrix}$, $M=\begin{bmatrix} M_1 \\ M_2 \end{bmatrix}$ and $\xi(t) = \left[x^T(t)\; x^T(t-d(t))\right]^T$

Adding (2.118) into (2.117), then carrying out algebraic manipulations, using bounding techniques (discussed in Theorem 2.11 and applying Schur-complement one can obtain,

$$\dot{V}(t) = \zeta^T(t)\Phi\zeta(t)$$

where $\zeta(t) = [x(t)^T, x(t-d(t)^T, x(t-d_l)^T, x(t-d_u)^T]^T$. The LMI Φ is already defined above. If $\Phi < 0$ then the system Σ_2 is guaranteed to be asymptotically stable.

Remark 2.17 It is possible to obtain delay-dependent stability condition from this theorem by setting $d_l = 0$ in the LK functional (2.113) thus reducing the double integral term to $\int_{-d_u}^{0}\int_{t+\theta}^{t} \dot{x}^T(s)(Z_1+Z_2)\dot{x}(s)dsd\theta$ and setting $T = 0$ the single integral term becomes zero while rest of the term appears as in (2.113).

Considering $N = [N_1^T, N_2^T, 0]^T$, $S = [S_1^T, S_2^T, 0]^T$ and $M = [0]$ in (2.109) of Theorem 2.13 one can obtain the delay-dependent condition of Corollary 3 of [3].

Theorem 2.13 and corollary 3 of [3] both are applicable for unknown $\mu \geq 1$ due to the presence of free matrices in the (2,2) element of the LMI conditions, so separate condition need not to be derived for this case.

Further improvements of delay-range-dependent stability criteria as well as delay-dependent stability criteria with less number of matrix variables applicable for both slow and fast varying time-delay (i.e., satisfying the conditions (2.7) and (2.8)) have been proposed in [4] and [5]. Both the stability criteria are presented below.

Theorem 2.16 ([4]) *The system Σ_2 is asymptotically stable for any time-delay $d(t) \in [-d_u, 0]$ satisfying (2.6), (2.7) and (2.8), if there exist symmetric positive definite matrices, P, T, Q, R, Z_i $(j = 1, 2)$ such that following LMI holds:*

$$\begin{bmatrix} \gamma_{11} & PA_d & Z_1 & 0 & d_l A^T Z_1 & d_{lu} A^T Z_2 \\ \star & \gamma_{22} & Z_2 & Z_2 & d_l A_d^T Z_1 & d_{lu} A_d^T Z_2 \\ \star & \star & \gamma_{33} & 0 & 0 & 0 \\ \star & \star & \star & -R - Z_2 & 0 & 0 \\ \star & \star & \star & \star & -Z_1 & 0 \\ \star & \star & \star & \star & \star & -Z_2 \end{bmatrix} < 0 \tag{2.119}$$

where, $\gamma_{11}{=}PA + A^T P + Q + T + R - Z_1$, $\gamma_{22}{=}-(1-\mu)Q - 2Z_2$ and $\gamma_{33}{=}-T - Z_1 - Z_2$

Proof The LK functional candidate is selected as,

$$\begin{aligned} V(t) = {} & x^T(t)Px(t) + \int_{-d_l}^{0}\int_{t+\theta}^{t} d_l \dot{x}^T(s)Z_1\dot{x}(s)dsd\theta \\ & + \int_{-d_u}^{-d_l}\int_{t+\theta}^{t} d_{lu}\dot{x}^T(s)Z_2\dot{x}(s)dsd\theta + \int_{t-d(t)}^{t} x^T(s)Qx(s)ds \\ & + \int_{t-d_u}^{t} x^T(s)Rx(s)ds + \int_{t-d_l}^{t} x^T(s)Tx(s)ds \end{aligned} \tag{2.120}$$

Considering time-derivative of (2.120) and substituting the value of $\dot{x}(t) = Ax(t) + A_d x(t - d(t))$ one can obtain

$$\begin{aligned} \dot{V}(t) \leq {} & 2x^T(t)P(Ax(t) + A_d x(t - d(t))) + x^T(Q + R + T)x(t) \\ & -(1-\mu)x^T(t - d(t))Qx(t - d(t)) \\ & -x^T(t - d_u)Rx(t - d_u) - x^T(t - d_l)Tx(t - d_l) \\ & +(Ax(t) + A_d x(t - d(t)))^T(d_l^2 Z_1 + d_{lu}^2 Z_2)(Ax(t) + A_d x(t - d(t))) \\ & -\int_{t-d_l}^{t} d_l\dot{x}^T(s)Z_1\dot{x}(s)ds - \int_{t-d_u}^{t-d_l} d_{lu}\dot{x}^T(s)Z_2\dot{x}(s)ds \end{aligned} \tag{2.121}$$

Applying Lemma 2.5 (Jensen's integral inequality), the integral terms in (2.121) are approximated as,

$$-\int_{t-d_l}^{t} d_l\dot{x}(s)^T Z_1\dot{x}(s)ds \leq -(x(t)-x(t-d_l))^T Z_1(x(t)-x(t-d_l)) \tag{2.122}$$

and,

$$\begin{aligned}-\int_{t-d_u}^{t-d_l} d_{lu}\dot{x}(s)^T Z_2\dot{x}(s)ds \leq &-\int_{t-d_u}^{t-d(t)} (d_u-d(t))\dot{x}(s)^T Z_2\dot{x}(s)ds\\ &-\int_{t-d(t)}^{t-d_l} (d(t)-d_l)\dot{x}(s)^T Z_2\dot{x}(s)ds\end{aligned} \tag{2.123}$$

further one can write (2.123) as,

$$\begin{aligned}-\int_{t-d_u}^{t-d_l} d_{lu}\dot{x}(s)^T Z_2\dot{x}(s)ds \leq &-(x(t-d(t))-x(t-d_u))^T Z_2(x(t-d(t))-x(t-d_u))\\ &-(x(t-d_l)-x(t-d(t)))^T Z_2(x(t-d_l)-x(t-d(t)))\end{aligned} \tag{2.124}$$

Using (2.121)–(2.124) and with algebraic manipulations, one can easily obtain the following expression,

$$\dot{V}(t) \leq \zeta^T\Gamma\zeta(t) \tag{2.125}$$

where, $\zeta(t)=\left[x^T(t)\ x^T(t-d(t))\ x^T(t-d_l)\ x^T(t-d_u)\right]^T$

$$\Gamma=\begin{bmatrix}\gamma_{11} & PA_d & Z_1 & 0\\ \star & \gamma_{22} & Z_2 & Z_2\\ \star & \star & \gamma_{33} & 0\\ \star & \star & \star & -R-Z_2\end{bmatrix}+\left[A\ A_d\ 0\ 0\right]^T(d_l^2 Z_1+d_{lu}^2 Z_2)\left[A\ A_d\ 0\ 0\right]$$

γ_{11}, γ_{22} and γ_{33} are defined in (2.119), if $\Gamma < 0$ in (2.125) then the system is asymptotically stable.

Remark 2.18 When $d_l = 0$, the above Theorem reduces to delay-dependent stability condition, which is given in corollary 1 of [4] and it is valid for $\mu < 1$.

When $\mu > 1$ then Theorem 2.16 is not applicable, thus setting $Q = 0$ in Theorem 2.16, one can easily obtain corollary 2 of [4]

The modifications made in Theorem 2.16 compared to Theorem 2.15 are (i) selection of different LK functional (ii) use of Jensens integral to approximate the quadratic integral terms arising in the LK functional derivative unlike introducing free matrices via Newton-Leibniz formula in [3]. It is clear from the above theorem that, even without using free matrices it is possible to derive an LMI condition which can provide feasible solution for both $\mu < 1$ and $\mu > 1$, with lower matrix variables due to the use of Jensen integral inequality.

As only weighting matrices of LK functional is involved in the LMI of Theorem 2.16 thus the computational burden of this theorem is much lesser then that of [3], as the latter method involves lot of free weighting matrices.

Recently in [5] another improved delay-range-dependent stability analysis has been reported with a tight bounding of the following integral terms,

$$-\int_{t-d_u}^{t-d(t)} d_{lu}\dot{x}(s)^T Z_2 \dot{x}(s)ds$$
$$and -\int_{t-d(t)}^{t-d_l} d_{lu}\dot{x}(s)^T Z_2 \dot{x}(s)ds$$

The bounding of the above integral terms carried out in [5] and it is expressed as,

$$\begin{aligned}
-\int_{t-d_u}^{t-d_l} d_{lu}\dot{x}(s)^T Z_2 \dot{x}(s)ds &= -\int_{t-d_u}^{t-d(t)} d_{lu}\dot{x}(s)^T Z_2 \dot{x}(s)ds \\
&\quad -\int_{t-d(t)}^{t-d_l} d_{lu}\dot{x}(s)^T Z_2 \dot{x}(s)ds \\
-\int_{t-d_u}^{t-d_l} d_{lu}\dot{x}(s)^T Z_2 \dot{x}(s)ds &= -\int_{t-d_u}^{t-d(t)} (d_u - d(t))\dot{x}(s)^T Z_2 \dot{x}(s)ds \\
&\quad -\int_{t-d_u}^{t-d(t)} (d(t) - d_l)\dot{x}(s)^T Z_2 \dot{x}(s)ds \\
&\quad -\int_{t-d(t)}^{t-d_l} (d(t) - d_l)\dot{x}(s)^T Z_2 \dot{x}(s)ds \\
&\quad -\int_{t-d(t)}^{t-d_l} (d_u - d(t))\dot{x}(s)^T Z_2 \dot{x}(s)ds \quad (2.126)
\end{aligned}$$

Defining $\beta = (d(t) - d_l)/d_{lu}$, so $1 - \beta = (d_u - d(t))/d_{lu}$, thus following will be true

$$\begin{aligned}
-\int_{t-d_u}^{t-d(t)} (d(t) - d_l)\dot{x}(s)^T Z_2 \dot{x}(s)ds &= -\beta \int_{t-d_u}^{t-d(t)} d_{lu}\dot{x}(s)^T Z_2 \dot{x}(s)ds \\
&\leq -\beta \int_{t-d_u}^{t-d(t)} (d_u - d(t))\dot{x}(s)^T Z_2 \dot{x}(s)ds \\
and, \quad -\int_{t-d(t)}^{t-d_l} (d_u - d(t))\dot{x}(s)^T Z_2 \dot{x}(s)ds &= (1-\beta) \int_{t-d(t)}^{t-d_l} d_{lu}\dot{x}(s)^T Z_2 \dot{x}(s)ds \\
&\leq -(1-\beta) \int_{t-d(t)}^{t-d_l} (d(t) - d_l)\dot{x}(s)^T Z_2 \dot{x}(s)ds
\end{aligned}$$

Incorporating the above modifications, the quadratic integral term $-\int_{t-d_u}^{t-d_l} d_{lu}\dot{x}(s)^T Z_2\dot{x}(s)ds$ is approximated using Lemma 2.5 (Jensens Integral Inequality) as

$$\begin{aligned}-\int_{t-d_u}^{t-d_l} d_{lu}\dot{x}(s)^T Z_2\dot{x}(s)ds \leq & -(x(t-d(t))-x(t-d_u))^T Z_2(x(t-d(t))-x(t-d_u))\\ & -(x(t-d_l)-x(t-d(t)))^T Z_2(x(t-d_l)-x(t-d(t)))\\ & -\beta(x(t-d(t))-x(t-d_u))^T Z_2(x(t-d(t))-x(t-d_u))\\ & -(1-\beta)(x(t-d_l)-x(t-d(t)))^T Z_2(x(t-d_l)-x(t-d(t)))\end{aligned} \tag{2.127}$$

Considering similar LK functional as in [4] and applying bounding technique of (2.127) results a stability condition which is stated below.

Theorem 2.17 ([5]) *The system Σ_2 is asymptotically stable for any time-delay $d(t) \in [-d_u, 0]$ satisfying (2.6), (2.7), if there exist symmetric positive definite matrices, $P, T, Q, R, Z_i (j = 1, 2)$ such that following LMI holds:*

$$\Phi_1 = \Phi - \begin{bmatrix} 0 & -I & I & 0 \end{bmatrix}^T Z_2 \begin{bmatrix} 0 & -I & I & 0 \end{bmatrix} < 0 \tag{2.128}$$

and

$$\Phi_2 = \Phi - \begin{bmatrix} 0 & I & 0 & -I \end{bmatrix}^T Z_2 \begin{bmatrix} 0 & I & 0 & -I \end{bmatrix} < 0 \tag{2.129}$$

$$\text{where,}\Phi = \begin{bmatrix} \gamma_{11} & PA_d & Z_1 & 0 \\ \star & \gamma_{22} & Z_2 & Z_2 \\ \star & \star & \gamma_{33} & 0 \\ \star & \star & \star & -R-Z_2 \end{bmatrix} + \begin{bmatrix} A & A_d & 0 & 0 \end{bmatrix}^T (d_l^2 Z_1 + d_{lu}^2 Z_2) \begin{bmatrix} A & A_d & 0 & 0 \end{bmatrix}$$

$\gamma_{11} = PA + A^T P + Q + T + R - Z_1$, $\gamma_{22} = -(1-\mu)Q - 2Z_2$ and

$\gamma_{33} = -T - Z_1 - Z_2$

Proof Considering thc similar LK functional candidate as in Theorem 2.15 and incorporating the integral term approximation in (2.127) one can easily obtain,

$$\dot{V}(t) \leq \zeta(t)^T[(1-\beta)\Phi_1 + \beta\Phi_2]\zeta(t)$$

where, $\zeta(t) = \begin{bmatrix} x(t)^T & x(t-d(t))^T & x(t-d_l)^T & x(t-d_u)^T \end{bmatrix}^T$. One can observe that the above $\dot{V}(t)$ expression is a convex combination of the matrices Φ_1 and Φ_2. The negativity of $\dot{V}(t)$ is ensured if both Φ_1 and Φ_2 are negative definite, which in turn guarantees the asymptotic stability of the system Σ_2. The details derivation can be found in [5].

Remark 2.19 The improvement on delay upper bound result compared to the Theorem 2.17 is expected as the over bounding of the integrals (2.122)–(2.124) has been avoided in the present theorem.

Delay-dependent stability condition is a special case of delay-range-dependent stability condition when $d_l = 0$.

Delay-dependent stability conditions in LMI framework have been reviewed extensively that are directly relevant to the present work of the thesis. Initially, some basic delay-independent stability condition are recalled. Next, four basic approaches have been discussed in length for developing delay-dependent stability conditions in LMI framework. they are namely: Model transformation approach, Bounding techniques, Descriptor system approach and Free-Weighting matrix approach. Finally the recent results on additive delay and delay-range-dependent stability conditions have been discussed in previous sections.

For convenience of the discussion of the main results of this chapter, some preliminaries including few definitions, basic theorems on stability of time-delay systems which are related to the main results are presented in previous sections.

The main and improved results on delay-dependent stability analysis are proposed in this chapter and presented below.

2.5 Main Results on Stability Analysis of Time-Delay System

In this section, delay-dependent-stability analysis of nominal time-delay systems are presented with an objective to (i) obtain less conservative estimate of the delay upper bound result and (ii) obtain a stability condition in an LMI framework (without sacrificing the conservatism) that can be further extended to derive improved robust stability and/(or) stabilization conditions. The necessary condition for the delay-dependent stability of the time-delay systems is that the matrix $(A + A_d)$ must be Hurwitz.

The results of the proposed stability conditions are compared with existing methods by considering several numerical examples.

2.5.1 Stability Analysis of TDS with Single Time Delay

Consider the system Σ_2 in (2.3), satisfying the condition (2.5) and (2.7), the stability condition is presented by considering a new LK functional and replacing the quadratic integral inequalities arising in the LK derivative with more exact expressions as suggested in [3]. The proposed method is described in the form of theorem below.

Theorem 2.18 *The system Σ_2 satisfying the condition (2.5) for $0 < \mu < 1$, is asymptotically stable if there exist matrices $R_i = R_i^T > 0$, (i=1,2) and any free* *matrices* L_i, $(i = 1, ..3)$, M_i $(i = 1, ..3)$, G_i, $(i = 1, 2)$ *such that following LMI's are satisfied*

$$P = \begin{bmatrix} P_{11} & P_{12} \\ P_{12}^T & P_{22} \end{bmatrix} > 0, \tag{2.130}$$

$$Q = \begin{bmatrix} Q_{11} & 0 \\ 0 & Q_{22} \end{bmatrix} > 0 \tag{2.131}$$

and,

$$\begin{bmatrix} \Omega & \mu\tilde{P} & d_u M \\ \star & -\mu T & 0 \\ \star & 0 & -Q_{22} \end{bmatrix} < 0 \tag{2.132}$$

$$\begin{bmatrix} \Omega & \mu\tilde{P} & d_u L \\ \star & -\mu T & 0 \\ \star & 0 & -Q_{22} \end{bmatrix} < 0 \tag{2.133}$$

where,

$$\Omega = \begin{bmatrix} \Omega_{11} & \Omega_{12} & \Omega_{13} & \Omega_{14} & 0 & \Omega_{16} \\ \star & \Omega_{22} & \Omega_{23} & \Omega_{24} & 0 & \Omega_{26} \\ \star & \star & \Omega_{33} & 0 & 0 & 0 \\ \star & \star & 0 & -Q_{11} & 0 & \Omega_{46} \\ 0 & 0 & 0 & 0 & -Q_{11} & 0 \\ \star & \star & 0 & \star & 0 & \Omega_{66} \end{bmatrix} \tag{2.134}$$

and, $L = \begin{bmatrix} L_1^T & L_2^T & L_3^T & 0 & 0 & 0 \end{bmatrix}^T$, $M = \begin{bmatrix} M_1^T & M_2^T & M_3^T & 0 & 0 & 0 \end{bmatrix}^T$

$\tilde{P} = \begin{bmatrix} P_{12}^T & 0 & 0 & P_{22} & 0 & 0 \end{bmatrix}^T$

where, $\Omega_{11}=d_u^2 Q_{11} + G_1^T A + A^T G_1 + P_{12} + P_{12}^T + R_1 + R_2 + d_u(L_1 + L_1^T)$

$\Omega_{12}=G_1^T A_d - P_{12} + d_u(-L_1 + L_2^T + M_1)$, $\Omega_{13}=d_u(L_3^T - M_1)$, $\Omega_{14}=P_{22}$,

$\Omega_{16}=P_{11} - G_1^T + A^T G_2$ $\Omega_{22}=-(1-\mu)R_1 + \mu T + d_u(-L_2 - L_2^T + M_2 + M_2^T)$

$\Omega_{23}=d_u(-M_2 + M_3^T - L_3^T)$, $\Omega_{24}=-P_{22}$, $\Omega_{26}=A_d^T G_2$, $\Omega_{33}=d_u(-M_3 - M_3^T) - R_2$,

$\Omega_{66}=-G_2 - G_2^T + d_u^2 Q_{22}$

Proof The LK functional is selected as,

$$V(t) = V_1(t) + V_2(t) + V_3(t)$$

where,

$$V_1(t) = \varphi^T(t)P\varphi(t) \tag{2.135}$$

with, $\varphi(t) = \left[x^T(t) \left(\int_{t-d(t)}^{t} x(s)ds \right)^T \right]^T$

$$V_2(t) = \int_{t-d(t)}^{t} x^T(s)R_1x(s)ds + \int_{t-d_u}^{t} x^T(s)R_2x(s)ds \tag{2.136}$$

$$V_3(t) = \int_{-d_u}^{0}\int_{t+\theta}^{t} d_u\gamma^T(s)Q\gamma(s)dsd\theta \tag{2.137}$$

where, $\gamma(s) = \left[x^T(s)\ \dot{x}^T(s) \right]^T$. The time-derivative of (2.135) along the solution of (2.3), one can obtain

$$\dot{V}_1(t) = 2\varphi^T(t)P\left(\begin{bmatrix} I & 0 & 0 \\ 0 & I & -I \end{bmatrix} \varphi_1(t) + \dot{d}(t)\begin{bmatrix} 0 \\ I \end{bmatrix} x(t-d(t)) \right) \tag{2.138}$$

where,

$$\varphi_1(t) = \left[\dot{x}^T(t)\ x^T(t)\ x^T(t-d(t)) \right]^T$$

Defining the augmented state vector as, $\eta(t) = \left[x^T(t)\ x^T(t-d(t))\ x^T(t-d_u) \left(\int_{t-d(t)}^{t} x(s)ds \right)^T \left(\int_{t-d_u}^{t-d(t)} x(s)ds \right)^T \dot{x}^T(t) \right]^T$

One can rewrite the second term of (2.138) as,

$$2\dot{d}(t)\eta^T\tilde{P}\left[0\ I\ 0\ 0\ 0\ 0 \right]\eta(t) = 2\dot{d}(t)a^T(.)b(.) \tag{2.139}$$

where, $a(.) = \tilde{P}^T\eta(t)$, $b(.) = \left[0\ I\ 0\ 0\ 0\ 0 \right]\eta(t)$ and $\tilde{P} = \left[P_{12}^T\ 0\ 0\ P_{22}\ 0\ 0 \right]^T$. Using bounding lemma (Lemma 2.1) on (2.139) one can have,

$$\begin{aligned} 2\dot{d}(t)a^T(.)b(.) &\leq \mu\{a^T(.)T^{-1}a(.) + b^T(.)Tb(.)\} \\ &\leq \mu\eta^T(t)\left(\tilde{P}T^{-1}\tilde{P}^T + \begin{bmatrix} 0 \\ I \\ 0 \\ 0 \\ 0 \\ 0 \end{bmatrix} T\left[0\ I\ 0\ 0\ 0\ 0 \right] \right)\eta(t) \end{aligned} \tag{2.140}$$

Substituting (2.140) on (2.138) and upon expansion of (2.138) one can have,

$$\dot{V}_1(t) \leq \eta^T(t)\left(\begin{bmatrix} P_{12}+P_{12}^T & -P_{12} & 0 & P_{22} & 0 & P_{11} \\ \star & \mu T & 0 & -P_{22} & 0 & 0 \\ 0 & 0 & 0 & 0 & 0 & 0 \\ \star & \star & 0 & 0 & 0 & P_{12}^T \\ 0 & 0 & 0 & 0 & 0 & 0 \\ \star & 0 & 0 & \star & 0 & 0 \end{bmatrix} + \tilde{P}\mu T^{-1}\tilde{P}^T\right)\eta(t) \quad (2.141)$$

Finding the time-derivative of (2.136) and (2.137), one can obtain

$$\begin{aligned} \dot{V}_2(t) \leq\ & x^T(t)(R_1+R_2)x(t) - (1-\mu)x^T(t-d(t))R_1x(t-d(t)) \\ & -x^T(t-d_u)R_2x(t-d_u) \end{aligned} \quad (2.142)$$

$$\dot{V}_3(t) \leq d_u^2\gamma^T(t)Q\gamma(t) - d_u\int_{t-d_u}^{t}\gamma^T(s)Q\gamma(s)ds \quad (2.143)$$

Express the first quadratic term of (2.143) into the following form,

$$d_u^2\gamma^T(t)Q\gamma(t) = d_u^2\eta^T(t)\Theta\,\eta(t) \quad (2.144)$$

where,

$$\Theta = \begin{bmatrix} d_u^2Q_{11} & 0 & 0 & 0 & 0 & 0 \\ 0 & 0 & 0 & 0 & 0 & 0 \\ 0 & 0 & 0 & 0 & 0 & 0 \\ 0 & 0 & 0 & 0 & 0 & 0 \\ 0 & 0 & 0 & 0 & 0 & 0 \\ 0 & 0 & 0 & 0 & 0 & d_u^2Q_{22} \end{bmatrix}$$

The integral term in (2.143) can be written as,

$$\begin{aligned} -d_u\int_{t-d_u}^{t}\gamma^T(s)Q\gamma(s)ds = & -d_u\int_{t-d(t)}^{t}\gamma^T(s)Q\gamma(s)ds \\ & -d_u\int_{t-d_u}^{t-d(t)}\gamma^T(s)Q\gamma(s)ds \end{aligned} \quad (2.145)$$

Expanding (2.145), one can write

$$-d_u\int_{t-d_u}^{t}\gamma^T(s)Q\gamma(s)ds=-\int_{t-d(t)}^{t}d_ux^T(s)Q_{11}x(s)ds-\int_{t-d_u}^{t-d(t)}d_ux^T(s)Q_{11}x(s)ds$$
$$-\int_{t-d(t)}^{t}d_u\dot{x}^T(s)Q_{22}\dot{x}(s)ds-\int_{t-d_u}^{t-d(t)}d_u\dot{x}^T(s)Q_{22}\dot{x}(s)ds \tag{2.146}$$

Treating first two integral terms of (2.146) using Jensen's integral inequality (Lemma 2.5), one can write

$$-d_u\int_{t-d_u}^{t}\gamma^T(s)Q\gamma(s)ds\le-\left(\int_{t-d(t)}^{t}x(s)ds\right)^TQ_{11}\int_{t-d(t)}^{t}x(s)ds$$
$$-\left(\int_{t-d_u}^{t-d(t)}x(s)ds\right)^TQ_{11}\int_{t-d_u}^{t-d(t)}x(s)ds$$
$$-\int_{t-d_u}^{t-d(t)}d_u\dot{x}^T(s)Q_{22}\dot{x}(s)ds-\int_{t-d(t)}^{t}d_u\dot{x}^T(s)Q_{22}\dot{x}(s)ds \tag{2.147}$$

The last two integral terms in (2.147) are eliminated using Newton-Leibniz formula by introducing free matrices L and M and they have the following forms,

$$2\eta^T(t)\left[L_1^T\ L_2^T\ L_3^T\ 0\ 0\ 0\right]^T[x(t)-x(t-d(t)-\int_{t-d(t)}^{t}\dot{x}(s)ds]=0 \tag{2.148}$$

$$2\eta^T(t)\left[M_1^T\ M_2^T\ M_3^T\ 0\ 0\ 0\right]^T[x(t-d(t))-x(t-d_u)-\int_{t-d_u}^{t-d(t)}\dot{x}(s)ds]=0 \tag{2.149}$$

Expanding (2.148) one can obtain,

$$0=\eta^T(t)\begin{bmatrix}L_1+L_1^T & -L_1+L_2^T & L_3^T & 0\,0\,0\\ \star & -L_2-L_2^T & -L_3^T & 0\,0\,0\\ \star & \star & 0 & 0\,0\,0\\ 0 & 0 & 0 & 0\,0\,0\\ 0 & 0 & 0 & 0\,0\,0\\ 0 & 0 & 0 & 0\,0\,0\end{bmatrix}\eta(t)$$
$$-2\int_{t-d(t)}^{t}\eta^T(t)\left[L_1^T\ L_2^T\ L_3^T\ 0\ 0\ 0^T\right]\dot{x}(s)ds \tag{2.150}$$

Now applying bounding lemma (Lemma 2.1) and with algebraic manipulations one can obtain,

$$-\int_{t-d(t)}^{t} d_u \dot{x}^T(s) Q_{22} \dot{x}(s) ds \leq \eta^T(t) \left\{ d_u \begin{bmatrix} L_1 + L_1^T & -L_1 + L_2^T & L_3^T & 0\,0\,0 \\ \star & -L_2 - L_2^T & -L_3^T & 0\,0\,0 \\ \star & \star & 0 & 0\,0\,0 \\ 0 & 0 & 0 & 0\,0\,0 \\ 0 & 0 & 0 & 0\,0\,0 \\ 0 & 0 & 0 & 0\,0\,0 \end{bmatrix} + d(t) L d_u Q_{22}^{-1} L^T \right\} \eta(t) \quad (2.151)$$

Similarly the simplification of (2.149) will yield,

$$-\int_{t-d_u}^{t-d(t)} d_u \dot{x}^T(s) Q_{22} \dot{x}(s) ds \leq \eta^T(t) \left\{ d_u \begin{bmatrix} 0 & M_1 & -M_1 & 0\,0\,0 \\ \star & M_2 + M_2^T & -M_2 + M_3^T & 0\,0\,0 \\ \star & \star & -M_3 - M_3^T & 0\,0\,0 \\ 0 & 0 & 0 & 0\,0\,0 \\ 0 & 0 & 0 & 0\,0\,0 \\ 0 & 0 & 0 & 0\,0\,0 \end{bmatrix} + (d_u - d(t)) M d_u Q_{22}^{-1} M^T \right\} \eta(t) \quad (2.152)$$

Substituting the values of integrals from (2.151) and (2.152) in (2.147) and rearranging the terms one can write

$$-d_u \int_{t-d_u}^{t} \gamma^T(s) Q \gamma(s) ds \leq \eta^T(t) \{\Delta + d(t) L (d_u Q_{22}^{-1}) L^T + (d_u - d(t)) M (d_u Q_{22}^{-1}) M^T \} \eta(t) \quad (2.153)$$

where,

$$\Delta = \begin{bmatrix} d_u(L_1 + L_1^T) & d_u(-L_1 + L_2^T + M_1) & d_u(L_3^T - M_1) & 0 & 0 & 0 \\ \star & d_u(-L_2 - L_2^T + M_2 + M_2^T) & d_u(-L_3^T - M_2 + M_3^T) & 0 & 0 & 0 \\ \star & \star & d_u(-M_3 - M_3^T) & 0 & 0 & 0 \\ 0 & 0 & 0 & -Q_{11} & 0 & 0 \\ 0 & 0 & 0 & 0 & -Q_{11} & 0 \\ 0 & 0 & 0 & 0 & 0 & 0 \end{bmatrix},$$

Substituting (2.144) and (2.153) in (2.143) one can obtain,

$$\dot{V}_3(t) \leq \eta^T(t)\left\{\begin{bmatrix} \Upsilon_{11} & \Upsilon_{12} & \Upsilon_{13} & 0 & 0 & 0 \\ \star & \Upsilon_{22} & \Upsilon_{23} & 0 & 0 & 0 \\ \star & \star & \Upsilon_{33} & 0 & 0 & 0 \\ 0 & 0 & 0 & -Q_{11} & 0 & 0 \\ 0 & 0 & 0 & 0 & -Q_{11} & 0 \\ 0 & 0 & 0 & 0 & 0 & d_u^2 Q_{22} \end{bmatrix}\right.$$
$$\left. + d(t)L(d_u Q_{22}^{-1})L^T + (d_u - d(t))M(d_u Q_{22}^{-1})M^T\right\}\eta(t) \quad (2.154)$$

where, $\Upsilon_{11}{=}d_u^2 Q_{11} + d_u(L_1 + L_1^T)$, $\Upsilon_{12}{=}d_u(-L_1 + L_2^T + M_1)$

$\Upsilon_{13}{=}d_u(L_3^T - M_1)$, $\Upsilon_{22}{=}d_u(-L_2 - L_2^T + M_2 + M_2^T)$

$\Upsilon_{23}{=}d_u(-L_3^T - M_2 + M_3^T)$, $\Upsilon_{33}{=}d_u(-M_3 - M_3^T)$

Now for any matrices G_1 and G_2 the equation shown below is satisfied,

$$2[x^T(t)G_1^T + \dot{x}^T(t)G_2^T]\,[Ax(t) + A_d x(t - d(t)) - \dot{x}(t)] = 0 \quad (2.155)$$

Expanding (2.155) one can get,

$$\eta^T(t)\begin{bmatrix} G_1^T A + A^T G_1 & G_1^T A_d & 0 & 0 & 0 & -G_1^T + A^T G_2 \\ \star & 0 & 0 & 0 & 0 & A_d^T G_2 \\ 0 & 0 & 0 & 0 & 0 & 0 \\ 0 & 0 & 0 & 0 & 0 & 0 \\ 0 & 0 & 0 & 0 & 0 & 0 \\ \star & \star & 0 & 0 & 0 & -G_2 - G_2^T \end{bmatrix}\eta(t) = 0 \quad (2.156)$$

Now, adding all the three derivative terms in (2.141), (2.142) and (2.154), along with (2.156) one can obtain;

$$\dot{V}(t) \leq \eta^T(t)\left\{\Omega + \tilde{P}\mu T^{-1}\tilde{P}^T + d(t)L(d_u Q_{22}^{-1})L^T + (d_u - d(t))M(d_u Q_{22}^{-1})M^T\right\}\eta(t) \quad (2.157)$$

where, the matrix Ω is defined in (2.134). Since in (2.157) the term $d(t)L(d_u Q_{22}^{-1})L^T + (d_u - d(t))M(d_u Q_{22}^{-1})M^T$ is a convex combination of the matrices $L(d_u Q_{22}^{-1})L^T$ and $M(d_u Q_{22}^{-1})M^T$ on $d(t)$ so one can express (2.157) by two equivalent LMI conditions, one for $d(t) = 0$ and another for $d(t) = d_u$ that are described in (2.132) and (2.133) respectively.

Hence, to guarantee the asymptotic stability of the time-delay system Σ_2 in (2.3) the LMI conditions (2.132) and (2.133) need to be satisfied.

When the time-varying delay is not differentiable (or equivalently called fast time-varying delay), then by setting $P_{12} = 0$, $P_{22} = 0$, $Q_{11} = 0$ and $R_1 = 0$ in Theorem 2.18 one can obtain the delay-dependent stability condition for $\mu \geq 1$ which is presented in the following corollary.

Corollary 2.2 *The system Σ_2 for $\mu \geq 1$, satisfying the condition (2.5), is asymptotically stable if there exist matrices $R_2 = R_2^T > 0$, $P_{11} = P_{11}^T > 0$, $Q_{22} = Q_{22}^T > 0$ and any free matrices L_i, $(i = 1, ..3)$, $M_i(i = 1, ..3)$ such that following LMI's are satisfied:*

$$\begin{bmatrix} \tilde{\Omega} & d_u M \\ \star & -Q_{22} \end{bmatrix} < 0 \tag{2.158}$$

$$\begin{bmatrix} \tilde{\Omega} & d_u L \\ \star & -Q_{22} \end{bmatrix} < 0 \tag{2.159}$$

$$\tilde{\Omega} = \begin{bmatrix} \tilde{\Omega}_{11} & \tilde{\Omega}_{12} & \tilde{\Omega}_{13} & \Omega_{14} \\ \star & \tilde{\Omega}_{22} & \tilde{\Omega}_{23} & \tilde{\Omega}_{24} \\ \star & \star & \tilde{\Omega}_{33} & 0 \\ \star & \star & 0 & \tilde{\Omega}_{44} \end{bmatrix} < 0 \tag{2.160}$$

where, $\tilde{\Omega}_{11}=G_1^T A + A^T G_1 + R_2 + d_u(L_1 + L_1^T)$; $\tilde{\Omega}_{12}=G_1^T A_d + d_u(-L_1 + L_2^T + M_1)$

$\tilde{\Omega}_{13}=d_u(L_3^T - M_1)$, $\Omega_{14}=P_{11} - G_1^T + A^T G_2$

$\tilde{\Omega}_{22}=d_u(-L_2 - L_2^T + M_2 + M_2^T)$ $\tilde{\Omega}_{23}=d_u(-M_2 + M_3^T - L_3^T)$, $\Omega_{24}=A_d^T G_2$

$\tilde{\Omega}_{33}=d_u(-M_3 - M_3^T) - R_2$, $\Omega_{44}=-G_2^T - G_2 + d_u^2 Q_{22}$

Corollary 2.3 *The system Σ_2 satisfying the condition (2.5) for $\mu = 0$ (i.e., constant delay), is asymptotically stable if there exist matrices $R_i = R_i^T > 0$, $(i = 1,2)$ and any free matrices L_i, $(i = 1, ..3)$, $M_i(i = 1, ..3)$ such that following LMIs are satisfied*

$$P = \begin{bmatrix} P_{11} & P_{12} \\ P_{12}^T & P_{22} \end{bmatrix} > 0 \tag{2.161}$$

$$Q = \begin{bmatrix} Q_{11} & 0 \\ 0 & Q_{22} \end{bmatrix} > 0 \tag{2.162}$$

and,

$$\begin{bmatrix} \Omega & d_u M \\ \star & -Q_{22} \end{bmatrix} < 0 \tag{2.163}$$

$$\begin{bmatrix} \Omega & d_u L \\ \star & -Q_{22} \end{bmatrix} < 0 \tag{2.164}$$

where, $L = \begin{bmatrix} L_1^T & L_2^T & L_3^T & 0 & 0 & 0 \end{bmatrix}^T$, $M = \begin{bmatrix} M_1^T & M_2^T & M_3^T & 0 & 0 & 0 \end{bmatrix}^T$ *and*

$$\Omega = \begin{bmatrix} \Omega_{11} & \Omega_{12} & \Omega_{13} & \Omega_{14} & 0 & \Omega_{16} \\ \star & \Omega_{22}|_{\mu=0} & \Omega_{23} & \Omega_{24} & 0 & \Omega_{26} \\ \star & \star & \Omega_{33} & 0 & 0 & 0 \\ \star & \star & 0 & -Q_{11} & 0 & 0 \\ \star & \star & 0 & 0 & -Q_{11} & 0 \\ \star & \star & 0 & 0 & 0 & \Omega_{66} \end{bmatrix} < 0 \tag{2.165}$$

where, $\Omega_{11} = d_u^2 Q_{11} + G_1^T A + A^T G_1 + P_{12} + P_{12}^T + R_1 + R_2 + d_u(L_1 + L_1^T)$
$\Omega_{12} = G_1^T A_d - P_{12} + d_u(-L_1 + L_2^T + M_1)$
$\Omega_{13} = d_u(L_3^T - M_1)$, $\Omega_{14} = P_{22}$, $\Omega_{16} = P_{11} - G_1^T + A^T G_2$
$\Omega_{22} = -R_1 + d_u(-L_2 - L_2^T + M_2 + M_2^T)$
$\Omega_{23} = d_u(-M_2 + M_3^T - L_3^T)$, $\Omega_{24} = -P_{22}$, $\Omega_{26} = A_d^T G_2$
$\Omega_{33} = d_u(-M_3 - M_3^T) - R_2$, $\Omega_{66} = -G_2 - G_2^T + d_u^2 Q_{22}$

Remark 2.20 The proposed delay-dependent stability method has been obtained in an LMI framework by combining the method of [14] and [36]. The proposed method selects augmented LK functional of the type in [14] by introducing a new LK functional term ($\int_{t-d_u}^{t} x^T(s) R_2 x(s) ds$) and finally obtaining a convex combination of LMIs as in [36].

The proposed method differs with the method in [14] and its extension in [37] withe the fact that, the augmented state vector does not include the term $\int_{(s)} \dot{x}(s) ds$ and in sequel the matrix Q_{12} in $V_3(t)$ is not considered. Further due to the introduction of new delay-independent functional, the term $\int_{t-d_u}^{t-d(t)} x(s) ds$ is included in the augmented state vector and subsequently with the use of Jensen's integral inequality in the derivative of LK functional concerning integral of $x(s)$ can be taken inside the LMI condition, whereas the quadratic integral terms associated with the vector $\dot{x}(s)$ is eliminated using Newton-Leibniz formula. Unlike the stability condition in [14], the proposed method can estimate the delay upper bound for $\mu \geq 1$ which is due to the modification as suggested in the selection of LK functional followed by different techniques adopted for bounding the quadratic integral terms appearing in the LK functional derivative.

The proposed formulation can assess the stability of (i) TDS with constant time-delay, (ii) TDS with slow varying time-delay ($\mu < 1$) and (iii) TDS with fast varying time-delay ($\mu \geq 1$).

The proposed delay-dependent stability theorem and corollaries for different delay-derivatives by considering the numerical example 2.1 have been tested. The results of the delay upper bound are tabulated in Table 2.3 and 2.4.

Table 2.3 (d_u) results of Example 2.1 for $\mu = 0$ and $\mu < 1$

Stability methods	$\mu = 0$	$\mu = 0.5$	$\mu = 0.9$
[9, 27]	4.4721	2.00	1.18
[38]	4.4721	2.00	1.18
[39]	4.472	2.00	1.18
[40]	4.472	2.00	1.18
Cor. 3 [32]	4.4721	2.04	1.37
Cor. 1 [4]	4.4721	2.04	1.3789
Cor. 1 [5]	4.4721	2.07	1.5304
Cor. 3 [36]	4.472	2.337	1.873
Cor. 1 [14]	4.472	2.0100	1.1801
[31]	4.4721	2.0083	1.1801
Proposed method	4.4721 (Cor. 2.3)	2.3372 (§ 2.18)	1.8731 (§ 2.18)

Table 2.4 (d_u) results of Example 2.1 for $\mu \geq 1$

Stability methods	Unknown μ
Cor. 1 [4]	1.3454
Cor. 1 [5]	1.3454
Cor. 3 [36]	1.868
[40]	Not applicable
[14]	Not applicable
[39]	0.999
Cor. 2.2	1.868

Numerical Example 2.3 ([41]) *Consider the system Σ_2 with the following constant matrices*

$$A = \begin{bmatrix} -1 & 13.5 & -1 \\ -3 & -1 & -2 \\ -2 & -1 & -4 \end{bmatrix}, A_d = \begin{bmatrix} -5.9 & 7.1 & -70.3 \\ 2 & -1 & 5 \\ 2 & 0 & 6 \end{bmatrix}$$

The eigenvalues of the matrix $[A + A_d]$ is Hurwitz and the eigenvalues of the matrix $[A - A_d]$ is unstable, thus the given system is delay-dependently stable, i.e., the system is asymptotically stable for certain finite delay value. The exact delay upper bound value for this system is $d_u = 0.1624$ for $\mu = 0$ ([19] and [41]).

The delay upper bound estimate of Numerical Example 2.3 for $\mu = 0$ and $\mu < 1$ using proposed DDS condition are presented in Table 2.6.

Remark 2.21 The stability results presented in Tables 2.3 and 2.4 shows that the proposed stability method provides better estimate of delay upper bound compared to [14] for increasing delay derivatives, whereas the results of the proposed method is

Table 2.5 Comparison of decision variables and LMIs for Example 2.1

Stability methods	Decision variables	No. of LMIs
Cor. 3 [36]	$17n^2 + 9n$	6
Cor. 2.2	$15n^2 + 5n$	4

Table 2.6 (d_u) results of Example 2.3 for $\mu = 0$ and $\mu < 1$

Stability methods	$\mu = 0$	$\mu = 0.5$
Cor. 3 [32]	0.0751	–
Cor. 1 [4]	0.0751	–
Cor. 1 [5]	0.0751	–
Cor. 1 [14]	0.1091	0.0723
[36]	–	0.0736
Present method	0.0803 (§ 2.18)	0.0736 (Cor. 2.3)

found to be same when compared with the results obtained in [4, 5, 32, 36] for $\mu = 0$. Note that the stability condition in [36] is not in a conventional LMI framework. The advantage of the proposed method over [36] is indicated in Table 2.5.

Remark 2.22 The LK functional considered in Theorem 2.18 can be treated as generalized one as other choices of functional can be obtained from (2.135)–(2.137) with following choices of the matrices described below:

1. setting $P_{12} = P_{22} = 0,\ P_{11} = P,\ Q_{11} = Q_{12} = 0$ and $Q_{22} = \frac{Q}{d_u}$ in proposed LK functional yields the LK function of [32].
2. $P_{12} = P_{22} = 0,\ P_{11} = P,\ Q_{11} = Q_{12} = 0$ and $Q_{22} = Z_2, Q_3 = R_1$ and $Q_2 = R_2$ in proposed LK functional yields LK functional of [5] and [4].

2.5.2 Stability Analysis of TDS with Two Additive Time-Varying Delays

Consider the system Σ_4 described in (2.11) satisfying the conditions (2.13), the stability condition for this system is presented in the following theorem.

Theorem 2.19 ([42]) *The system Σ_4 described in (2.11) satisfying the conditions (2.13)–(2.14) is asymptotically stable if there exist $P = P^T > 0, Q_1 = Q_1^T > 0, Q_2 = Q_2^T > 0, Q_3 = Q_3^T > 0, R_1 = R_1^T > 0, R_2 = R_2^T > 0, R_3 = R_3^T > 0$ and $G_1, G_2, G_3, G_4, M_1, M_2, M_3, M_4, N_1, N_2, N_3$ and N_4 are free matrices with $Q_2 \geq Q_3$. satisfying following LMI:*

$$\begin{bmatrix} \Omega_{11} & \Omega_{12} & \Omega_{13} & \Omega_{14} & L_1 & M_1 & N_1 \\ \star & \Omega_{22} & \Omega_{23} & \Omega_{24} & L_2 & M_2 & N_2 \\ \star & \star & \Omega_{33} & \Omega_{34} & L_3 & M_3 & N_3 \\ \star & \star & \star & \Omega_{44} & L_4 & M_4 & N_4 \\ \star & \star & \star & \star & -\frac{1}{d_u}R_1 & 0 & 0 \\ \star & \star & \star & \star & \star & -\frac{1}{d_{u1}}R_2 & 0 \\ \star & \star & \star & \star & \star & \star & -\frac{1}{d_{u2}}R_3 \end{bmatrix} < 0 \tag{2.166}$$

where, $\Omega_{11}=Q_1 + Q_2 + G_1 A + A^T G_1^T + L_1 + L1^T + M_1 + M_1^T$
$\Omega_{12}=A^T G_2^T + L_2^T - M_1 + M_2^T + N_1$, $\Omega_{13}=G_1 A_d + A_T G_3^T - L_1 + L_3^T + M_3^T - N_1$
$\Omega_{14}=P - G_1 + A^T G_4^T + L_4^T + M_4^T$, $\Omega_{22}=-(1 - d_{u1})(Q_2 - Q_3) - M_2 - M_2^T + N_2 + N_2^T$
$\Omega_{23}=G_2 A_d - L_2 - M_3^T - N_2 + N_3^T$, $\Omega_{24}=-G_2 - M_4^T + N_4^T$
$\Omega_{33}=-(1 - d_{u1} - d_{u2})(Q_1 + Q_3) + G_3 A_d + A_d^T G_3^T - L_3 - L_3^T - N_3 - N_3^T$
$\Omega_{34}=-G_3 + A_d^T G_4^T - L_4 - N_4^T$, and $\Omega_{44}=d_u R_1 + d_{u1} R_2 + d_{u2} R_3 - G_4 - G_4^T$

Proof Considering the LK functional candidate as

$$V(t) = V_1(t) + V_2(t) + V_3(t) \tag{2.167}$$

$$V_1(t) = x^T(t) P x(t) \tag{2.168}$$

$$V_2(t) = \int_{t-d_1(t)-d_2(t)}^{t} x^T(s) Q_1 x(s) ds + \int_{t-d_1(t)}^{t} x^T(s) Q_2 x(s) ds + \int_{t-d_1(t)-d_2(t)}^{t-d_1(t)} x^T(s) Q_3 x(s) ds \tag{2.169}$$

$$V_3(t) = \int_{t-d_{u1}-d_{u2}}^{t} \int_{\theta}^{t} \dot{x}^T(s) R_1 \dot{x}(s) ds d\theta + \int_{t-d_{u1}}^{t} \int_{\theta}^{t} \dot{x}^T(s) Q_2 \dot{x}(s) ds d\theta + \int_{t-d_{u1}-d_{u2}}^{t-d_{u1}} \int_{\theta}^{t} \dot{x}^T(s) Q_3 \dot{x}(s) ds d\theta \tag{2.170}$$

Finding the time-derivative of (2.168)–(2.170). The time derivative of $\dot{V}_3(t)$ will be,

$$\begin{aligned}
\dot{V}_3(t) &= d_u \dot{x}^T(t) R_1 \dot{x}(t) - \int_{t-d_u}^{t} \dot{x}^T(s) R_1 \dot{x}(s) ds \\
&+ d_{u1} \dot{x}^T(t) R_2 \dot{x}(t) - \int_{t-d_{u1}}^{t} \dot{x}^T(s) R_2 \dot{x}(s) ds \\
&+ d_{u2} \dot{x}^T(t) R_3 \dot{x}(t) - \int_{t-d_u}^{t-d_{u1}} \dot{x}^T(s) R_3 \dot{x}(s) ds
\end{aligned} \tag{2.171}$$

As $\dot{V}_3(t)$ contains integral terms so to formulate quadratic conditions we need to replace them. For any symmetric positive definite matrices R_1, R_2 and R_3, the following inequalities are satisfied,

$$\begin{aligned}
-\int_{t-d_u}^{t} \dot{x}^T(s) R_1 \dot{x}(s) ds &\le -\int_{t-d(t)}^{t} \dot{x}^T(s) R_1 \dot{x}(s) ds \\
-\int_{t-d_{u1}}^{t} \dot{x}^T(s) R_2 \dot{x}(s) ds &\le -\int_{t-d_1(t)}^{t} \dot{x}^T(s) R_2 \dot{x}(s) ds \\
-\int_{t-d_u}^{t-d_{u1}} \dot{x}^T(s) R_3 \dot{x}(s) ds &\le -\int_{t-d(t)}^{t-d_1(t)} \dot{x}^T(s) R_3 \dot{x}(s) ds
\end{aligned} \tag{2.172}$$

thus (2.171) can be written as,

$$\begin{aligned}
\dot{V}_3(t) &\le \dot{x}^T(t)(d_u R_1 + d_{u1} R_2 + d_{u2} R_3)\dot{x}(t) - \int_{t-d(t)}^{t} \dot{x}^T(s) R_1 \dot{x}(s) ds \\
&- \int_{t-d_1(t)}^{t} \dot{x}^T(s) R_2 \dot{x}(s) ds - \int_{t-d(t)}^{t-d_1(t)} \dot{x}^T(s) R_3 \dot{x}(s) ds
\end{aligned} \tag{2.173}$$

Removing the integral terms from (2.173) with the help of Newton-Leibniz formula and introduction of free matrices. Free matrices $L_i, i = 1, 2, 3, 4$, $M_i, i = 1, 2, 3, 4$ and $N_i, i = 1, 2, 3, 4$ are introduced in the second, third and fourth integral terms respectively in (2.173). The following identities with free matrices will satisfy

$$\begin{aligned}
\Sigma &= 2[x^T L_1 + x^T(t - d_1(t)) L_2 + x^T(t - d(t)) L_3 + \dot{x}(t) L_4] \\
&\times [x(t) - x(t - d(t)) - \int_{t-d(t)}^{t} \dot{x}(s) ds] = 0
\end{aligned} \tag{2.174}$$

$$\begin{aligned}
0 &= 2[x^T M_1 + x^T(t - d_1(t)) M_2 + x^T(t - d(t)) M_3 + \dot{x}^T(t) M_4] \\
&\times [x(t) - x(t - d_1(t)) - \int_{t-d_1(t)}^{t} \dot{x}(s) ds]
\end{aligned} \tag{2.175}$$

$$0 = 2[x^T N_1 + x^T(t - d_1(t))N_2 + x^T(t - d(t))N_3 + \dot{x}(t)N_4] \times [x(t - d_1(t)) - x(t - d(t)) - \int_{t-d(t)}^{t-d_1(t)} \dot{x}(s)ds] \quad (2.176)$$

Simple algebraic manipulations along with the application of Lemma 2.1 one can obtain following,

$$-\int_{t-d(t)}^{t} \dot{x}^T(s)R_1\dot{x}(s)ds \leq \xi^T \begin{bmatrix} L_1 + L_1^T & L_2^T & -L_1 + L_3^T & L_4^T \\ \star & 0 & -L_2 & 0 \\ \star & \star & -L_3 - L_3^T & -L_4^T \\ \star & \star & \star & 0 \end{bmatrix} \xi(t) + \xi^T(t)d_u \begin{bmatrix} L_1 \\ L_2 \\ L_3 \\ L_4 \end{bmatrix} R_1^{-1} \begin{bmatrix} L_1 \\ L_2 \\ L_3 \\ L_4 \end{bmatrix}^T \xi(t) \quad (2.177)$$

$$-\int_{t-d_1(t)}^{t} \dot{x}^T(s)R_2\dot{x}(s)ds \leq \xi^T \begin{bmatrix} M_1 + M_1^T & -M_1 + M_2^T & M_3^T & M_4^T \\ \star & -M_2 - M_2^T & -M_3 & -M_4^T \\ \star & \star & 0 & 0 \\ \star & \star & \star & 0 \end{bmatrix} \xi(t) + \xi^T d_{u1} \begin{bmatrix} M_1 \\ M_2 \\ M_3 \\ M_4 \end{bmatrix} R_2^{-1} \begin{bmatrix} M_1 \\ M_2 \\ M_3 \\ M_4 \end{bmatrix}^T \xi(t) \quad (2.178)$$

$$-\int_{t-d(t)}^{t-d_1(t)} \dot{x}^T(s)R_2\dot{x}(s)ds \leq \xi^T \begin{bmatrix} 0 & N_1 & -N_1^T & 0 \\ \star & N_2 + N_2^T & -N_2 + N_3^T & N_4^T \\ \star & \star & -N_3 - N_3^T & -N_4^T \\ \star & \star & \star & 0 \end{bmatrix} \xi(t) + \xi^T d_{u2} \begin{bmatrix} N_1 \\ N_2 \\ N_3 \\ N_4 \end{bmatrix} R_3^{-1} \begin{bmatrix} N_1 \\ N_2 \\ N_3 \\ N_4 \end{bmatrix}^T \xi(t) \quad (2.179)$$

In this formulation as system dynamics $\dot{x}(t)$ is retained in the formulation, so introducing free matrices $G_i, i = 1, 2, 3, 4$ system matrices are introduced in the LMI expression as,

$$\Sigma = 2[x^T G_1 + x^T(t - d_1(t))G_2 + x^T(t - d(t))G_3 + \dot{x}(t)G_4] \times [-\dot{x}(t) + Ax(t) + A_d x(t - d(t))] = 0 \quad (2.180)$$

Simplifying (2.180) one can write

$$\xi^T(t)\begin{bmatrix} G_1A+A^TG_1^T & A^TG_2^T & G_1A_d+A^TG_3^T & -G_1+A^TG_4^T \\ \star & 0 & G_2A_d & -G_2 \\ \star & \star & G_3A_d+A_d^TG_3^T & -G_3 \\ \star & \star & \star & -G_4-G_4^T \end{bmatrix}\xi(t)=0 \quad (2.181)$$

where, $\xi(t)=\left[x(t)^T\ x(t-d_1(t))^T\ x(t-d(t))^T\ \dot{x}(t)^T\right]^T$

Substituting (2.177), (2.178) and (2.179) into (2.173) and then finally adding all the derivative terms one obtains

$$\dot{V}(t)\le\xi^T(t)\begin{bmatrix} \Omega_{11} & \Omega_{12} & \Omega_{13} & \Omega_{14} & L_1 & M_1 & N_1 \\ \star & \Omega_{22} & \Omega_{23} & \Omega_{24} & L_2 & M_2 & N_2 \\ \star & \star & \Omega_{33} & \Omega_{34} & L_3 & M_3 & N_3 \\ \star & \star & \star & \Omega_{44} & L_4 & M_4 & N_4 \\ \star & \star & \star & \star & -\frac{1}{d_u}R_1 & 0 & 0 \\ \star & \star & \star & \star & \star & -\frac{1}{d_{u1}}R_2 & 0 \\ \star & \star & \star & \star & \star & \star & -\frac{1}{d_{u2}}R_3 \end{bmatrix}\xi(t) \quad (2.182)$$

The LMI (2.182) is defined in (2.166). This completes the proof of the theorem. □

Remark 2.23 The numerical example 2.1 is considered for validating the result of this theorem. The delay-derivatives are assumed to be $\dot{d}_1(t)\le 0.1$ and $\dot{d}_2(t)\le 0.8$. Here the delay upper bound d_{u1} or d_{u2} is calculated, when the value of either one is known. By combining the two delay factors the results of some existing stability theorems have been provided in the Tables 2.7 and 2.8 respectively. The results validate the fact that the formulation of the stability conditions in LMI framework by single delay approach for ascertaining the stability of time-delay systems provides conservative results compared to independent treatment of the delays.

Table 2.7 Computed delay bound d_{u2} for a given d_{u1} with $\mu_1=0.1$ and $\mu_2=0.8$

Stability methods	$d_{u1}=1$	$d_{u1}=1.2$	$d_{u1}=1.5$	Remarks
Theorem 2.19	0.5188	0.4528	0.3777	†
[43]	0.512	0.406	0.283	†
[33]	0.415	0.376	0.283	†
[9, 13, 44]	0.180	0.080	Infeasible	‡
[45]	Infeasible	Infeasible	Infeasible	‡
[18]	Infeasible	Infeasible	Infeasible	‡

'†' – $d_1(t)$ and $d_2(t)$ treated separately
'‡'- $d_1(t)$ and $d_2(t)$ treated combinedly

Table 2.8 Computed delay bound d_{u1} for a given d_{u2} with $\mu_1 = 0.1$ and $\mu_2 = 0.1$

Stability methods	$d_{u2} = 0.1$	$d_{u2} = 0.2$	$d_{u3} = 0.3$	Remarks
Theorem 2.19	2.9182	2.3304	1.8324	†
[43]	2.300	1.779	1.453	†
[33]	2.263	1.696	1.324	†
[9, 13, 44]	1.080	0.980	0.880	‡
[45]	0.098	Infeasible	Infeasible	‡
[18]	0.074	Infeasible	Infeasible	‡

'†' - $d_1(t)$ and $d_2(t)$ treated separately
'‡'- $d_1(t)$ and $d_2(t)$ treated combinedly

2.5.3 *Stability Analysis of TDS with Interval Time Varying Delay*

Consider the system Σ_2 described in (2.3) satisfying following conditions

$$0 \leq d_l \leq d(t) \leq d_u \tag{2.183}$$

$$d_{lu} = d_u - d_l \tag{2.184}$$

$$0 \leq \dot{d}(t) \leq \mu \tag{2.185}$$

To establish the stability condition using LK functional approach, the integral inequalities arising in the LK functional derivatives need to be approximated by tight bounding inequality to achieve less conservative delay upper bound estimate. To do so, two different types of bounding inequalities have been proposed depending upon the nature of the limits of integration in the integral inequalities and they are discussed below.

- If the limit of integral inequality is certain, then the approximation can be done by following bounding inequality,

$$-\int_{t-\beta}^{t-\alpha} \dot{x}^T(\theta)R\dot{x}(\theta)d\theta \leq \gamma^{-1} \begin{bmatrix} x(t-\alpha) \\ x(t-\beta) \end{bmatrix}^T \begin{bmatrix} -R & R \\ \star & -R \end{bmatrix} \begin{bmatrix} x(t-\alpha) \\ x(t-\beta) \end{bmatrix} \tag{2.186}$$

- If the limit of integral inequality is uncertain, then the following bounding inequality that can be used is,

$$-\int_{t-\beta}^{t-\alpha}\dot{x}^T(\theta)R\dot{x}(\theta)d\theta \le \begin{bmatrix} x(t-\alpha) \\ x(t-\beta) \end{bmatrix}^T \begin{bmatrix} M+M^T & -M+N^T \\ \star & -N-N^T \end{bmatrix} \begin{bmatrix} x(t-\alpha) \\ x(t-\beta) \end{bmatrix}$$
$$+\gamma \begin{bmatrix} x(t-\alpha) \\ x(t-\beta) \end{bmatrix}^T \begin{bmatrix} M \\ N \end{bmatrix} R^{-1} \begin{bmatrix} M \\ N \end{bmatrix}^T \begin{bmatrix} x(t-\alpha) \\ x(t-\beta) \end{bmatrix} \quad (2.187)$$

where, $\gamma = \beta - \alpha$. The main result of the delay-range-dependent stability of the time-delay system (Σ_2) is presented below in the form of theorem.

Theorem 2.20 *System Σ_2 is asymptotically stable satisfying the conditions (2.183)–(2.185) if there exist symmetric matrices $P > 0$, $Q_i > 0, i = 1, 2, \ldots 4$, $R_j > 0$, and any arbitrary matrices of appropriate dimensions M_j, N_j, $j = 1, 2$ satisfying the following LMIs:*

$$\begin{bmatrix} \Theta & \Phi_1 \\ \star & -R_2 \end{bmatrix} < 0, \quad (2.188)$$

$$\begin{bmatrix} \Theta & \Phi_2 \\ \star & -R_2 \end{bmatrix} < 0, \quad (2.189)$$

where, $\Phi_1 = \begin{bmatrix} 0 \\ M_1 \\ N_1 \\ 0 \end{bmatrix}, \Phi_2 = \begin{bmatrix} 0 \\ 0 \\ M_2 \\ N_2 \end{bmatrix}, \Theta = \begin{bmatrix} \Theta_{11} & R_1 & \Theta_{13} & 0 \\ \star & \Theta_{22} & \Theta_{23} & 0 \\ \star & \star & \Theta_{33} & \Theta_{34} \\ \star & \star & \star & \Theta_{44} \end{bmatrix}$

$\Theta_{11}=PA + A^T P + \sum_{i=1}^{3} Q_i + A^T(d_l^2 R_1 + R_2)A - R_1,$

$\Theta_{13}=PA_d + A^T(d_l^2 R_1 + R_2)A_d + R_1$, $\Theta_{22}=Q_4 - Q_1 - R_1 + d_{lu}^{-1}(M_1 + M_1^T),$

$\Theta_{23}=d_{lu}^{-1}(-M_1 + N_1^T),$

$\Theta_{33}=-(1-\mu)(Q_3 + Q4) + A_d^T(d_l^2 R_1 + R_2)A_d + d_{lu}^{-1}(M_2 + M_2^T - N_1 - N_1^T),$

$\Theta_{34}=d_{lu}^{-1}(-M_2 + N_2^T)$, $\Theta_{44}=-Q_2 - d_{lu}^{-1}(N_2 + N_2^T)$

Proof Selecting the LK functional candidate as

$$V(t) = x^T(t)Px(t) + \int_{t-d_l}^{t} x^T(\theta)Q_1x(\theta)d\theta$$
$$+ \int_{t-d_u}^{t} x^T(\theta)Q_2x(\theta)d\theta + \int_{t-d(t)}^{t} x^T(\theta)Q_3x(\theta)d\theta$$
$$+ \int_{t-d(t)}^{t-d_l} x^T(\theta)Q_4x(\theta)d\theta + d_l\int_{t-d_l}^{t}\int_{\theta}^{t} \dot{x}^T(s)R_1\dot{x}(s)dsd\theta$$

$$
+d_{lu}^{-1}\int_{t-d_u}^{t-d_l}\int_{\theta}^{t}\dot{x}^T(s)R_2\dot{x}(s)d\theta \tag{2.190}
$$

Finding the time-derivative of (2.190) one can obtain

$$
\begin{aligned}
\dot{V}(t) &= 2x^T(t)P\dot{x}(t) + x^T(t)(\sum_{i=1}^{3} Q_i)x(t) \\
&\quad -(1-\dot{d}(t))x^T(t-d(t))(\sum_{i=3}^{4} Q_i)x(t-d(t)) \\
&\quad -x^T(t-d_l)(Q_1-Q_4)x(t-d_l) - x^T(t-d_u)Q_2x(t-d_u) \\
&\quad +\dot{x}^T(t)(d_l^2R_1+R_2)\dot{x}(t) - d_l\int_{t-d_l}^{t}\dot{x}^T(\theta)R_1\dot{x}(\theta)d\theta \\
&\quad -d_{lu}^{-1}\int_{t-d_u}^{t-d_l}\dot{x}^T(\theta)R_2\dot{x}(\theta)d\theta
\end{aligned} \tag{2.191}
$$

Replacing $\dot{x}(t)$ using (2.3) in (2.191) one gets

$$
\begin{aligned}
\dot{V}(t) &= 2x^T(t)PAx(t) + 2x^T(t)PA_dx(t-d(t)) + x^T(t)(\sum_{i=1}^{3} Q_i)x(t) \\
&\quad -(1-\dot{d}(t))x^T(t-d(t))(\sum_{i=3}^{4} Q_i)x(t-d(t)) \\
&\quad -x^T(t-d_l)(Q_1-Q_4)x(t-d_l) - x^T(t-d_u)Q_2x(t-d_u) \\
&\quad +(Ax(t)+A_dx(t-d(t)))^T\{d_l^2R_1+R_2\}(Ax(t)+A_dx(t-d(t))) \\
&\quad -d_l\int_{t-d_l}^{t}\dot{x}^T(\theta)R_1\dot{x}(\theta)d\theta - d_{lu}^{-1}\int_{t-d_u}^{t-d_l}\dot{x}^T(\theta)R_2\dot{x}(\theta)d\theta
\end{aligned} \tag{2.192}
$$

Now, using (2.186) the first integral term in (2.192) is approximated as,

$$
-d_l\int_{t-d_l}^{t}\dot{x}^T(\theta)R_1\dot{x}(\theta)d\theta \le \begin{bmatrix} x(t) \\ x(t-d_l) \end{bmatrix}^T \begin{bmatrix} -R_1 & R_1 \\ \star & -R_1 \end{bmatrix} \begin{bmatrix} x(t) \\ x(t-d_l) \end{bmatrix} \tag{2.193}
$$

Next, the last integral term in (2.192) can be written as,

$$-d_{lu}^{-1}\int_{t-d_u}^{t-d_l}\dot{x}^T(\theta)R_2\dot{x}(\theta)d\theta = -d_{lu}^{-1}\int_{t-d(t)}^{t-d_l}\dot{x}^T(\theta)R_2\dot{x}(\theta)d\theta$$
$$-d_{lu}^{-1}\int_{t-d_u}^{t-d(t)}\dot{x}^T(\theta)R_2\dot{x}(\theta)d\theta \quad (2.194)$$

Now, using (2.187), one may approximate the two integral terms of (2.194) as,

$$-d_{lu}^{-1}\int_{t-d(t)}^{t-d_l}\dot{x}^T(\theta)R_2\dot{x}(\theta)d\theta \le \begin{bmatrix} x(t-d_l) \\ x(t-d(t)) \end{bmatrix}^T \Big\{d_{lu}^{-1}$$
$$\times \begin{bmatrix} M_1^T + M_1^T & -M_1 + N_1^T \\ \star & -N_1 - N_1^T \end{bmatrix} + \varrho \begin{bmatrix} M_1 \\ N_1 \end{bmatrix} R_2^{-1} \begin{bmatrix} M_1 \\ N_1 \end{bmatrix}^T \Big\}$$
$$\times \begin{bmatrix} x(t-d_l) \\ x(t-d(t)) \end{bmatrix} \quad (2.195)$$

and,

$$-d_{lu}^{-1}\int_{t-d_u}^{t-d(t)}\dot{x}^T(\theta)R_2\dot{x}(\theta)d\theta \le \begin{bmatrix} x(t-d(t)) \\ x(t-d_u) \end{bmatrix}^T \Big\{d_{lu}^{-1}$$
$$\times \begin{bmatrix} M_2^T + M_2^T & -M_2 + N_2^T \\ \star & -N_2 - N_2^T \end{bmatrix}$$
$$+ (1-\varrho) \begin{bmatrix} M_2 \\ N_2 \end{bmatrix} R_2^{-1} \begin{bmatrix} M_2 \\ N_2 \end{bmatrix}^T \Big\}$$
$$\times \begin{bmatrix} x(t-d(t)) \\ x(t-d_u) \end{bmatrix} \quad (2.196)$$

where,

$$\varrho = \frac{d(t)-d_l}{d_{lu}}, 0 \le \varrho \le 1 \quad (2.197)$$

In view of (2.185) one may replace the uncertain $\dot{d}(t)$ by μ in (2.192). Substituting (2.195) and (2.196) into (2.194) and substituting the integral term (2.193) into (2.192) one can get

$$\dot{V}(t) \le \xi^T(t)(\Theta + \varrho\Phi_1 R_2^{-1}\Phi_1^T + (1-\varrho)\Phi_2 R_2^{-1}\Phi_2^T)\xi(t) \quad (2.198)$$

where, Θ, Φ_1 and Φ_2 are already defined above and
$\xi(t) = \left[x^T(t)\ x^T(t-d_l)\ x^T(t-d(t))\ x^T(t-d_u)\right]^T$

To ensure the asymptotic stability of (2.3) the matrix $\Theta + \varrho\Phi_1 R_2^{-1}\Phi_1^T + (1-\varrho)\Phi_2 R_2^{-1}\Phi_2^T)$ must be negative definite which further can be simplified in the following form,

$$\varrho(\Theta + \Phi_1 R_2^{-1}\Phi_1^T) + (1-\varrho)(\Theta + \Phi_2 R_2^{-1}\Phi_2^T) < 0 \tag{2.199}$$

The above matrix inequality holds equivalently can be written as the following two LMIs.

$$\Theta + \Phi_1 R_2^{-1}\phi_1^T < 0 \tag{2.200}$$

$$\Theta + \Phi_2 R_2^{-1}\phi_2^T < 0 \tag{2.201}$$

Finally using Schur-complement on (2.200) and (2.201) one can obtain the stability condition stated in (2.188) and (2.189).

For the case of $d_l = 0$, one may set $Q_1 = Q_4 = R_1 = 0$ in (2.190). Then Theorem 2.20 reduces to following corollary.

Corollary 2.4 *System Σ_2 with $d_l = 0$ is asymptotically stable satisfying the conditions (2.183)–(2.185) if there exist symmetric matrices $P > 0$, $Q_i > 0, i = 2, 3$ $R_2 > 0$, and any arbitrary matrices of appropriate dimensions $M_j, N_j, j = 1, 2$ satisfying the following LMIs:*

$$\begin{bmatrix} \Sigma & \Xi_1 \\ \star & -R_2 \end{bmatrix} < 0, \tag{2.202}$$

$$\begin{bmatrix} \Sigma & \Xi_2 \\ \star & -R_2 \end{bmatrix} < 0, \tag{2.203}$$

where, $\Xi_1 = \begin{bmatrix} M_1 \\ N_1 \\ 0 \end{bmatrix}$, $\Xi_2 = \begin{bmatrix} 0 \\ M_2 \\ N_2 \end{bmatrix}$, $\Sigma = \begin{bmatrix} \Sigma_{11} & \Sigma_{12} & 0 \\ \star & \Sigma_{22} & \Sigma_{23} \\ \star & \star & \Sigma_{33} \end{bmatrix}$

$\Sigma_{11} = PA + A^T P + \sum_{i=2}^{3} Q_i + A^T R_2 A + d_u^{-1}(M_1 + M_1^T)$,
$\Sigma_{12} = PA_d + A^T R_2 A_d + d_u^{-1}(-M_1 + N_1^T)$,
$\Sigma_{22} = -(1-\mu)Q_3 + A_d^T R_2 A_d + d_u^{-1}(M_2 + M_2^T - N_1 - N_1^T)$
$\Sigma_{23} = d_u^{-1}(-M_2 + N_2^T)$, $\Sigma_{33} = -Q_2 - d_u^{-1}(N_2 + N_2^T)$

Numerical Example 2.4 *Consider the system* Σ_2 *with the following constant matrices*

$$A = \begin{bmatrix} 0 & 1 \\ -1 & -2 \end{bmatrix}, A_d = \begin{bmatrix} 0 & 0 \\ -1 & -1 \end{bmatrix}$$

For a given delay lower bound (for different μ) the delay upper bound estimates are computed using the obtained LMI condition for Numerical Example 2.1 and Numerical Example 2.4 that are tabulated in Tables 2.9 and 2.10 respectively. The computed results are compared with the existing results.

Remark 2.24 The proposed delay-range-dependent stability analysis is a modification over the work in [5]. It is observed in [5] that, to reduce the conservatism in the delay upper bound estimate, a convex combination of LMIs are derived by approximating the uncertain factor $\gamma = \beta - \alpha$. However, approximation of this γ could not be fully avoided in [5] as the terms $d(t) - d_l$ and $d_u - d(t)$ are assigned to zero.

To avoid the approximation on uncertain factor γ stability analysis is derived using both the integral inequalities (2.186) and (2.187), instead of using single bounding inequality as in Lemma 1 of [5]. To implement the bounding inequality in (2.187) along with (2.186) a new LK functional is proposed. The inequality in (2.187) is used when the limit of integral is uncertain, else (2.186). This is an important feature and in addition helps to improve the delay upper bound estimate.

Remark 2.25 Note that the present method emphasizes the less conservativeness of the derived stability criterion based on the LK functional that belongs to the class of (2.190). However, there exists LK functional that is based on delay partitioning approach (where the delay intervals are further divided into sub intervals) as in [35] and [47]. It appears that the use of such LK functional along with the present bounding inequalities proposed in (2.186) and (2.187) may leads to further improvement in the estimate of the delay upper bound estimate.

Remark 2.26 In order to derive less conservative stability condition a separate class of LK functional was considered that consists of triple integral as well as single-integral terms with $t - d_l$ as the upper limit of integration in the latter term [46]. But it is observed from the results presented in Tables 2.9, 2.10 and 2.11 that the suggested modifications cannot effectively and consistently reduce the conservatism in the delay bound estimate compared to the results obtained by the proposed method.

Table 2.9 Computed delay bound d_u for different μ (Example 2.1)

d_l	[35]	[3]	[5]	[46]	Present result
$\mu = 0.5$					
0	–	2.04	2.0723	–	2.2594
1	–	2.07	2.1277	–	2.3303
2	–	2.43	2.5049	–	2.6127
3	–	3.22	3.2591	–	3.3147
4	–	4.07	4.0744	–	4.0900
$\mu = 0.9$					
0	–	1.37	1.5305	–	1.8502
1	–	1.74	1.8736	–	2.0550
2	–	2.43	2.5049	–	2.6127
3	–	3.22	3.2591	–	3.3147
4	–	4.07	4.0744	–	4.0900
$\mu \geq 1$					
0	1.01	1.34	1.5295	–	1.8497
1	1.64	1.7424	1.8737	–	2.0550
2	2.39	2.4328	2.5049	2.5663	2.6127
3	3.20	3.2234	3.2591	3.3408	3.3147
4	4.06	4.0644	4.0744	4.1690	4.0900

Table 2.10 Computed delay bound d_u for $\mu = 0.3$ (Example 2.4)

d_l	[3]	[4]	[5]	[46]	Present result
0	2.19	2.1959	2.2161	–	2.3369
1	2.2125	2.2128	2.2474	2.3167 - Theorem 2	2.4043
2	2.4091	2.4179	2.4798	–	2.5871
3	3.3342	3.3382	3.3893	–	3.4766
4	4.2799	4.2819	4.3250	–	4.3978
5	5.2393	5.2403	5.2773	–	5.3394

Table 2.11 Computed delay bound d_u for $\mu \geq 1$ (Example 2.4)

d_l	[35]	[3]	[4]	[5]	[46]	Present result
0	0.67	0.77	0.7744	0.8714	–	1.0420
0.3	0.91	0.9431	0.9860	1.0715	–	1.2301
0.5	1.07	1.0991	1.1325	1.2191		1.3713
0.8	1.33	1.3476	1.3733	1.4539	–	1.5960
1	1.50	1.5187	1.5401	1.6169	1.6198	1.7523
2	2.39	2.4000	2.4100	2.4798	2.4884	2.5871

2.6 Robust Stability Analysis of Time-Delay System

The robust stability problem of time-delay systems considers parametric uncertainties in the system matrices. The uncertainty arises in the system model due to following reasons:

1. System parameters are often not known accurately while modeling, rather the ranges are known.
2. Due to the limitation on the part of availability of the mathematical tools, one tends to create simple mathematical models that approximate a practical systems, thus some aspects of system dynamics are ignored, which is known as un-modeled dynamics.
3. Some control strategies are required to operate the systems under different operating conditions.

2.6.1 *Characteristic of Structured Uncertainties [7]*

To include these parametric uncertainties, a bounding set containing all possible uncertainties in the set is considered that makes the mathematical analysis and synthesis simpler. In the present discussion, the structured uncertainties are considered for robust stability analysis. Consider a single time-delay system,

$$\dot{x}(t) = A(t)x(t) + A_d(t)x(t - d(t)) \tag{2.204}$$

where, $A(t)$ and $A_d(t)$ are uncertain system matrices, and defined within a compact uncertain set Π as,

$$(A(t), A_d(t)) \in \Pi, \forall t \geq 0 \tag{2.205}$$

The various characterization of the structured uncertainties [7] are as follows,

(i) **Polytopic uncertainty**: In practice the parameters of the system are not completely known and may vary between lower and upper limits, and these uncertain parameters are found to vary linearly in the system matrices. Thus, the collection of all possible system matrices form a polytopic uncertainty set. Say there exist n_p uncertain parameters, then the number of uncertain elements in the set Π is $n_m = 2^{n_p}$, as the parameters vary between upper and lower limits,

$$\pi^k = [A^k, A_d^k], k = 1, 2, \ldots n_m$$

n_m is known as vertices. The uncertainty set Π is expressed as the convex hull of these vertices as given by,

$$\Pi = \left\{ \sum_{k=1}^{n_m} \alpha_k \pi^k | \alpha_k \geq 0, k = 1, 2, \ldots n_m; \sum_{k=1}^{n_m} \alpha_k = 1 \right\}$$

(ii) **Sub-polytopic uncertainty**: The subpolytopic uncertainty is more general than polytopic uncertainty. In this case the uncertainty set π possesses n_m vertices and the uncertainty set Π is contained in the convex hull of the vertices

$$\Pi \subset co\left\{\pi^i, i = 1, 2, \ldots n_m\right\}$$

where, $\pi = \sum_{i=1}^{n_m} \beta_i \pi_i$, for some scalar $\beta_i \geq 0$ and $\sum_{i=1}^{n_m} \beta_i = 1$

(iii) **Norm-bounded uncertainty**: Here the uncertain system matrices $\pi = (A(t), A_d(t))$ in (2.204) is decomposed into two parts, nominal part $\pi_n = (A, A_d)$ and the uncertain part $\Delta\pi = (\Delta A, \Delta A_d)$, thus $\pi = \pi_n + \Delta\pi$. The uncertain part can be further decomposed as [12],

$$\Delta A = D_a F_a(t) E_a$$
$$\Delta A_d = D_d F_d(t) E_d$$

where, $F_a(t)$ and $F_d(t)$ are unknown real time-varying matrices with Lebesgue measurable elements satisfying

$$\| F_a(t) \| \leq 1$$
$$\| F_d(t) \| \leq 1$$

and, D_a, D_d, E_a and E_d are known real constant matrices that characterizes how the uncertain parameters in $F_a(t)$ and $F_d(t)$ enter the nominal system matrices (A, A_d).

The delay-dependent robust stability conditions are generally obtained by directly extending the stability conditions of nominal time-delay systems. To do so, the nominal system A and A_d are replaced by $A(t)$ and $A_d(t)$ in the stability conditions derived for nominal systems. The uncertain matrices/parameters that now appear in the formulation of stability analysis are eliminated with the help of appropriate bounding inequality lemma along with the condition $\| F_a(t) \| \leq 1$ and $\| F_d(t) \| \leq 1$ to get the robust stability condition.

Note: Here, the robust stability analysis of an uncertain time-delay systems is discussed for norm-bounded type uncertainty structure only. The time-delay system with norm-bounded uncertainty structure is presented in details in Sect. 2.2.2 from (2.15)–(2.22).

Following lemmas are useful for deriving robust stability condition of an uncertain time-delay systems with norm-bounded uncertainties.

Lemma 2.6 ([12, 13]) *Let D, E and F be real matrices of appropriate dimensions with $\| F \| \leq 1$, then for any scalar $\epsilon > 0$,*

$$DFE + E^T F^T D^T \leq \frac{1}{\epsilon} DD^T + \epsilon E^T E \tag{2.206}$$

Lemma 2.7 ($\mathcal{S}$-Procedure for quadratic forms and strict inequalities [24]). *Let $T_0, \ldots, T_p \in \mathcal{R}^{n \times n}$ be symmetric matrices. Considering the following conditions on $T_0, \ldots, T_p$*

$$\zeta^T T_0 \zeta > 0, \forall \zeta \neq 0 \tag{2.207}$$

such that,

$$\zeta^T T_i \zeta \geq 0, i = 1, \ldots p \tag{2.208}$$

If there exists some scalars, $\tau_1 \geq 0, \ldots \tau_p \geq 0$, such that

$$T_0 - \sum_{i=1}^{i=p} \tau_i T_i > 0 \tag{2.209}$$

then, (2.207) and (2.208) holds.

2.6.2 Delay-Dependent Robust Stability Analysis

In this subsection, some relevant existing results on delay-dependent robust stability analysis of time-delay systems considering norm-bounded uncertainties in the system matrices are discussed to understand the application of lemmas 2.6 and 2.7 for eliminating the uncertain matrix from the derived stability condition in order to obtain desired robust stability condition.

Several literature on delay-dependent robust stability analysis considering polytopic type model uncertainties can be found in [9, 28, 29, 48], and references cited therein and are not discussed here as the present work focuses on norm-bounded uncertainty structure only.

Theorem 2.21 ([25]) *If there exist matrices $P = P^T > 0, Q = Q^T > 0, X > 0, Y > 0$, any matrix Z and scalars $\epsilon_1 > 0$ and $\epsilon_2 > 0$ such that following LMIs hold:*

$$\begin{bmatrix} Y_{11} & -Y + PA_d & d_u A^T Z & PD_a & PD_d \\ \star & -Q + \epsilon_2 E_d^T E_d & d_u A_d^T Z & 0 & 0 \\ \star & \star & -d_u Z & d_u Z D_a & d_u Z D_d \\ \star & \star & \star & -\epsilon_1 I & 0 \\ \star & \star & \star & \star & -\epsilon_2 I \end{bmatrix} < 0 \tag{2.210}$$

$$\begin{bmatrix} X & Y \\ \star & Z \end{bmatrix} \geq 0 \tag{2.211}$$

where, $Y_{11} = A^T P + PA + d_u X + Y + Y^T + Q + \epsilon_1 E_a^T E_a$, *then the system (2.15)–(2.16) is asymptotically stable for any time-delay satisfying* $0 \leq d \leq d_u$ *and all admissible uncertainties as defined in (2.19)–(2.20).*

Proof The robust stability condition in (2.210) can be obtained as a direct extension of the stability results in Theorem 2.5. Note that, the stability condition in Theorem 2.5 was obtained for constant delay d.

Replace A and A_d in (2.52) with $A(t)$[1] and $A_d(t)$[2] respectively, which gives,

$$\begin{bmatrix} (1,1) & (1,2) & (1,3) \\ \star & -Q & (2,3) \\ \star & \star & -d_u Z \end{bmatrix} \tag{2.212}$$

where, $(1,1)$=$A^T P + PA + d_u X + Y + Y^T + Q + E_a^T F_a^T D_a^T P + P D_a F_a E_a$,

$(1,2)$=$-Y + PA_d + PD_d F_d E_d$, $(1,3)$=$d_u(E_a^T F_a^T D_a^T Z + A^T Z)$,

$(2,3)$=$d_u(E_d^T F_d^T D_d^T Z + A_d^T Z)$

Now, multiplying both the sides of (2.212) by any vector y_i $(i = 1, 2, 3)$ and its transpose one gets,

$$\begin{bmatrix} y_1 \\ y_2 \\ y_3 \end{bmatrix}^T \begin{bmatrix} (1,1) & (1,2) & (1,3) \\ \star & -Q & (2,3) \\ \star & \star & -d_u Z \end{bmatrix} \begin{bmatrix} y_1 \\ y_2 \\ y_3 \end{bmatrix} \tag{2.213}$$

Expanding (2.213) one can get following cross terms involving the product of vectors y_i and the uncertain matrices,

$$y_1^T(E_a^T F_a^T D_a^T P + P D_a F_a E_a)y_1, \; y_1^T P D_d F_d E_d y_2$$
$$d_u y_1^T E_a^T F_a^T D_a^T Z y_3 \; and \; d_u y_2^T E_d^T F_d^T D_d^T Z y_3$$

Defining,

$$p = F_a(t) E_a y_1, \; q = F_d(t) E_d E_d y_2$$

Thus one can rewrite (2.213) in view of above as,

---

[1] $A(t) = A + \Delta A(t)$, where, $\Delta A(t) = D_a F_a(t) E_a$.
[2] $A_d(t) = A + \Delta A_d(t)$, where, $\Delta A_d(t) = D_d F_d(t) E_d$.

$$\begin{bmatrix} y_1 \\ y_2 \\ y_3 \\ p \\ q \end{bmatrix}^T \begin{bmatrix} X_{11} & -Y + PA_d & d_u A^T Z & PD_a & PD_d \\ \star & -Q & d_u A_d^T Z & 0 & 0 \\ \star & \star & -d_u Z & d_u Z D_a & d_u Z D_d \\ \star & \star & \star & 0 & 0 \\ \star & \star & \star & \star & 0 \end{bmatrix} \begin{bmatrix} y_1 \\ y_2 \\ y_3 \\ p \\ q \end{bmatrix} \tag{2.214}$$

where, $X_{11} = A^T P + PA + d_u X + Y + Y^T + Q$

It may be noted that as the elements (5,5) and (6,6) in (2.214) are zero, so it cannot be solved for negative definiteness. To overcome this problem one can have,

$$\begin{aligned} \epsilon_1 p^T p &\leq \epsilon_1 y_1^T E_a^T E_a y_1, \ with \parallel F_a(t) \parallel \leq 1 \\ \epsilon_2 q^T q &\leq \epsilon_2 y_2^T E_d^T E_d y_2, \ with \parallel F_d(t) \parallel \leq 1 \end{aligned} \tag{2.215}$$

where, $\epsilon_1 > 0$ and $\epsilon_2 > 0$.
Now, applying Lemma 2.7 ($\mathcal{S} - procedure$), one can combine (2.214) and (2.215) to obtain the LMI condition of (2.210). As, the LMI condition in (2.211) do not contain any uncertain terms so it remains unchanged as in (2.53) (see Theorem 2.5).

Theorem 2.22 ([13]) *Given scalars $d_u > 0$ and $\mu < 1$, satisfying the conditions $0 \leq d(t) \leq d_u$ and $\dot{d}(t) \leq \mu < 1$, the uncertain system (2.15) is robustly stable if there exist symmetric positive definite matrices $P = P^T > 0$, $Q = Q^T > 0$ and $Z = Z^T > 0$, a symmetric-semi-positive-definite matrix $X = \begin{bmatrix} X_{11} & X_{12} \\ X_{12}^T & X_{22} \end{bmatrix} \geq 0$, any matrices Y and T such that following LMIs hold:*

$$\phi = \begin{bmatrix} \phi_{11} + \epsilon E_a^T E_a & \phi_{12} + \epsilon E_a^T E_d & d_u A^T Z & PD \\ \star & \phi_{22} + \epsilon E_d^T E_d & d_u A_d^T Z & 0 \\ \star & \star & -d_u Z & d_u Z D \\ \star & \star & \star & -\epsilon I \end{bmatrix} < 0 \tag{2.216}$$

$$\begin{bmatrix} X_{11} & X_{12} & Y \\ \star & X_{22} & T \\ \star & \star & Z \end{bmatrix} \geq 0 \tag{2.217}$$

where,ϕ_{11}=$PA + A^T P + Y + Y^T + Q + d_u X_{11}$
ϕ_{12}=$PA_d - Y + Y^T + d_u X_{12}$ and ϕ_{22}=$-T - T^T - (1-\mu)Q + d_u X_{22}$

Remark 2.27 The proof of this theorem follows directly from the stability condition derived in Theorem 2 of [13] by replacing the A and A_d matrices with $A + DF(t)E_a$ and $A_d + DF(t)E_d$. In this method to establish the the robust stability condition in an LMI framework the uncertain matrix $F(t)$ has been eliminated using Lemma 2.6 instead of using ($\mathcal{S} - procedure$) Lemma 2.7.

Theorem 2.23 ([14]) *Consider the uncertain systems (2.15). Given the scalars* $d_u > 0$ *and* $\mu > 0$*, the system described in (2.15) is robustly asymptotically stable for any time-delay satisfying the conditions* $0 \le d(t) \le d_u$ *and* $\dot{d}(t) \le \mu < 1$*, with the admissible uncertainties (2.19)–(2.20) satisfying (2.21)–(2.22), if there exist symmetric positive definite matrices,* P, Q, R, T*, matrices* S_i $(i = 1, 2 \ldots, 4)$ *with appropriate dimensions and scalars* ϵ_i $(i = 1, 2)$*, then following LMIs hold:*

$$\begin{bmatrix} \Sigma_{11} & \Sigma_{12} & \Sigma_{13} & \Sigma_{14} & \Sigma_{15} & \Sigma_{16} & P_{11}D_a & P_{11}D_d & \mu P_{12} \\ \star & \Sigma_{22} & \Sigma_{23} & \Sigma_{24} & \Sigma_{25} & \Sigma_{26} & 0 & 0 & 0 \\ \star & \star & -Q_{11} & \Sigma_{34} & 0 & 0 & P_{12}^T D_a & P_{12}^T D_d & \mu P_{22} \\ \star & \star & \star & \Sigma_{44} & 0 & 0 & 0 & 0 & 0 \\ \star & \star & \star & \star & -Q_{11} & -Q_{12} & \Sigma_{57} & \Sigma_{58} & 0 \\ \star & \star & \star & \star & \star & -Q_{22} & \Sigma_{67} & \Sigma_{68} & 0 \\ \star & \star & \star & \star & \star & \star & -\epsilon_1 I & 0 & 0 \\ \star & \star & \star & \star & \star & \star & \star & -\epsilon_2 I & 0 \\ \star & \star & \star & \star & \star & \star & \star & \star & -\mu T \end{bmatrix} < 0 \quad (2.218)$$

$$P = \begin{bmatrix} P_{11} & P_{12} \\ \star & P_{22} \end{bmatrix} \ge 0, \; with \;\; P_{11} > 0 \quad (2.219)$$

$$Q = \begin{bmatrix} Q_{11} & Q_{12} \\ \star & Q_{22} \end{bmatrix} \ge 0 \quad (2.220)$$

where, $\Sigma_{11} = A^T P_{11} + P_{11} A + R + S_1 + S_1^T + \epsilon_1 E_a^T E_a$, $\Sigma_{12} = P_{11} A_d - S_1^T + S_2$
$\Sigma_{13} = A^T P_{12} + S_3$, $\Sigma_{14} = P_{12} - S_1^T + S_4$, $\Sigma_{15} = d_u(Q_{11} + A^T Q_{12}^T)$
$\Sigma_{16} = d_u(Q_{12} + A_T Q_{22})$, $\Sigma_{22} = -(1 - \mu)R + \mu T - S_2^T - S_2 + \epsilon_2 E_d^T E_d$
$\Sigma_{23} = A_d^T P_{12} - S_3$, $\Sigma_{24} = -S_2^T - S_4$, $\Sigma_{25} = d_u A_d^T Q_{12}^T$, $\Sigma_{26} = d_u A_d^T Q_{22}$
$\Sigma_{34} = P_{22} - Q_{12} - S_3^T$, $\Sigma_{44} = -Q_{22} - S_4^T - S_4$
$\Sigma_{57} = d_u Q_{12} D_a$, $\Sigma_{58} = d_u Q_{12} D_d$, $\Sigma_{67} = d_u Q_{22} D_a$, and $\Sigma_{68} = d_u Q_{22} D_d$

Proof This theorem can be proved by directly extending the stability condition (2.95) (see Theorem 2.12). To accomplish this one can replace A and A_d by $A(t)$ and $A_d(t)$ in (2.95) and subsequently the resulting matrix inequality is decomposed into nominal and uncertain matrices as,

$$\Sigma_n + \Sigma_{un}^T + \Sigma_{un} < 0 \quad (2.221)$$

where the uncertain matrices are defined as,

$$\Sigma_{un} = D_1 F_a(t) E_1 + D_2 F_d(t) E_2$$

and, $D_1 = [D_a^T P_{11}\ 0\ D_a^T P_{12}\ 0\ d_u D_a^T Q_{12}^T\ d_u D_a^T Q_{22}\ 0]^T$, $E_1 = [E_a\ 0\ 0\ 0\ 0\ 0\ 0]$, $D_2 = [D_d^T P_{11}\ 0\ D_d^T P_{12}\ 0\ d_u D_d^T Q_{12}^T\ d_u D_d^T Q_{22}\ 0]^T$ and $E_2 = [0\ E_d\ 0\ 0\ 0\ 0\ 0]$.

Employing Lemma 2.6 and Schur's complement on (2.221) one can obtain the LMI condition (2.218). If the uncertainties in (2.19) and (2.20) are given as

$$D_a = D_d = D \quad F_a(t) = F_d(t) = F(t)$$

then the LMI condition in (2.218) reduces to the following LMI condition,

$$\begin{bmatrix} \Sigma_{11} & \Sigma_{12} & \Sigma_{13} & \Sigma_{14} & \Sigma_{15} & \Sigma_{16} & P_{11}D & \mu P_{12} \\ \star & \Sigma_{22} & \Sigma_{23} & \Sigma_{24} & \Sigma_{25} & \Sigma_{26} & 0 & 0 \\ \star & \star & -Q_{11} & \Sigma_{34} & 0 & 0 & P_{12}^T D & \mu P_{22} \\ \star & \star & \star & \Sigma_{44} & 0 & 0 & 0 & 0 \\ \star & \star & \star & \star & -Q_{11} & -Q_{12} & \Sigma_{57} & 0 \\ \star & \star & \star & \star & \star & -Q_{22} & \Sigma_{67} & 0 \\ \star & \star & \star & \star & \star & \star & -\epsilon I & 0 \\ \star & \star & \star & \star & \star & \star & \star & -\mu T \end{bmatrix} < 0 \tag{2.222}$$

where, $\Sigma_{11}= A^T P_{11} + P_{11} A + R + S_1 + S_1^T + \epsilon E_a^T E_a$, $\Sigma_{12}=P_{11} A_d - S_1^T + S_2 + \epsilon E_a^T E_d$
$\Sigma_{13}=A^T P_{12} + S_3$, $\Sigma_{14}=P_{12} - S_1^T + S_4$, $\Sigma_{15}=d_u(Q_{11} + A^T Q_{12}^T)$
$\Sigma_{16}=d_u(Q_{12} + A_T Q_{22})$, $\Sigma_{22}=-(1-\mu)R + \mu T - S_2^T - S_2 + \epsilon E_d^T E_d$
$\Sigma_{23}=A_d^T P_{12} - S_3$, $\Sigma_{24}=-S_2^T - S_4$, $\Sigma_{25}=d_u A_d^T Q_{12}^T$, $\Sigma_{26}=d_u A_d^T Q_{22}$
$\Sigma_{34}=P_{22} - Q_{12} - S_3^T$, $\Sigma_{44}=-Q_{22} - S_4^T - S_4$ $\Sigma_{57}=d_u Q_{12} D$, $\Sigma_{67}=d_u Q_{22} D$

In [36] a delay-dependent robust stability condition has been proposed using new type of LK functional and integral inequality lemma of [25]. The uncertain time-delay system in (2.15) for $t \geq 0$ with an assumption that $D_a = D_d = D$ and $F_a(t) = F_d(t) = F(t)$ is expressed as,

$$\begin{aligned} \dot{x}(t) &= (A + DF(t)E_a)x(t) + (A_d + DF(t)E_d)x(t-d(t)) \\ &= Ax(t) + A_d x(t-d(t)) + D[F(t)E_a x(t) + F(t)E_d x(t-d(t))] \\ &= Ax(t) + A_d x(t-d(t)) + Dp(t), \ \forall\, t \in [-d_u, 0] \end{aligned} \tag{2.223}$$

where, $p(t) = F(t)q(t)$ and $q(t) = E_a x(t) + E_d x(t-d(t))$ and $F(t)^T F(t) \leq \gamma^{-2} I$. The LMI stability condition for the uncertain system described in (2.223) is presented below.

Theorem 2.24 ([36]) *For a given γ, the delayed uncertain system (2.223) with $\dot{d}(t) \leq \mu$ is asymptotically stable if there exist symmetric matrices $P > 0$, $Q_0 > 0$, $Q_1 > 0$, $S_0 > 0$, $S_1 > 0$, Y_{11}, Y_{12}, Y_{22}, Z_{11}, Z_{12}, Z_{22} and Σ such that the following conditions hold,*

$$
\begin{aligned}
0 > {} & \Sigma(Ae_1^T + A_d e_2^T - e_4^T + De_5^T) + Y_{12}(e_1 - e_2)^T + d_u Y_{11} \\
& +(e_1 A^T + e_2 A_d^T - e_4 + e_5 D^T)\Sigma^T + (e_1 - e_2)Y_{12}^T \\
& +(e_1 E_a^T + e_2 E_d^T)(E_a e_1^T + E_d e_2^T) + Z_{12}(e_2 - e_3)^T \\
& +(e_2 - e_3)Z_{12}^T + e_4 P e_1^T + e_1 P e_4^T - (1-\mu)e_2 Q_1 e_2^T \\
& +d_u e_4 (S_0 + S_1)e_4^T + e_1(Q_0 + Q_1)e_1^T - e_3 Q_0 e_3^T - \gamma^2 e_5 e_5^T
\end{aligned} \tag{2.224}
$$

$$
\begin{aligned}
0 > {} & \Sigma(Ae_1^T + A_d e_2^T - e_4^T + De_5^T) + Y_{12}(e_1 - e_2)^T + d_u Z_{11} \\
& +(e_1 A^T + e_2 A_d^T - e_4 + e_5 D^T)\Sigma^T + (e_1 - e_2)Y_{12}^T \\
& +(e_1 E_a^T + e_2 E_d^T)(E_a e_1^T + E_d e_2^T) + Z_{12}(e_2 - e_3)^T \\
& +(e_2 - e_3)Z_{12}^T + e_4 P e_1^T + e_1 P e_4^T - (1-\mu)e_2 Q_1 e_2^T \\
& +d_u e_4 S_0 e_4^T + e_1(Q_0 + Q_1)e_1^T - e_3 Q_0 e_3^T - \gamma^2 e_5 e_5^T
\end{aligned} \tag{2.225}
$$

$$
\begin{bmatrix} Y_{11} & Y_{12} \\ \star & Y_{22} \end{bmatrix} \geq 0, \; \begin{bmatrix} Z_{11} & Z_{12} \\ \star & Z_{22} \end{bmatrix} \geq 0, \; S_0 \geq Z_{22}, \; S_0 + (1-\mu)S_1 \geq Y_{22} \tag{2.226}
$$

Remark 2.28 The stability method in Theorem 2.24 uses a new LK functional of the form,

$$
\begin{aligned}
V(t) = {} & x^T(t)Px(t) + \int_{t-d_u}^{t} x^T(s)Q_0 x(s)ds + \int_{t-d(t)}^{t} x^T(s)Q_1 x(s)ds \\
& + \int_{-d_u}^{0}\int_{t+s}^{t} \dot{x}^T(\alpha)S_0\dot{x}(\alpha)d\alpha ds + \int_{-d(t)}^{0}\int_{t+s}^{t} \dot{x}^T(\alpha)S_1\dot{x}(\alpha)d\alpha ds
\end{aligned} \tag{2.227}
$$

with Q_0, Q_1, S_0 and S_1 are symmetric positive-definite matrices. One may note that, the delay-dependent functional (double integral) term in (2.227) contains uncertain limit of integration which is usually not used in any LK functional, in this regard the chosen LK functional is new. The time-derivative of the functional in (2.227) yields two quadratic integral terms of the form,

$$
-\int_{t-d_u}^{t} \dot{x}^T(s)S_0\dot{x}(s)ds, \; \textit{and} \; (1-\dot{d}(t))\int_{t-d(t)}^{t} \dot{x}^T(s)S_1\dot{x}(s)ds
$$

The information of the system matrices are incorporated into the formulation by introducing one more matrix variables to satisfy the following,

$$
0 = 2\chi^T(t)\Sigma(Ae_1^T + A_d e_2^T - e_4^T + De_5^T)\chi(t) \tag{2.228}
$$

where, $\chi(t) = [x^T(t)\; x^T(t-d(t))\; x^T(t-d_u)\; \dot{x}(t)\; p^T(t)]^T$ and e_i $(i = 1, ..5)$ are corresponding block identity matrix. For handling the uncertainty, one more constraint is introduced which is of the form,

$$0 \leq q^T(t)q(t) - \gamma^2 p^t(t)p(t) \tag{2.229}$$

this can further be written as,

$$0 \leq \chi^T(t)\{(e_1 E_a^T + e_2 E_d^T)(E_a e_1^T + E_d e_2^T) - \gamma^2 p^t(t)p(t)\}\chi(t) \tag{2.230}$$

The integral terms shown above are approximated using integral bounding lemma (Lemma 2.3) which is further added with the other derivative terms of LK functional along with the terms of (2.228), the resulting quadratic expression is finally combined with (2.230) using $\mathcal{S} - Procedure$ lemma to get $\dot{V}(t)$ expression of the form,

$$\dot{V}(t) \leq \chi^T(t)\{d(t)\Gamma + (d_u - d(t))\Pi + \Omega\}\chi(t) - \int_{s_1} (.)^T \Phi_1(.)ds$$
$$- \int_{s_2} (.)^T \Phi_2(.)ds \tag{2.231}$$

where, Γ, Π, Ω, Φ_1 and Φ_2 are matrices of compatible dimensions. The detailed structure of the matrices can be found in [36]. The above expression is a convex combination of matrices Γ and Π on $d(t)$ that can be further expressed by two matrix inequality conditions as,

1. When $d(t) = 0$ one can write

$$d_u\Pi + \Omega - \int_{s_1} (.)^T \Phi_1(.)ds \quad - \int_{s_2} (.)^T \Phi_2(.)ds < 0$$

2. When $d(t) = d_u$ one can write

$$d_u\Gamma + \Omega - \int_{s_1} (.)^T \Phi_1(.)ds \quad - \int_{s_2} (.)^T \Phi_2(.)ds < 0$$

In order to guarantee the negativity of $\dot{V}(t)$ in (2.231), first one has to impose two matrix inequality constraints, $\Phi_1 \geq 0$ and $\Phi_2 \geq 0$ such that the quadratic integral terms in the above expression remains semi-positive definite. In addition to these constraints two more LMI constraints due to the use of bounding lemma (Lemma 2.3) need to be satisfied (refer (2.226)).

Note that, the number of LMIs used are six and number of matrix variables involved are 12, so this theorem has still room for improvement in terms of reducing the number of LMI constraints and matrix variables by using efficient bounding inequalities as proposed in the stability condition in Sect. 2.5.3.

For detailed proof of Theorem 2.24 one can refer [36].

For convenience of the discussion of the main results of this chapter, some preliminaries including few definitions, basic theorems on robust stability of time-delay systems which are related to the main results on delay-dependent robust stability analysis are presented in preceding sections.

2.7 Main Results on Delay-Dependent Robust Stability Analysis of TDS

In this section, delay-dependent robust stability conditions for uncertain time-delay systems in an LMI framework are presented. The additive uncertainties are assumed to be of norm-bounded type (explained in Sect. 2.6.1). For robust stability analysis, an uncertain time-delay system described in (2.15)–(2.20) satisfying the conditions (2.21) and (2.22) (refer Sect. 2.2.2 of the chapter) is considered.

The delay-dependent robust stability conditions are direct extension of the stability analysis of TDS as discussed in Sect. 2.5.1. The nominal matrices A and A_d in the stability condition are replaced by the uncertain matrices $A(t)$ and $A_d(t)$. With the proper choice of lemma (i.e., 2.6 or 2.7) the uncertain matrices are eliminated.

2.7.1 Delay-Dependent Robust Stability Analysis of TDS with Single Time Delay

Considering an uncertain time-delay system (2.15) with the structure as described in (2.17)–(2.20), satisfying the condition (2.21)–(2.22). The robust stability condition is presented in the following theorem by constructing a new LK functional and subsequently using improved bounding inequalities for the integral quadratic terms arising in the LK functional derivative which, in turn, yields convex combination of LMIs.

The main contribution of the proposed method have already been discussed in Sect. 2.5.1 (see Remarks 2.20–2.22).

Theorem 2.25 *For given scalars $d_u > 0$ and ϵ_i $(i = 1, 2) > 0$, the system (2.15) is robustly asymptotically stable satisfying the conditions (2.5) and $0 < \mu < 1$, for the admissible uncertainties (2.19)–(2.20) satisfying (2.21)–(2.22), if there exist symmetric positive definite matrices P, Q, R_i $(i = 1, 2)$, any free matrices L_i $(i = 1, 2, 3)$, M_i $(i = 1, 2, 3)$ and $G_i (i = 1, 2)$ such that, $P = \begin{bmatrix} P_{11} & P_{12} \\ P_{12}^T & P_{22} \end{bmatrix} > 0$, $Q = \begin{bmatrix} Q_{11} & 0 \\ 0 & Q_{22} \end{bmatrix} > 0$, then the following LMIs hold:*

$$\begin{bmatrix} \Xi & \mu\tilde{P} & d_u M & D_1 & D_2 \\ \star & -\mu T & 0 & 0 & 0 \\ \star & 0 & -Q_{22} & 0 & 0 \\ \star & 0 & 0 & -\epsilon_1 I & 0 \\ \star & 0 & 0 & 0 & -\epsilon_2 I \end{bmatrix} < 0 \quad (2.232)$$

$$\begin{bmatrix} \Xi & \mu\tilde{P} & d_u L & D_1 & D_2 \\ \star & -\mu T & 0 & 0 & 0 \\ \star & 0 & -Q_{22} & 0 & 0 \\ \star & 0 & 0 & -\epsilon_1 I & 0 \\ \star & 0 & 0 & 0 & -\epsilon_2 I \end{bmatrix} < 0 \quad (2.233)$$

where, $L = \begin{bmatrix} L_1^T \; L_2^T \; L_3^T \; 0\,0\,0 \end{bmatrix}^T$, $M = \begin{bmatrix} M_1^T \; M_2^T \; M_3^T \; 0\,0\,0 \end{bmatrix}^T$,
$D_1 = \begin{bmatrix} D_a^T G_1 \; 0\,0\,0\,0 \; D_a^T G_2 \end{bmatrix}^T$, $D_2 = \begin{bmatrix} D_d^T G_1 \; 0\,0\,0\,0 \; D_d^T G_2 \end{bmatrix}^T$, $\tilde{P}$ *is defined in (2.132) and*

$$\Xi = \begin{bmatrix} \Xi_{11} & \Xi_{12} & \Xi_{13} & \Xi_{14} & 0 & \Xi_{16} \\ \star & \Xi_{22} & \Xi_{23} & \Xi_{24} & 0 & \Xi_{16} \\ \star & \star & \Xi_{33} & 0 & 0 & 0 \\ \star & \star & 0 & -Q_{11} & 0 & \Xi_{46} \\ 0 & 0 & 0 & 0 & -Q_{11} & 0 \\ \star & \star & 0 & \star & 0 & \Xi_{66} \end{bmatrix} \quad (2.234)$$

where,
$\Xi_{11} = \Omega_{11} + \epsilon_1 E_a^T E_a$, $\Xi_{12} = \Omega_{12}$, $\Xi_{13} = \Omega_{13}$, $\Xi_{14} = \Omega_{14}$, $\Xi_{16} = \Omega_{16}$
$\Xi_{22} = \Omega_{22} + \epsilon_2 E_d^T E_d$, $\Xi_{23} = \Omega_{23}$, $\Xi_{24} = \Omega_{24}$, $\Xi_{26} = \Omega_{26}$
$\Xi_{33} = \Omega_{23}$, $\Xi_{46} = \Omega_{46}$, $\Xi_{66} = \Omega_{66}$
the elements of matrix Ω ($\Omega_{i,j}$, $i = 1, ..6$; $j = 1, ..6$) are already defined in stability Theorem 2.18.

Proof The proof of this theorem follows directly from the stability condition in (2.132)–(2.134). The nominal matrices A and A_d appearing in Ω matrix in (2.132) are replaced by time-varying matrices $A(t)$ and $A_d(t)$ which are defined as,

$$A(t) = A + \Delta_a(t)$$

$$A_d(t) = A_d + \Delta_d(t)$$

where $\Delta_a(t) = D_a F_a(t) E_a$ and $\Delta_d(t) = D_d F_d(t) E_d$. With these replacement in Ω matrix one can get resulting matrix as,

$$\Xi = \begin{bmatrix} \Omega_{11}|_\Delta & \Omega_{12}|_\Delta & \Omega_{13} & \Omega_{14} & 0 & \Omega_{16}|_\Delta \\ \star & \Omega_{22} & \Omega_{23} & \Omega_{24} & 0 & \Omega_{26}|_\Delta \\ \star & \star & \Omega_{33} & 0 & 0 & 0 \\ \star & \star & 0 & -Q_{11} & 0 & \Omega_{46} \\ 0 & 0 & 0 & 0 & -Q_{11} & 0 \\ \star & \star & 0 & \star & 0 & \Omega_{66} \end{bmatrix} \quad (2.235)$$

where, $\Omega_{11}|_\Delta = d_u^2 Q_{11} + G_1^T A(t) + A^T(t) G_1 + P_{12} + P_{12}^T + R_1 + R_2 + d_u(L_1 + L_1^T)$
$\Omega_{12}|_\Delta = G_1^T A_d(t) - P_{12} + d_u(-L_1 + L_2^T + M_1)$
$\Omega_{13} = d_u(L_3^T - M_1)$, $\Omega_{14} = P_{22}$, $\Omega_{16}|_\Delta = P_{11} - G_1^T + A^T(t) G_2$
$\Omega_{22} = -(1-\mu) R_1 - \mu T + d_u(-L_2 - L_2^T + M_2 + M_2^T)$
$\Omega_{23} = d_u(-M_2 + M_3^T - L_3^T)$, $\Omega_{24} = -P_{22}$, $\Omega_{26} = A_d^T G_2$
$\Omega_{33} = d_u(-M_3 - M_3^T) - R_2$, $\Omega_{66} = -G_2 - G_2^T + d_u^2 Q_{22}$

The matrix in (2.235) can be further decomposed in the form,

$$\Xi = \Omega_{nom} + \Omega_{unc} \quad (2.236)$$

where, $\Omega_{nom} = \Omega$ and Ω_{unc} can be represented as,

$$\Omega_{unc} = \Sigma + \Sigma^T \quad (2.237)$$

where,

$$\Sigma = D_1 F_a(t) E_1 + D_2 F_d(t) E_d \quad (2.238)$$

$$D_1 = \left[D_a^T G_1 \; 0 \; 0 \; 0 \; 0 \; D_a^T G_2 \right]^T,$$

$$D_2 = \left[D_d^T G_1 \; 0 \; 0 \; 0 \; 0 \; D_d^T G_2 \right]^T,$$

$$E_1 = \left[E_a \; 0 \; 0 \; 0 \; 0 \; 0 \right],$$

$$E_2 = \left[0 \; E_d \; 0 \; 0 \; 0 \; 0 \right].$$

Using Lemma 2.6 one can write (2.237) as,

$$\Sigma + \Sigma^T \leq \epsilon_1 E_1^T E_1 + \epsilon_2 E_2^T E_2 + \epsilon_1^{-1} D_1 D_1^T + \epsilon_2^{-1} D_2 D_2^T \tag{2.239}$$

So, in view of (2.239) and with the application of Schur-complement one can obtain the LMI conditions (2.232)–(2.233), while remaining conditions $P > 0$ and $Q > 0$ remains same as in stability Theorem 2.18 as they do not contain A and A_d matrices. This completes the proof. □

If the uncertainties in (2.19) and (2.20) are defined as,

$$\begin{aligned} \Delta A(t) &= DF(t)E_a \\ \Delta A_d(t) &= DF(t)E_d, \ F^T(t)F(t) \leq I \end{aligned} \tag{2.240}$$

then with $D_a = D_d = D$, Theorem 2.25 can be modified in the form of following corollary.

Corollary 2.5 *Let $d_u > 0$ and $\epsilon > 0$ be given scalars, the system (2.15) with (2.5) and $0 \leq \mu < 1$ is robustly asymptotically stable for the admissible uncertainties (2.19)–(2.20) satisfying (2.240), if there exist symmetric positive definite matrices $P, \ Q, \ R_i \ (i = 1, 2)$, and any free matrices $L_i \ (i = 1, 2, 3)$, $M_i \ (i = 1, 2, 3)$ and $G_i \ (i = 1, 2)$ such that, $P = \begin{bmatrix} P_{11} & P_{12} \\ P_{12}^T & P_{22} \end{bmatrix} > 0$, $Q = \begin{bmatrix} Q_{11} & 0 \\ 0 & Q_{22} \end{bmatrix} > 0$, the following LMIs hold:*

$$\begin{bmatrix} \tilde{\Xi} & \mu\tilde{P} & d_u M & \bar{D} \\ \star & -\mu T & 0 & 0 \\ \star & 0 & -Q_{22} & 0 \\ \star & 0 & 0 & -\epsilon I \end{bmatrix} < 0 \tag{2.241}$$

$$\begin{bmatrix} \tilde{\Xi} & \mu\tilde{P} & d_u L & \bar{D} \\ \star & -\mu T & 0 & 0 \\ \star & 0 & -Q_{22} & 0 \\ \star & 0 & 0 & -\epsilon I \end{bmatrix} < 0 \tag{2.242}$$

where, $\bar{D} = \begin{bmatrix} D^T G_1 \ 0 \ 0 \ 0 \ 0 \ D^T G_2 \end{bmatrix}^T$, *and*

$$\tilde{\Xi} = \begin{bmatrix} \tilde{\Xi}_{11} & \tilde{\Xi}_{12} & \tilde{\Xi}_{13} & \tilde{\Xi}_{14} & 0 & \tilde{\Xi}_{16} \\ \star & \tilde{\Xi}_{22} & \tilde{\Xi}_{23} & \tilde{\Xi}_{24} & 0 & \tilde{\Xi}_{16} \\ \star & \star & \tilde{\Xi}_{33} & 0 & 0 & 0 \\ \star & \star & 0 & -Q_{11} & 0 & \tilde{\Xi}_{46} \\ 0 & 0 & 0 & 0 & -Q_{11} & 0 \\ \star & \star & 0 & \star & 0 & \tilde{\Xi}_{66} \end{bmatrix} < 0 \tag{2.243}$$

where, $\tilde{\Xi}_{11} = \Omega_{11} + \epsilon E_a^T E_a$, $\tilde{\Xi}_{12} = \Omega_{12} + \epsilon E_a^T E_d$,

$$\tilde{\Xi}_{13}=\Omega_{13},\ \ \tilde{\Xi}_{14}=\Omega_{14},\ \ \tilde{\Xi}_{16}=\Omega_{16}$$
$$\tilde{\Xi}_{22}=\Omega_{22}+\epsilon_2 E_d^T E_d,\ \tilde{\Xi}_{23}=\Omega_{23},\ \ \tilde{\Xi}_{24}=\Omega_{24},\ \ \tilde{\Xi}_{26}=\Omega_{26}$$
$$\tilde{\Xi}_{33}=\Omega_{23},\ \ \tilde{\Xi}_{46}=\Omega_{46},\ \ \tilde{\Xi}_{66}=\Omega_{66}$$

The the elements of matrix Ω ($\Omega_{i,j},\ i=1,..6;\ j=1,..6$) are defined in stability Theorem 2.18.

If $\mu=0$ (i.e., constant time-delay case) and the uncertainty structure is as defined in 2.240) then the robust stability condition for an uncertain system satisfying the condition $F^T(t)F(t)\leq I$, will follow directly from Corollary 2.5 by substituting $\mu=0$.

Corollary 2.6 *Let $d_u>0$ and $\epsilon>0$ be given scalars, an uncertain time-delay systems is robustly asymptotically stable for $\mu=0$ and admissible uncertainties described in (2.19)–(2.20), if there exist symmetric positive definite matrices $P,\ Q,\ R_i\ (i=1,2)$, and any free matrices $L_i\ (i=1,2,3)$, $M_i\ (i=1,2,3)$ and $G_i\ (i=1,2)$ such that, $P=\begin{bmatrix} P_{11} & P_{12} \\ P_{12}^T & P_{22}\end{bmatrix}>0$ $Q=\begin{bmatrix} Q_{11} & 0 \\ 0 & Q_{22}\end{bmatrix}>0$, the following LMIs hold:*

$$\begin{bmatrix} \tilde{\Xi} & d_u M & \bar{D} \\ \star & -Q_{22} & 0 \\ \star & 0 & -\epsilon I \end{bmatrix}<0 \tag{2.244}$$

$$\begin{bmatrix} \tilde{\Xi} & d_u L & \bar{D} \\ \star & -Q_{22} & 0 \\ \star & 0 & -\epsilon I \end{bmatrix}<0 \tag{2.245}$$

where, $\bar{D}$ is defined in corollary 2.5. and

$$\tilde{\Xi}=\begin{bmatrix} \tilde{\Xi}_{11} & \tilde{\Xi}_{12} & \tilde{\Xi}_{13} & \tilde{\Xi}_{14} & 0 & \tilde{\Xi}_{16} \\ \star & \tilde{\Xi}_{22}|_{\mu=0} & \tilde{\Xi}_{23} & \tilde{\Xi}_{24} & 0 & \tilde{\Xi}_{16} \\ \star & \star & \tilde{\Xi}_{33} & 0 & 0 & 0 \\ \star & \star & 0 & -Q_{11} & 0 & \tilde{\Xi}_{46} \\ 0 & 0 & 0 & 0 & -Q_{11} & 0 \\ \star & \star & 0 & \star & 0 & \tilde{\Xi}_{66} \end{bmatrix}<0 \tag{2.246}$$

where, $\tilde{\Xi}_{22}=-R_1+d_u(-L_2-L_2^T+M_2+M_2^T)+\epsilon E_d^T E_d$ and rest of the elements of $\tilde{\Xi}$ are same as in Corollary 2.5.

To illustrate the effectiveness of the proposed theorem three numerical examples are considered and the results are compared in tabular form with the existing robust stability methods.

Numerical Example 2.5 ([14, 36]) *Consider the uncertain time-delay system (2.15) with the following constant matrices*

$$A = \begin{bmatrix} -2 & 0 \\ 0 & -1 \end{bmatrix}, \quad A_d = \begin{bmatrix} -1 & 0 \\ -1 & -1 \end{bmatrix}$$

$$E_a = \begin{bmatrix} 1.6 & 0 \\ 0 & 0.05 \end{bmatrix}, \quad E_d = \begin{bmatrix} 0.1 & 0 \\ 0 & 0.3 \end{bmatrix}, \quad D_a = D_d = D = \begin{bmatrix} 1 & 0 \\ 0 & 1 \end{bmatrix}$$

The analytical value of the delay upper bound d_u considering $F(t) = I$ is found to be 1.3771 [14] for $\mu = 0$, so it cannot be more than 1.3771 for any given uncertainties defined in (2.240) as $F^T(t)F(t) \leq I$. The results obtained for different delay-derivatives μ are presented in Table 2.12.

Numerical Example 2.6 ([14]) *Consider the uncertain time-delay system (2.15) with the following constant matrices*

$$A = \begin{bmatrix} -0.5 & -2 \\ 1 & -1 \end{bmatrix}, \quad A_d = \begin{bmatrix} -0.5 & -1 \\ -1 & 0.6 \end{bmatrix}$$

$$E_a = \begin{bmatrix} 0.2 & 0 \\ 0 & 0.2 \end{bmatrix}, \quad = E_d, \text{ and } D_a = D_d = D = I$$

The results obtained for this Example using Theorem 2.25 for different delay-derivatives (μ) are presented in Table 2.13.

Numerical Example 2.7 ([14]) *Consider the uncertain time-delay system (2.15) with the following constant matrices*

Table 2.12 (d_u) results of Example 2.5 for $0 \leq \mu < 1$, $d_l = 0$

Stability methods	$\mu = 0$	$\mu = 0.4$	$\mu = 0.5$	$\mu = 0.6$	$\mu = 0.8$	$\mu = 0.9$
[12]	0.2086	*	*	*	*	*
[18]	0.2299	–	0.1758	–	–	0.0557
[25]	Infeas.	*	*	*	*	*
[27]	1.1490	–	0.9247	–	–	0.6710
[40]	1.1490	0.973	0.9247	0.873	0.760	0.6954
[38]	1.03	0.61	0.40		0.18	–
[14]	1.1623	–	0.9273	–	–	0.6954
[36]	1.149	1.077	–	1.070	1.068	–
Cor. 2.6 and Cor. 2.5	1.1606	1.0778	1.0733	1.0708	1.0686	1.0686

'–' means result is not available in reference, '*' means improper

Table 2.13 (d_u) results of Example 2.6 for $0 \leq \mu < 1, d_l = 0$

Stability methods	$\mu = 0$	$\mu = 0.5$	$\mu = 0.9$
[12]	0.3010	*	*
[18]	0.3513	0.2587	0.0825
[25]	0.5799	*	*
[27]	0.6812	0.1820	0.1820
[40]	0.8435	0.2433	0.2420
[14]	1.8542	0.3459	0.2542
Cor. 2.6 and Cor. 2.5	0.8435	0.3972	0.3972

'–' means result is not available in reference, '*' means improper

$$A = \begin{bmatrix} -2 & 0 \\ 0 & -0.9 \end{bmatrix}, \quad A_d = \begin{bmatrix} -1 & 0 \\ -1 & -1 \end{bmatrix}$$

$$D_a = D_d = D = \begin{bmatrix} 0.2 & 0 \\ 0 & 0.2 \end{bmatrix}, \quad and\ E_a = E_d = E = I$$

The delay upper bound estimate obtained (i) using Corollary 2.6 is $d_u = 2.4317$ *for* $\mu = 0$ *(ii) using Corollary 2.5 are* $d_u = 1.5276$ *and* $d_u = 1.2658$ *for* $\mu = 0.5$ *and* 0.9 *respectively, whereas the corresponding results obtained in [14] for* $\mu = 0$ *is* $d_u = 2.4390$, *and* $d_u = 1.3214$ *and* 0.7938 *for* $\mu = 0.5$ *and* 0.9 *respectively.*

Remark 2.29 It may be noted from Tables 2.12 and 2.13 that the proposed method gives less conservative estimate of delay upper bound compared to the [14] for non zero delay derivatives for all the systems. It may be emphasized here that the stability method adopted in [14] leads to better delay upper bound only for $\mu = 0$.

2.7.2 Robust Stability Analysis of TDS with Delay Varying in Ranges

Theorem 2.26 *Given a system (2.15) for* $\mu \geq 0$ *with the admissible uncertainties (2.19)–(2.20), satisfying the conditions (2.21)–(2.22) and (2.183)–(2.184) is robustly asymptotically stable, if there exist symmetric matrices* $P,\ Q_i > 0, i = 1, 2, \ldots 4,\ R_j > 0$, *any free matrices of appropriate dimensions* $M_j,\ N_j, j = 1, 2$ *and the scalars* $\epsilon_i > 0\ (i = 1, 2)$ *such that the following LMIs hold:*

$$\begin{bmatrix} \Psi & \Phi_1 & D_1 & D_2 \\ \star & -R_2 & 0 & 0 \\ \star & 0 & -\epsilon_1 I & 0 \\ \star & 0 & 0 & -\epsilon_2 I \end{bmatrix} < 0, \tag{2.247}$$

$$\begin{bmatrix} \Psi & \Phi_2 & D_1 & D_2 \\ \star & -R_2 & 0 & 0 \\ \star & 0 & -\epsilon_1 I & 0 \\ \star & 0 & 0 & -\epsilon_2 I \end{bmatrix} < 0, \tag{2.248}$$

where, $\Phi_1 = \left[0\ M_1^T\ N_1^T\ 0\,0\,0\right]^T$, $\Phi_2 = \left[0\,0\ M_2^T\ N_2^T\ 0\,0\right]$,

$$D_1 = \left[D_a^T P\ 0\,0\,0\ d_l D_a^T R_1\ D_a^T R_2\right]^T,\ D_2 = \left[D_d^T P\ 0\,0\,0\ d_l D_d^T R_1\ D_d^T R_2\right]^T$$

and

$$\Psi = \begin{bmatrix} \Theta_{11}|_{\Delta=0} + \epsilon_1 E_a^T E_a & R_1 & \Theta_{13}|_{\Delta=0} & 0 & \Theta_{15}|_{\Delta=0} & \Theta_{16}|_{\Delta=0} \\ \star & \Theta_{22} & \Theta_{23} & 0 & 0 & 0 \\ \star & \star & \Theta_{33}|_{\Delta=0} + \epsilon_2 E_d^T E_d & \Theta_{34} & \Theta_{35}|_{\Delta=0} & \Theta_{36}|_{\Delta=0} \\ \star & \star & \star & \Theta_{44} & 0 & 0 \\ \star & 0 & \star & 0 & -R_1 & 0 \\ \star & 0 & \star & 0 & 0 & -R_2 \end{bmatrix}$$

$\Theta_{11}|_{\Delta=0}=PA + A^T P + \sum_{i=1}^{3} Q_i - R_1$, $\Theta_{13}|_{\Delta=0}=PA_d + R_1$, $\Theta_{15}|_{\Delta=0}=d_l A^T R_1$,

$\Theta_{16}|_{\Delta=0}=A^T R_2$, $\Theta_{22}=Q_4 - Q_1 - R_1 + d_{lu}^{-1}(M_1 + M_1^T)$, $\Theta_{23}=d_{lu}^{-1}(-M_1 + N_1^T)$,

$\Theta_{33}=-(1-\mu)(Q_3 + Q4) + d_{lu}^{-1}(M_2 + M_2^T - N_1 - N_1^T)$,

$\Theta_{34}=d_{lu}^{-1}(-M_2 + N_2^T)$, $\Theta_{35}|_{\Delta=0}=d_l A_d^T R_1$, $\Theta_{36}|_{\Delta=0}=A_d^T R_2$, $\Theta_{44}=-Q_2 - d_{lu}^{-1}(N_2 + N_2^T)$

Proof The proof of this theorem follows directly from the stability conditions derived in Theorem 2.20 for nominal time-delay systems. The nominal matrices A and A_d appearing in Θ matrix of (2.188) and (2.189) are replaced by time-varying matrices $A(t)$ and $A_d(t)$ respectively, and then using Schur-complement one can rewrite matrix Θ as,

$$\Theta(t) = \begin{bmatrix} \Theta_{11} & R_1 & \Theta_{13} & 0 & \Theta_{15} & \Theta_{16} \\ \star & \Theta_{22} & \Theta_{23} & 0 & 0 & 0 \\ \star & \star & \Theta_{33} & \Theta_{34} & \Theta_{35} & \Theta_{36} \\ \star & \star & \star & \Theta_{44} & 0 & 0 \\ \star & 0 & \star & 0 & -R_1 & 0 \\ \star & 0 & \star & 0 & 0 & -R_2 \end{bmatrix} \tag{2.249}$$

$\Theta_{11}=PA(t) + A^T(t)P + \sum_{i=1}^{3} Q_i - R_1$, $\Theta_{13}=PA_d(t) + R_1$, $\Theta_{15}=d_l A^T(t)R_1$,

$\Theta_{16}=A^T(t)R_2$, $\Theta_{22}=Q_4 - Q_1 - R_1 + d_{lu}^{-1}(M_1 + M_1^T)$, $\Theta_{23}=dlu^{-1}(-M_1 + N_1^T)$,

$\Theta_{33}=-(1-\mu)(Q_3 + Q4) + d_{lu}^{-1}(M_2 + M_2^T - N_1 - N_1^T)$,

$\Theta_{34}=d_{lu}^{-1}(-M_2 + N_2^T)$, $\Theta_{35}=d_l A_d(t)^T R_1$, $\Theta_{36}=A_d(t)^T R_2$, $\Theta_{44}=-Q_2 - d_{lu}^{-1}(N_2 + N_2^T)$

Now, $A(t)$ and $A_d(t)$ are substituted in (2.249) with the values as defined in (2.17) and (2.18) respectively and subsequently separating nominal and uncertain matrices, one can rewrite (2.249) as,

$$\Theta(t) = \Theta|_{\Delta=0} + \Delta \tag{2.250}$$

where,

$$\Delta = \Pi + \Pi^T \tag{2.251}$$

and

$$\Pi = D_1 F_a(t) E_1 + D_2 F_d(t) E_d \tag{2.252}$$

where,
$D_1 = \left[\, D_a^T P \; 0 \; 0 \; 0 \; d_l D_a^T R_1 \; D_a^T R_2 \,\right]^T$, $D_2 = \left[\, D_d^T P \; 0 \; 0 \; 0 \; d_l D_d^T R_1 \; D_d^T R_2 \,\right]^T$,

$E_1 = \left[\, E_a \; 0 \; 0 \; 0 \; 0 \; 0 \,\right]$, $E_2 = \left[\, 0 \; 0 \; E_d \; 0 \; 0 \; 0 \,\right]$.

Using Lemma 2.6 for eliminating the uncertain matrix, one can write (2.252) as,

$$\Pi + \Pi^T \le \epsilon_1 E_1^T E_1 + \epsilon_2 E_2^T E_2 + \epsilon_1^{-1} D_1 D_1^T + \epsilon_2^{-1} D_2 D_2^T \tag{2.253}$$

So, using (2.253) in (2.250) and applying Schur-complement one can obtain the LMI conditions (2.243) and (2.248).

If $F_a(t) = F_d(t) = F(t)$ and $D_a = D_d = D$, then the above theorem is stated in the following corollary.

Corollary 2.7 *System (2.15) for $\mu \ge 0$ is robustly asymptotically stable with the admissible uncertainties defined in (2.240) satisfying the conditions $F(t)^T F(t) \le I$, if there exist symmetric matrices $P > 0$, $Q_i > 0$, $i = 1, 2, \ldots 4$, $R_j > 0$, any matrices of appropriate dimensions M_j, N_j, $j = 1, 2$ and the scalars $\epsilon_i > 0$ $(i = 1, 2)$ such that following LMIs hold:*

$$\begin{bmatrix} \Psi & \Phi_1 & D \\ \star & -R_2 & 0 \\ \star & 0 & -\epsilon I \end{bmatrix} < 0, \tag{2.254}$$

$$\begin{bmatrix} \Psi & \Psi_2 & D \\ \star & -R_2 & 0 \\ \star & 0 & -\epsilon I \end{bmatrix} < 0, \tag{2.255}$$

where,

$$D = \begin{bmatrix} D^T P \; 0 \; 0 \; 0 \; d_l D^T R_1 \; D^T R_2 \end{bmatrix}^T,$$

and,

$$\Psi = \begin{bmatrix} \Theta_{11}|_{\Delta=0} + \epsilon E_a^T E_a & R_1 & \Theta_{13}|_{\Delta=0} + \epsilon E_a^T E_d & 0 & \Theta_{15}|_{\Delta=0} & \Theta_{16}|_{\Delta=0} \\ \star & \Theta_{22} & \Theta_{23} & 0 & 0 & 0 \\ \star & \star & \Theta_{33}|_{\Delta=0} + \epsilon E_d^T E_d & \Theta_{34} & \Theta_{35}|_{\Delta=0} & \Theta_{36}|_{\Delta=0} \\ \star & \star & \star & \Theta_{44} & 0 & 0 \\ \star & 0 & \star & 0 & -R_1 & 0 \\ \star & 0 & \star & 0 & 0 & -R_2 \end{bmatrix}$$

Note: Ψ, Φ_1 and Φ_2 are defined in Theorem 2.26.

If $d_l = 0$ and the uncertainties are as defined in (2.240) then the Corollary 2.4 (stability condition of nominal TDS) can be extended for obtaining the corresponding delay-range-dependent robust stability condition which is presented below.

Corollary 2.8 *Given a system (2.15) with $d_l = 0$ for $\mu \geq 0$ is robustly asymptotically stable with the admissible uncertainties defined in (2.240), if there exist symmetric matrices $P > 0$, $Q_i > 0$, $i = 2, 3$ $R_2 > 0$, and any free matrices of appropriate dimensions M_j, N_j, $j = 1, 2$ and the scalar $\epsilon > 0$ such that the following LMIs hold:*

$$\begin{bmatrix} \Gamma & \Xi_1 & D \\ \star & -R_2 & 0 \\ \star & 0 & -\epsilon I \end{bmatrix} < 0, \tag{2.256}$$

$$\begin{bmatrix} \Gamma & \Xi_2 & D \\ \star & -R_2 & 0 \\ \star & 0 & -\epsilon I \end{bmatrix} < 0, \tag{2.257}$$

where, $\Gamma = \begin{bmatrix} \Gamma_{11} & \Gamma_{12} & 0 & A^T R_2 \\ \star & \Gamma_{22} & \Gamma_{23} & A_d^T R_2 \\ \star & \star & \Gamma_{33} & 0 \\ \star & \star & \star & -R_2 \end{bmatrix}$, and

$\Gamma_{11} = PA + A^T P + \sum_{i=2}^{3} Q_i + d_u^{-1}(M_1 + M_1^T) + \epsilon E_a^T E_a$,
$\Gamma_{12} = PA_d + d_u^{-1}(-M_1 + N_1^T) + \epsilon E_a^T E_d$,
$\Gamma_{22} = -(1-\mu)Q_3 + d_u^{-1}(M_2 + M_2^T - N_1 - N_1^T) + \epsilon E_d^T E_d$, $\Gamma_{23} = d_u^{-1}(-M_2 + N_2^T)$,
$\Gamma_{33} = -Q_2 - d_u^{-1}(N_2 + N_2^T)$

The matrices Ξ_1 *and* Ξ_2 are already defined in Corollary 2.4 while matrix D is defined as $D = \begin{bmatrix} D^T P \; 0 \; 0 \; D^T R_2 \; 0 \end{bmatrix}$.

The delay upper bound estimates using Corollary 2.7 and 2.8 for the numerical examples 2.6 and 2.7 are computed when (i) $d_l < d(t) < d_u$ (where d_l is specified) and (ii) $0 < d(t) < d_u$, for different μ (delay-derivative) values.

Table 2.14 d_u results of Example 2.6 for $0 \leq \mu < 1$ for specified d_l

Stability methods	μ	d_l	d_u
[37]	0.5	0	0.31
		0.5	–
[37]	0.9	0	0.31
		0.5	–
Corollary 2.8	0.5	0	0.3972
Corollary 2.7		0.1	0.2783
Corollary 2.7		0.2	0.3688
Corollary 2.7		0.5	0.6076
Corollary 2.8	0.9	0	0.3972
Corollary 2.7		0.1	0.2783
Corollary 2.7		0.2	0.3688
Corollary 2.7		0.5	0.6076

Table 2.15 d_u results of Example 2.7 for $0 \leq \mu < 1$ for specified d_l

Stability methods	μ	d_l	d_u
Corollary 2.8	0	0	2.3970
	0.5	0	1.4818
Corollary 2.7	0.5	0.1	1.4952
		0.5	1.5234
		1	1.5458
		2	2.1277
		2.2	2.2612
		2.38	2.3851
Corollary 2.7	0.9	0.1	1.2526
		0.5	1.3199
		1	1.5391
		2	2.1279
		2.2	2.2612
		2.38	2.3851

The numerical results presented above suggests that the proposed method improves the results over the published paper [37] mainly due to the (i) new LK functional (ii) tighter bounding conditions (2.193) and (2.194) used in the robust stability analysis.

2.8 Delay-Range-Dependent Stability Analysis of Uncertain TDS by Delay Partitioning Approach

In this section, a robust delay-range-dependent stability method in the frame work of delay partitioning approach is considered by adopting the LK functional in [47] (here the delay range between d_l to d_u is divided into two equal intervals $\delta = \frac{d_l + d_u}{2}$) and using the proposed bounding inequality discussed in Theorem 2.20 (see Sect. 2.5.3). The stability analysis is presented below in the form of theorem.

Theorem 2.27 *Given a system (2.15) for $0 \le d_l \le d(t) \le d_u$, $0 \le \mu < 1$ is robustly asymptotically stable with the admissible uncertainties defined in (2.240) satisfying the conditions $F(t)^T F(t) \le I$, if there exist symmetric matrices $P > 0$, $Q_i > 0$, $i = 1, 2, 3$, $R_j > 0$, and any free matrices of appropriate dimensions M_j, N_j, $j = 1, 2$ and the scalar $\epsilon > 0$ such that the following LMIs hold:*

$$\begin{bmatrix} \Phi_{11}(a) & \Phi_{12} \\ \star & -\epsilon I \end{bmatrix} < 0 \tag{2.258}$$

$$\begin{bmatrix} \Phi_{11}(b) & \Phi_{12} \\ \star & -\epsilon I \end{bmatrix} < 0 \tag{2.259}$$

where,

$$\Phi_{11}(a) = \begin{bmatrix} \Theta_{11}(a) & \Theta_{12} \\ \star & \Theta_{22} \end{bmatrix}, \; \Phi_{11}(b) = \begin{bmatrix} \Theta_{11}(b) & \Theta_{12} \\ \star & \Theta_{22} \end{bmatrix}, \tag{2.260}$$

$$\Phi_{12} = \left[D^T P \; 0 \; 0 \; 0 \; 0 \; \delta D^T R_1 \; D^T R_2 \right]^T, \text{and}$$

$$\Theta_{11}(a) = \begin{bmatrix} \theta_{11(0)} & \theta_{12(0)} & 0 & R_1 & 0 \\ \star & \theta_{22(0)} & \theta_{23} & \theta_{24} & N_1 \\ \star & \star & \theta_{33} & \theta_{34} & 0 \\ \star & \star & \star & \theta_{44} & M_1 \\ \star & \star & \star & \star & -R_2 \end{bmatrix} \tag{2.261}$$

$$\Theta_{11}(b) = \begin{bmatrix} \theta_{11(0)} & \theta_{12(0)} & 0 & R_1 & 0 \\ \star & \theta_{22(0)} & \theta_{23} & \theta_{24} & M_2 \\ \star & \star & \theta_{33} & \theta_{34} & N_2 \\ \star & \star & \star & \theta_{44} & 0 \\ \star & \star & \star & \star & -R_2 \end{bmatrix} \tag{2.262}$$

$$\Theta_{12}=\begin{bmatrix}\delta A^TR_1 & A^TR_2\\ \delta A_d^TR_1 & A_d^TR_2\\ 0 & 0\\ 0 & 0\\ 0 & 0\end{bmatrix},\quad \Theta_{22}=\begin{bmatrix}-R_1 & 0\\ 0 & -R_2\end{bmatrix}$$

The elements of the $\Theta_{11}(a)$ *and* $\Theta_{11}(b)$ *are as follows,*
$\theta_{11(0)}{=}PA+A^TP+\Sigma_{i=1}^{3}Q_i-R_1$, $\theta_{12(0)}{=}PA_d$,

$\theta_{22(0)}{=}-(1-\mu)Q_2+d_{lu}^{-1}(M_2+M_2^T-N_1-N_1^T)$, $\theta_{23}{=}d_{lu}^{-1}(-M_2+N_2^T)$,

$\theta_{24}{=}d_{lu}^{-1}(-M_1^T+N_1)$, $\theta_{33}{=}d_{lu}^{-1}(-N_2^T-N_2)-Q_3$,

$\theta_{44}{=}d_{lu}^{-1}(M_1^T+M_1)-Q_1-R_1$, $\delta{=}\frac{d_l+d_u}{2}$, $d_{lu}{=}(d_u-\delta)$

Proof Considering the similar type of LK functional as in [47],

$$V(t)=V_1(t)+V_2(t)+V_3(t)+V_4(t)+V_5(t)+V_6(t) \tag{2.263}$$

where the individual functionals are as follows,

$$V_1(t)=x^TPx(t),\ \ V_2(t)=\int_{t-\delta}^{t}x^T(s)Q_1x(s)ds,\ \ V_3(t)=\int_{t-d(t)}^{t}x^T(s)Q_2x(s)ds$$

$$V_4(t)=\int_{t-d_u}^{t}x^T(s)Q_3x(s)ds,\ \ V_5(t)=\delta\int_{-\delta}^{0}\int_{t+\theta}^{t}x^T(\theta)R_1x(\theta)d\theta ds$$

$$V_6(t)=d_{lu}^{-1}\int_{-d_u}^{-\delta}\int_{t+\theta}^{t}x^T(\theta)R_2x(\theta)d\theta ds$$

Note that, the $V_6(t)$ functional term selected here contains a factor d_{lu}^{-1} whereas in [47] it appears as d_{lu}, this modification is required in the functional in order to use the proposed bounding inequality described in Sect. 2.5.3.

$$\begin{aligned}\dot V(t)&=2\dot x^T(t)Px(t)+x^T(t)(\sum_{i=1}^{3}Q_i)x(t)-(1-\mu)x^T(t-d(t))Q_2x(t-d(t))\\&-x^T(t-d_u)Q_3x(t-d_u)-x^T(t-\delta)Q_3x(t-\delta)+\delta^2\dot x^T(t)R_1x(t)\\&+\dot x^T(t)R_2\dot x(t)-\delta\int_{t-\delta}^{t}\dot x^T(s)R_1\dot x(s)ds\\&-d_{lu}^{-1}\int_{t-d_u}^{t-\delta}\dot x^T(s)R_2\dot x(s)ds\end{aligned} \tag{2.264}$$

Define augmented state space vector as,

$$\zeta(t) = \left[x^T(t)\; x^T(t-d(t))\; x^T(t-d_u)\; x^T(t-\delta)\right]^T$$

Now, one can rearrange the terms in (2.264) in view of augmented state space vector as,

$$\dot{V}(t) \leq \zeta^T(t)\Xi\zeta(t) - d_{lu}^{-1}\int_{t-d_u}^{t-\delta} \dot{x}^T(s)R_2\dot{x}(s)ds$$
$$-\delta\int_{t-\delta}^{t} \dot{x}^T(s)R_1\dot{x}(s)ds$$

$$\dot{V}(t) \leq \zeta^T(t)\Xi\zeta(t) - \delta\int_{t-\delta}^{t} \dot{x}^T(s)R_1\dot{x}(s)ds$$
$$-d_{lu}^{-1}\int_{t-d(t)}^{t-\delta} \dot{x}^T(s)R_2\dot{x}(s)ds - d_{lu}^{-1}\int_{t-d_u}^{t-d(t)} \dot{x}^T(s)R_2\dot{x}(s)ds \quad (2.265)$$

where, $\Xi = \begin{bmatrix} \Xi_{11} & \Xi_{12} & 0 & 0 \\ \star & \Xi_{22} & 0 & 0 \\ \star & \star & -Q_3 & 0 \\ 0 & 0 & 0 & -Q_1 \end{bmatrix}$

The elements of 'Ξ' matrix are as follows:

$\Xi_{11} = PA(t) + A^T(t)P + \sum_{i=1}^{3} Q_i + A^T(t)(\delta^2 R_1 + R_2)A(t),$

$\Xi_{12} = PA_d(t) + A^T(t)(\delta^2 R_1 + R_2)A_d(t),$

$\Xi_{22} = -(1-\mu)Q_2 + A_d^T(t)(\delta^2 R_1 + R_2)A_d(t)$

The integral terms in (2.265) are approximated using the proposed tighter bounding inequality condition as discussed in Theorem 2.20. One can obtain the following expression as,

$$-\delta\int_{t-\delta}^{t} \dot{x}^T(s)R_1\dot{x}(s)ds \leq \begin{bmatrix} x(t) \\ x(t-\delta) \end{bmatrix}^T \begin{bmatrix} -R_1 & R_1 \\ \star & -R_1 \end{bmatrix} \begin{bmatrix} x(t) \\ x(t-\delta) \end{bmatrix} \quad (2.266)$$

$$-\delta\int_{t-\delta}^{t}\dot{x}^T(s)R_1\dot{x}(s)ds \le \zeta^T \begin{bmatrix} -R_1 & 0 & 0 & R_1 \\ 0 & 0 & 0 & 0 \\ 0 & 0 & 0 & 0 \\ \star & 0 & 0 & -R_1 \end{bmatrix} \zeta(t) \tag{2.267}$$

$$\begin{aligned} -d_{lu}^{-1}\int_{t-d(t)}^{t-\delta}\dot{x}^T(s)R_2\dot{x}(s)ds \le & \begin{bmatrix} x(t-\delta) \\ x(t-d(t)) \end{bmatrix}^T \\ & \left\{ d_{lu}^{-1}\begin{bmatrix} M_1+M_1^T & -M_1+N_1^T \\ \star & -N_1-N_1^T \end{bmatrix} \right. \\ & \left. \times\, \varrho \begin{bmatrix} M_1 \\ N_1 \end{bmatrix} R_2^{-1} \begin{bmatrix} M_1 \\ N_1 \end{bmatrix}^T \right\} \\ & \times \begin{bmatrix} x(t-\delta) \\ x(t-d(t)) \end{bmatrix} \end{aligned} \tag{2.268}$$

and,

$$\begin{aligned} -d_{lu}^{-1}\int_{t-d_u}^{t-d(t)}\dot{x}^T(s)R_2\dot{x}(s)ds \le & \begin{bmatrix} x(t-d(t)) \\ x(t-d_u) \end{bmatrix}^T \\ & \left\{ d_{lu}^{-1}\begin{bmatrix} M_2+M_2^T & -M_2+N_2^T \\ \star & -N_2-N_2^T \end{bmatrix} \right. \\ & \left. \times\, (1-\varrho) \begin{bmatrix} M_2 \\ N_2 \end{bmatrix} R_2^{-1} \begin{bmatrix} M_2 \\ N_2 \end{bmatrix}^T \right\} \\ & \times \begin{bmatrix} x(t-d(t)) \\ x(t-d_u) \end{bmatrix} \end{aligned} \tag{2.269}$$

where, $\varrho = \frac{d(t)-\delta}{d_{lu}}$ and $0 \le \varrho \le 1$. Substituting the value of integral (2.267), (2.268) and (2.269) in (2.265), and carrying out some algebraic manipulations yields resulting expression as,

$$\dot{V}(t) \le \zeta^T(t)\{\Gamma(t) + \varrho\Psi_1 R_2^{-1}\Psi_1^T + (1-\varrho)\Psi_2 R_2^{-1}\Psi_2^T\}\zeta(t) \tag{2.270}$$

where, $\Psi_1 = \begin{bmatrix} 0 & N_1^T & 0 & M_1^T \end{bmatrix}^T$, and $\Psi_2 = \begin{bmatrix} 0 & M_2^T & N_2^T & 0 \end{bmatrix}^T$.

One can observe convex combination of matrices $\Psi_1 R_2^{-1}\Psi_1^T$ and $\Psi_2 R_2^{-1}\Psi_2^T$ in (2.270). For asymptotic stability of the system (2.15), $\dot{V}(t)$ must be negative-definite for which one must have,

$$\Gamma(t) + \varrho\Psi_1 R_2^{-1}\Psi_1^T + (1-\varrho)\Psi_2 R_2^{-1}\Psi_2^T < 0 \tag{2.271}$$

The above expression can be further rewritten as,

$$\varrho(\Gamma(t) + \Psi_1 R_2^{-1} \Psi_1^T) + (1-\varrho)(\Gamma(t) + \Psi_2 R_2^{-1} \Psi_2^T) < 0 \tag{2.272}$$

In view of the condition $0 \leq \varrho \leq 1$, one may write the above inequality as,

$$\Gamma(t) + \Psi_1 R_2^{-1} \Psi_1^T < 0 \tag{2.273}$$

$$\Gamma(t) + \Psi_2 R_2^{-1} \Psi_2^T < 0 \tag{2.274}$$

where,

$$\Gamma(t) = \begin{bmatrix} \gamma_{11} & \gamma_{12} & 0 & R_1 \\ \star & \gamma_{22} & \gamma_{23} & \gamma_{24} \\ \star & \star & \gamma_{33} & 0 \\ \star & \star & \star & \gamma_{44} \end{bmatrix} \tag{2.275}$$

and,
$\gamma_{11} = PA(t) + A^T(t)P + \sum_{i=1}^{3} Q_i - R_1 + A^T(t)(\delta^2 R_1 + R_2)A(t)$,
$\gamma_{12} = PA_d(t) + A^T(t)(\delta^2 R_1 + R_2)A_d(t)$,
$\gamma_{22} = -(1-\mu)Q_2 + d_{lu}^{-1}(M_2 + M_2^T - N_1 - N_1^T) + A_d^T(t)(\delta^2 R_1 + R_2)A_d(t)$, $\gamma_{23} = d_{lu}^{-1}(-M_2 + N_2^T)$,
$\gamma_{24} = d_{lu}^{-1}(-M_1^T + N_1)$, $\gamma_{33} = d_{lu}^{-1}(-N_2^T - N_2) - Q_3$, $\gamma_{44} = d_{lu}^{-1}(M_1^T + M_1) - Q_1 - R_1$

Using Schur-Complement one can write (2.273) and (2.274) as

$$\Theta(t)(a) = \begin{bmatrix} \theta_{11} & \theta_{12} & 0 & R_1 & 0 \\ \star & \theta_{22} & \theta_{23} & \theta_{24} & N_1 \\ \star & \star & \theta_{33} & \theta_{34} & 0 \\ \star & \star & \star & \theta_{44} & M_1 \\ \star & \star & \star & \star & -R_2 \end{bmatrix} \tag{2.276}$$

$$\Theta(t)(b) = \begin{bmatrix} \theta_{11} & \theta_{12} & 0 & R_1 & 0 \\ \star & \theta_{22} & \theta_{23} & \theta_{24} & M_2 \\ \star & \star & \theta_{33} & \theta_{34} & N_2 \\ \star & \star & \star & \theta_{44} & 0 \\ \star & \star & \star & \star & -R_2 \end{bmatrix} \tag{2.277}$$

where, $\theta_{11} = \gamma_{11}$, $\theta_{12} = \gamma_{12}$, $\theta_{22} = \gamma_{22}$, $\theta_{23} = \gamma_{23}$, $\theta_{24} = \gamma_{24}$, $\theta_{33} = \gamma_{33}$, $\theta_{44} = \gamma_{44}$
Once again using Schur-complement on (2.276) and (2.277) one can get,

$$\Phi(t)_{11}(a) = \begin{bmatrix} \Theta(t)_{11}(a) & \Theta(t)_{12} \\ \star & \Theta_{22} \end{bmatrix}, \; \Phi(t)_{11}(b) = \begin{bmatrix} \Theta(t)_{11}(b) & \Theta(t)_{12} \\ \star & \Theta_{22} \end{bmatrix} \tag{2.278}$$

where,

$$\Theta(t)_{12} = \begin{bmatrix} \delta A^T(t)R_1 & A^T(t)R_2 \\ \delta A_d(t)^T R_1 & A_d^T(t)R_2 \\ 0 & 0 \\ 0 & 0 \\ 0 & 0 \end{bmatrix}, \quad \Theta_{22} = \begin{bmatrix} -R_1 & 0 \\ 0 & -R_2 \end{bmatrix}$$

$$\Theta(t)_{11}(a) = \begin{bmatrix} \theta_{11(0)} & \theta_{12(0)} & 0 & R_1 & 0 \\ \star & \theta_{22(0)} & \theta_{23} & \theta_{24} & N_1 \\ \star & \star & \theta_{33} & \theta_{34} & 0 \\ \star & \star & \star & \theta_{44} & M_1 \\ \star & \star & \star & \star & -R_2 \end{bmatrix}$$

$$\Theta(t)_{11}(b) = \begin{bmatrix} \theta_{11(0)} & \theta_{12(0)} & 0 & R_1 & 0 \\ \star & \theta_{22(0)} & \theta_{23} & \theta_{24} & M_2 \\ \star & \star & \theta_{33} & \theta_{34} & N_2 \\ \star & \star & \star & \theta_{44} & 0 \\ \star & \star & \star & \star & -R_2 \end{bmatrix}$$

where,

$\theta_{11(0)}=PA(t) + A^T(t)P + \Sigma_{i=1}^3 Q_i - R_1, \theta_{12(0)}=PA_d(t),$
$\theta_{22(0)}=-(1-\mu)Q_2 + M_2 + M_2^T - N_1 - N_1^T, \theta_{23}=d_{lu}^{-1}(-M_2 + N_2^T),$
$\theta_{24}=d_{lu}^{-1}(-M_1^T + N_1), \theta_{33}=d_{lu}^{-1}(-N_2^T - N_2) - Q_3,$
$\theta_{44}=d_{lu}^{-1}(M_1^T + M_1) - Q_1 - R_1$

Next replace the matrices $A(t)$ and $A_d(t)$ with the uncertain matrices in (2.17) and the structure of matrices $\Delta A(t)$ and $\Delta A_d(t)$ are decomposed as $DF(t)E_a$ and $DF(t)E_d$ respectively. Using Lemma 2.6 for elimination of uncertain matrices and separating nominal and uncertain matrices the LMIs (2.258) and (2.259) are obtained. This completes the proof. □

The estimated delay upper bound using the stability condition in Theorem 2.27 for the system described in Numerical Example 2.6 is illustrated in Table 2.16.

Remark 2.30 One can observe from the results presented in Table 2.16 that, the proposed bounding inequalities in conjunction with the delay partitioning method gives less conservative estimate of the delay upper bound compared to the existing results [47, 49, 50], for different cases of delay lower bounds.

Table 2.16 d_u results of Example 2.6 for $0 \leq \mu < 1$ for specified d_l

Stability methods	d_l	d_u
For $\mu = 0.5$		
Theorem 2.27	0	0.5563
[47]		0.4760
[49]		0.4243
[50]		0.4783
Theorem 2.27	0.1	0.5935
[49]		0.4767
Theorem 2.27	0.2	0.6294
[49]		0.5429
Theorem 2.27	0.3	0.6642
[49]		0.6059
Theorem 2.27	0.4	0.6982
[49]		0.6656
Theorem 2.27	0.5	0.7315
[49]		0.7238
For $\mu = 0.9$		
Theorem 2.27	0	0.5563
[47]		0.4760
[50]		0.4783
Theorem 2.27	0.1	0.5935
	0.2	0.6294
	0.3	0.6642
	0.4	0.6982
	0.5	0.7315

2.9 Conclusion

This chapter first deals with the review of existing literature on development of stability analysis of linear time-delay systems using LK functional approach in an LMI framework. The new stability conditions have been presented by introducing new LK functional, improved bounding inequalities and free weighting matrices. Unlike other methods, some useful terms (integral of quadratic form of $\dot{x}(t)$) in the derivative of LK functional are not ignored and their presence is taken into account by using tighter bounding of the integral term, this in turn, results less conservative results. The stability conditions for three classes of time-delay systems have been proposed:

(i) Delay-dependent stability condition for TDS with single time-varying delay
(ii) Delay-dependent stability condition for TDS with two-additive time-varying delays

(iii) Delay-range-dependent stability condition for TDS with time-varying delay.

The effectiveness of the proposed stability criteria are successfully verified by numerical examples. Tables 2.3, 2.4, 2.5, 2.6, 2.7, 2.8, 2.9 and 2.10 show that the proposed delay-dependent stability criteria provide less conservative results for delay upper bound estimate consistently for different delay derivatives compared to the existing methods.

The second part of this chapter deals with the robust stability analysis of an uncertain time-delay systems where the structure of the uncertainty is assumed to be of norm-bounded type. New and improved robust stability conditions for the following problems have been obtained by adopting the same procedure as discussed in the first part of this chapter,

(i) Delay-dependent robust stability condition for TDS with single time-varying delay
(ii) Delay-range-dependent robust stability condition for TDS with single time-varying delay
(iii) Delay-range-dependent robust stability condition for TDS with single time-varying delay using delay partitioning approach.

Tables 2.12, 2.13, 2.14, 2.15 and 2.16 show that the proposed delay-dependent robust stability methods give less conservative results for delay upper bound estimate compared to the existing methods.

References

1. J.P. Richard, Time-delay systems: an overview of some recent advances and open problems. Automatica **39**, 1667–1694 (2003)
2. X. Li, C.E. de Souza, Delay-dependent robust stability and stabilization of uncertain linear delay system:a linear matrix inequality approach. IEEE Trans. Autom. Control **42**, 1144–1148 (1997)
3. Y. He, Q.G. Wang, C. Lin, M. Wu, Delay-range-dependent stability for systems with time-varying delay. Automatica **43**, 371–376 (2007)
4. H. Shao, Improved delay-dependent stability criteria for systems with a delay varying in range. Automatica **44**, 3215–3218 (2008)
5. H. Shao, New delay-dependent stability criteria for systems with interval delay. Automatica **45**, 744–749 (2009)
6. Dey, R., Ghosh, S., Ray, G.: Delay-dependent stability and state feedback stabilization criterion for linear time delay system. In: International Conference on Modeling and Simulation, Coimbator, vol. 2, pp. 963 – 968 (2007)
7. K. Gu, V.L. Kharitonov, J. Chen, *Stability of Time-Delay Systems* (Birkhuser, Boston, 2003)
8. S. Xu, J. Lam, A survey of linear matrix inequalities in stability analysis of delay systems. Int. J. Syst. Sci. **39**, 1095–1113 (2008)
9. E. Fridman, U. Shaked, Delay-dependent stability and H_{∞} control: constant and time-varying delays. Int. J. Control **76**, 48–60 (2003)
10. V.B. Kolmanovoskii, J.P. Richard, Stability of some linear system with delay. IEEE Trans. Autom. Control **44**, 984–989 (1999)

11. Dey, R., Ghosh, S., Ray, G.: Delay dependent stability analysis of linear system with multiple state delays. In: *2nd IEEE International conference on Industrial and Information systems (ICIIS'07)*, pp. 255–260 (2007)
12. X. Li, C.E. de Souza, Criteria for robust stability and stabilization of uncertain linear systems with state delays. Automatica **33**, 1657–1662 (1997)
13. M. Wu, Y. He, J.H. She, G.P. Liu, Delay-dependent criteria for robust stability of time-varying delay systems. Automatica **40**, 1435–1439 (2004)
14. M.N.A. Parlakci, Improved robust stability criteria and design of robust stabilizing controller for uncertain linear time-delay system. Int. J. Robust Nonlinear Control **16**, 599–636 (2006)
15. R. Dey, S. Ghosh, G. Ray, A. Rakshit, State feedback stabilization of uncertain linear time-delay systems: a nonlinear matrix inequality appraoch. Numer. Linear Algebra Appl. **18**(3), 351–361 (2011)
16. Y.Y. Cao, Y.X. Sun, J. Lam, Delay-dependent robust H_∞ control for uncertain system with time-varying delays. IEE Proc. CTA **145**, 338–344 (1998)
17. V.B. Kolmanovoskii, On the liapunov-krasovskii functionals for stability analysis for linear time-delay systems. Int. J. Control **72**, 374–384 (1999)
18. J.H. Kim, Delay and its time-derivative robust stability of time delayed linear systems with uncertainty. IEEE Trans. Autom. Control **46**, 789–792 (2001)
19. K. Gu, S.I. Niculescu, Additional dynamics in transformed time-dealy systems. IEEE Trans. Autom. Control **45**, 00–00 (2000)
20. V. Kharitonov, D. Melchor-Aguilar, Additional dynamics for general class of time-delay systems. IEEE Trans. Autom. Control **48**, 1060–1064 (2003)
21. S.I. Niclescu, On delay-dependent stability under model transformations of some neutral linear systems. Int. J. Control **74**, 609–617 (2001)
22. Boyd, S., Vandenberghe, L.: Convex Optimization. Cambridge University Press, Cambridge (http://www.stanford.edu/boyd/cvxbook/bv_cvxbook.pdf) (2004)
23. P. Park, A delay dependent stability criterion for systems with uncertain time-invariant delays. IEEE Trans. Autom. Control **44**, 876–877 (1999)
24. S. Boyd, L. Ghaoui, E. Feron, V. Balakrishnan, *LMI in Systems and Control Theory* (SIAM, Philadelphia, 1994)
25. Y.S. Moon, P. Park, W.H. Kwon, Y.S. Lee, Delay-dependent robust stabilization of uncertain state delayed system. Int. J. Control **74**, 1447–1455 (2001)
26. E. Fridman, New-lyapunov-krasovskii functional for stability of linear retarded and neutral type. Syst. Control Lett. **43**, 309–319 (2001)
27. E. Fridamn, U. Shaked, A descriptor system approach to H_∞ control of linear time-delay systems. IEEE Trans. Autom. Control **47**, 253–270 (2002)
28. E. Fridman, U. Shaked, An improved stabilization method for linear time-delay system. IEEE Trans. Autom. Control **47**, 1931–1937 (2002)
29. V. Suplin, E. Fridman, U. Shaked, H_∞ control of linear uncertain time-delay systems-a projection approach. IEEE Trans. Autom. Control **31**, 680–685 (2006)
30. Y. He, M. Wu, J.H. She, G. Liu, Delay-dependent robust stability criteria for uncertain neutral system with mixed delays. Syst. Control Lett. **51**, 57–65 (2004)
31. Dey, R., Ghosh, S., Ray, G.: Delay-dependent stability analysis with time-varying state delay. In 4th IFAC conference on Management and Control of Production and Logistics, pp. 313–318 (2007)
32. Y. He, Q.-G. Wang, L. Xie, C. Lin, Further improvements of free weighting matrices technique for systems with time-varying delay. IEEE Trans. Autom. Control **52**, 293–299 (2007)
33. J. Lam, H. Gao, C. Wang, Stability analysis for continuous system with two additive time-varying delay components. Syst. Control Lett. **56**, 16–24 (2007)
34. P.L. Liu, T.J. Su, Robust stability of interval time-delay systems with delay-dependence. Syst. Control Lett. **33**, 231–239 (1998)
35. X.J. Jiang, Q.L. Han, On H_∞ control for linear systems withinterval time-varying delay. Automatica **41**, 2099–2106 (2005)

36. P. Park, J.W. Ko, Stability and robust stability for system with time-varying delay. Automatica **43**, 1855–1858 (2007)
37. T. Li, L. Guo, Y. Zhang, Delay-range-dependent robust stability and stabilization for uncertain systems with time-varying delay. Int. J. Robust Nonlinear Control **18**, 1372–1387 (2008)
38. Q.L. Han, On robust stability of neutral systems with time-varying discrete delay and norm-bounded uncertainty. Automatica **40**, 1087–1092 (2004)
39. Y. He, W. Min, J.H. She, G.P. Liu, Parameter dependent lyapunov functional for stability of time delay systems with polytopic uncertainties. Automatica **492**, 828–832 (2004)
40. M. Wu, Y. He, J.H. She, New delay-dependent stability criteria and stabilizing method for neutral system. IEEE Trans. Autom. Control **49**, 2266–2271 (2004)
41. N. Olgac, R. Sipahi, An exact method for the stability analysis of time-delayed linear time-invariant systems. IEEE Trans. Autom. Control **47**, 793–797 (2002)
42. R. Dey, G. Ray, S. Ghosh, A. Rakshit, Stability analysis for contious system with additive time-varying delays: a less conservative result. Appl. Math. Comput. **215**, 3740–3745 (2010)
43. H. Gao, T. Chen, J. Lam, A new delay system approach to network based control. Automatica **44**, 39–52 (2008)
44. X.L. Jing, D.L. Tan, Y.C. Wang, An lmi approach to stability of systems with severe time-delay. IEEE Trans. Autom. Control **49**, 1192–1195 (2004)
45. Lee, Y.S., Moon, Y.S., Kwoon, W.H., Lee, K.H.: Delay-dependent robust H_∞ control of uncertain system with time-varying state delay, pp. 3208–3213 (2001)
46. J. Sun, G.P. Liu, J. Chen, D. Rees, Improved delay-range-dependent stability criteria for linear systems with time-varying delays. Automatica **46**, 466–470 (2010)
47. C. Peng, Y.C. Tian, Improved delay-dependent robust stability criteria of uncertain systems with interval time-varying delay. IET Control Theor. Appl. **2**, 752–761 (2008)
48. E. Fridman, U. Shaked, Parameter dependent stability and stabilization of uncertain time-delay systems. IEEE Trans. Autom. Control **48**, 861–866 (2003)
49. C. Peng, Y.C. Tian, Delay-dependent robust stability criteria for uncertain systems with time-varying delay. J. Comput. Appl. Math. **214**, 480–494 (2008)
50. Ramakrishnan, K., Ray, G.: Delay-dependent robust stability criteria for linear uncertain systems with interval time-varying delay. In IEEE TENCON, Singapore, pp. 1–6 (2009)

Chapter 3
Stabilization of Time-Delay Systems

This chapter deals with the stabilization of a nominal and uncertain time-delay systems using state feedback control law. Next, Load-Frequency Control (LFC) of an interconnected power systems with communication delay based on two different control configurations (i) pure state feedback (one-term control) and (ii) pure state feedback as well as delayed state feedback (two-term control) is considered by exploring the H_∞ performance criterion in the design procedure.

Note that, the stabilization condition for time-delay systems is obtained by directly extending the results of delay-dependent stability (or robust stability) conditions of TDS. The results of new stabilization conditions are validated by considering the numerical examples and compared with existing methods.

3.1 Introduction

As discussed in Chap. 2, the stability analysis of time-delay systems has been proposed for developing delay-dependent results in LMI framework based on LK functional approach with a tighter bounding technique. A significant research attention has been devoted to the delay-dependent studies owing to the fact that, in the delay-independent stability notion there is no upper limit to the time-delay, so often results are regarded as conservative. In true sense an unbounded time-delay is not so realistic to physical or engineering systems. In sequel, the stabilization (or robust stabilization) conditions are derived in a delay-dependent framework.

The earlier results on robust stabilization (and/or stabilization) based on delay-independent as well as on Ricatti equation approach are recalled [1–4] and [5]. Some of the results on delay-dependent robust stabilization (and/or stabilization) in an LMI framework can be found in [6–12], and [13], note that the condition derived in [6] is based on Lyapunov-Razumikhin approach. In [7, 8] and [13], stabilizing conditions were derived using LK functional approach adopting first model transformation and

R. Dey et al., *Stability and Stabilization of Linear and Fuzzy Time-Delay Systems*,
Intelligent Systems Reference Library 141, https://doi.org/10.1007/978-3-319-70149-3_3

they are expected to give conservative result with or without uncertainties as model transformation introduces additional dynamics [14] and [15]. The stabilizing conditions obtained in [9, 11, 12] and [10] are all NLMI. In [12], and [9] the non-linear matrix (NLMI) conditions were solved using cone complementarity linearization algorithm [16], which is an iterative algorithm, while in [11] and [10] a fixed relaxation matrix is introduced to transform NLMI condition to LMI condition. Possibly it is one of the probable reason for conservativeness in the stabilizing results. The robust stabilizing (and/or stabilizing) conditions in [14] and [10] have been obtained for polytopic uncertain systems based on descriptor method.

In this chapter, an improved few significant delay-dependent robust stabilization (and/or stabilization) conditions for the system (3.1) (and/or (3.6)) in an LMI framework are presented in the form of theorems. The robust stabilization condition of an uncertain time-delay system (3.1) can be obtained from robust stability conditions by substituting $A = A + BK$, or one can obtain the robust stabilization condition from the derived stabilization condition depending upon the type of bounding inequalities to eliminate the uncertain time-varying matrices.

3.2 Problem Statement

Consider an uncertain linear time-delay systems described by the following state equations

$$\dot{x}(t) = [A + \Delta A(t)]x(t) + [A_d + \Delta A_d(t)]x(t - d(t)) + [B + \Delta B(t)]u(t) \tag{3.1}$$

$$x(t) = \phi(t), \quad \forall t \in [-d_u, 0], \tag{3.2}$$

where, $x(t) \in \mathcal{R}^n$ is the state vector, $u(t) \in \mathcal{R}^m$ is the control input, and $\phi(t)$ is the initial condition. The matrices A, A_d, B, C and D are known real constant matrices of appropriate dimensions which describe the nominal system of (3.1), and $\Delta A(t)$, $\Delta A_d(t)$ and $\Delta B(t)$ are real matrix function representing time-varying parameter uncertainties. The delay $d(t)$ is time-varying and satisfies following conditions.

$$0 \leq d(t) \leq d_u, \quad \dot{d}(t) \leq \mu < 1 \tag{3.3}$$

The parametric uncertainties are assumed to be norm bounded type of the form:

$$\Delta A(t) = D_a F_a(t) E_a, \quad \Delta A_d(t) = D_d F_d(t) E_d, \quad \Delta B(t) = D_b F_b(t) E_b \tag{3.4}$$

where, $F_a(t) \in \mathcal{R}^{m_a \times p_a}$, $F_b(t) \in \mathcal{R}^{m_b \times p_b}$ and $F_d(t) \in \mathcal{R}^{m_d \times p_d}$ are unknown real time-varying matrices with Lebesgue measurable elements satisfying the conditions:

$$\| F_a(t) \| \leq 1, \quad \| F_b(t) \| \leq 1, \quad \| F_d(t) \| \leq 1, \quad \forall\, t \tag{3.5}$$

and, D_a, D_d, E_a, E_b and E_d are known real constant matrices that characterize how the uncertain parameters in $F_a(t)$, $F_b(t)$ and $F_d(t)$ enter the nominal system and input matrices.

If the uncertainties $\Delta A(t) = 0$, $\Delta A_d(t) = 0$ and $\Delta B(t) = 0$, then the uncertain system (3.1) reduces to nominal time-delay system described as,

$$\dot{x}(t) = Ax(t) + A_d x(t - d(t)) + Bu(t) \tag{3.6}$$

Stabilization *Given a scalar $d_u > 0$, find a control law $u(t) = Kx(t)$ for the system (3.6) such that the closed loop system is asymptotically stable for any time-delay $d(t)$ satisfying $0 \leq d(t) \leq d_u$. This problem is known in the literature as stabilization problem.*

Robust Stabilization *[21] Given a scalar $d_u > 0$, find a control law $u(t) = Kx(t)$ for the system (3.1), such that the closed loop system is asymptotically stable for any time-delay $d(t)$ satisfying $0 \leq d(t) \leq d_u$. This problem is known in the literature as robust stabilization problem.*

3.3 Delay-Dependent Stabilization of Nominal TDS

In this section, some existing state feedback stabilization sufficient conditions for system (3.6) using LK approach are presented in the form of theorems.

Assumption 3.1 The necessary condition for delay-dependent stabilization is that, $(A + A_d,\ B)$ is stabilizable.

Theorem 3.1 *(Corollary 3.2 [7]) Consider the system (3.6) with a constant delay $d(t) \equiv d$, satisfying the condition $0 \leq d(t) \leq d_u$, the system is stabilizable with the control law $u(t) = YX^{-1}x(t)$, if there exist matrices $X = X^T > 0$, Y and a scalar $\beta > 0$ such that the following LMI holds:*

$$\begin{bmatrix} Q_c + d_u A_d A_d^T & d_u(AX + BY)^T & d_u X A_d^T \\ \star & -d_u\beta & 0 \\ \star & 0 & -d_u(1-\beta) \end{bmatrix} < 0 \tag{3.7}$$

where, $Q_c = X(A + A_d)^T + (A + A_d)X + BY + Y^T B^T$.

Remark 3.1 The condition has been derived using first model transformation, hence the transformed system becomes,

$$\dot{\zeta}(t) = (A + A_d)\zeta(t) - A_d \int_{-d}^{0} [A\zeta(t+\theta) + A_d\zeta(t+\theta-d)]d\theta$$

$$\zeta(\theta) = \psi(\theta), \quad \forall \in [-2d, 0]$$

where, $\zeta(t)$ is the new state variable of the transformed system. Any solution of the system (3.6) with $d(t) = d$ and $u(t) = 0$ is also the solution of the above equation [7]. Thus, the LK function is chosen in accordance with the transformed system which is of the form,

$$V(\zeta, t) = \zeta^T(t)P\zeta(t) + W(\zeta, t)$$

where, $W(\zeta, t) = \int_{-d}^{0}\{(1+\alpha^{-1})\int_{t+\theta}^{t} \| A\zeta(s) \|^2 ds + \int_{t+\theta-d}^{t} \| A_d\zeta(s) \|^2 ds\}d\theta$ Finding the time-derivative of $V(\zeta, t)$, using bounding lemma (Lemma 2.1) for the cross terms and Schwartz inequality for quadratic integral terms, and finally using the change of variables ($X = P^{-1}$ *and* $\beta = \frac{1}{1+\alpha^{-1}}$) in $\dot{V}(\zeta, t)$ one can obtain the stabilization condition in (3.7) with the use of Schur-complement.

The stability conditions (2.52) and (2.53) discussed in Theorem 2.5 is extended to obtain the stabilization condition which is presented below in the form of theorem.

Theorem 3.2 *([9]) If there exist* $L = L^T > 0, M, N, R, V$ *and* $W = W^T > 0$ *such that following LMI holds:*

$$\begin{bmatrix} (1,1) & -N + A_d L & (1,3) \\ \star & -W & d_u L A_d^T \\ \star & \star & -d_u R \end{bmatrix} < 0 \tag{3.8}$$

$$\begin{bmatrix} M & N \\ \star & LR^{-1}L \end{bmatrix} \geq 0 \tag{3.9}$$

Then the system (3.6) with the control law $u(t) = VL^{-1}x(t)$ *is asymptotically stable for any constant time-delay* $0 \leq d(t) \leq d_u$.

Proof Substituting $u(t) = Kx(t)$ in (3.6) gives closed-loop system,

$$\dot{x}(t) = A_c x(t) + A_d x(t - d(t))$$

where, $A_c = A + BK$. One can now replace A in (2.52)[1](corresponding stability condition) with $A + BK$, then pre- and post-multiplying (2.52) and (2.53) by $diag\{P^{-1}, P^{-1}, Z^{-1}\}$ and $diag\{P^{-1}, P^{-1}\}$ respectively and finally applying adopting following change of variables as indicated below,

$$L = P^{-1},\ M = P^{-1}XP^{-1},\ N = P^{-1}YP^{-1},\ R = Z^{-1},\ W = P^{-1}QP^{-1},\ and\ V = KP^{-1}$$

one can obtain the stabilization condition in (3.8) and (3.9) with standard algebraic manipulations.

Remark 3.2 It can be observed that the resulting condition is not an LMI due to the presence of the term $LR^{-1}L$ in (3.9), hence it is not possible to solve this condition using any standard solver of LMI toolbox of MATLAB for obtaining delay

[1]refer sub-section 2.3.2, Theorem 2.5.

bound d_u. However, this difficulty can be overcome by substituting $R = L$ in (3.8) and (3.9) which transforms it to LMI condition, but the estimate of delay bound will be conservative in this case. To obtain better delay bound estimate cone complementarity algorithm was introduced in [16] and it is adopted in [9] and [12]. The detailed discussion on iterative non-linear minimization problem can be found in [9]. In this theorem the time-delay is assumed to be constant (i.e, $d(t) = d$ in (3.6) which makes $\dot{d}(t) = \mu = 0$).

Theorem 3.3 *([10]) The state feedback control law $u(t) = Kx(t)$ asymptotically stabilizes the system (3.6) for all the delays satisfying the condition (3.3), if there exist a diagonal matrix $\epsilon_1 I \in \mathcal{R}^{n\times n}$, such that the following LMIs hold: $Q_1 = Q^T > 0, Q_2, Q_3, \bar{S} = \bar{S}^T, \bar{R} = \bar{R}^T > 0, \bar{Z} = \begin{bmatrix} \bar{Z_{11}} & \bar{Z_{12}} \\ \bar{Z_{12}^T} & \bar{Z_{13}} \end{bmatrix}$ and $\bar{Y}$ matrices with appropriate dimensions, that satisfy the following LMIs,*

$$\begin{bmatrix} (1,1) & (1,2) & 0 & Q_1 & d_u Q_2^T \\ \star & (2,2) & A_d(I_n - \epsilon_1)\bar{S} & 0 & d_u Q_3^T \\ 0 & \star & -(1-\mu)\bar{S} & 0 & 0 \\ \star & 0 & 0 & -\bar{S} & 0 \\ \star & \star & 0 & 0 & -\bar{R} \end{bmatrix} < 0 \tag{3.10}$$

$$\begin{bmatrix} \bar{R} & 0 & \bar{R}\epsilon_1 A_d^T \\ 0 & \bar{Z_{11}} & \bar{Z_{12}} \\ \star & \star & \bar{Z_{13}} \end{bmatrix} < 0 \tag{3.11}$$

where, $(1,1) = Q_2 + Q_2^T + d_u Z_{11}$,
$(1,2) = Q_3 - Q_2^T + Q_1 A^T + \epsilon_1 A_d^T + d_u Z_{12} + \bar{Y}^T B^T$,
$(2,2) = -Q_3 - Q_3^T + d_u Z_{13}$.

Proof The condition stated above is obtained by extending the stability theorem in [10], here a brief sketch of the formulation is presented as a part proof, the details can be found in [10].

Consider the system (3.6) with $u(t) = 0$ satisfying the condition (3.3). The system in descriptor form by substituting $x(t - d(t)) = x(t) - \int_{t-d(t)}^{t} \dot{x}(s)ds$ can be written as,

$$\begin{aligned} \dot{x}(t) &= y(t) \\ 0 &= -y(t) + (A + A_d)x(t) - A_d \int_{t-d(t)}^{t} y(s)ds \end{aligned} \tag{3.12}$$

this can also be expressed as

$$\begin{aligned} E\dot{\bar{x}}(t) &= \begin{bmatrix} \dot{x}(t) \\ 0 \end{bmatrix} \\ &= \begin{bmatrix} 0 & I \\ (A + A_d) & -I \end{bmatrix} \bar{x}(t) - \begin{bmatrix} 0 \\ A_d \end{bmatrix} \int_{t-d(t)}^{t} y(s)ds \end{aligned} \tag{3.13}$$

where, $\bar{x}(t) = \left[\, x^T(t)\ y^T(t)\,\right]^T$, $E = diag\{I,\ 0\}$.

Following LK functional is selected for the descriptor system in (3.13),

$$V(t) = \bar{x}(t)EP\bar{x}(t) + \int_{-d_u}^{0}\int_{t+\theta}^{t} y^T(s)Ry(s)ds + \int_{t-d(t)}^{t} x^T(s)Sx(s)ds \quad (3.14)$$

where, $P = \begin{bmatrix} P_1 & 0 \\ P_2 & P_3 \end{bmatrix}$, $P_1 > 0,\ EP = P^T E \geq 0$

Finding the time-derivative of (3.14), one can obtain the following

$$\begin{aligned}\dot{V}(t) = {} & \bar{x}(t)\{P^T\begin{bmatrix} 0 & I \\ (A+A_d) & -I \end{bmatrix} + \begin{bmatrix} 0 & (A+A_d)^T \\ I & -I \end{bmatrix} P\}\bar{x}(t) \\ & -2\bar{x}(t)P^T\begin{bmatrix} 0 \\ A_d \end{bmatrix}\int_{t-d(t)}^{t} y(s)ds + d_u y^T(t)Ry(t) \\ & +x^T(t)Sx(t) - (1-\mu)x^T(t-d(t))Sx(t-d(t)) \\ & -\int_{t-d(t)}^{T} y^T(s)Ry(s)ds \end{aligned} \quad (3.15)$$

Using bounding Lemma 2.3 (Moon's Bounding Lemma) for the cross term in (3.15), one can rewrite (3.15) as

$$\begin{aligned}\dot{V}(t) \leq {} & \bar{x}(t)\left\{P^T\begin{bmatrix} 0 & I \\ (A+A_d) & -I \end{bmatrix} + \begin{bmatrix} 0 & (A+A_d)^T \\ I & -I \end{bmatrix} P + d_u Z + \begin{bmatrix} S & 0 \\ 0 & d_u R \end{bmatrix}\right\}\bar{x}(t) \\ & -(1-\mu)x^T(t-d(t))Sx(t-d(t)) + 2\int_{t-d(t)}^{t} y^T(s)ds\left\{Y - \begin{bmatrix} 0 \\ A_d \end{bmatrix}^T P\right\}\bar{x}(t) \end{aligned} \quad (3.16)$$

To treat the last term of (3.16), substitute $\int_{t-d(t)}^{t} \dot{x}(s)^T ds = \{x^T(t) - x^T(t-d(t))\}$, applying the bounding Lemma 2.1 and with little algebraic manipulations one can obtain

$$\begin{aligned} 2\int_{t-d(t)}^{t} y^T(s)ds\left\{Y - \begin{bmatrix} 0 \\ A_d \end{bmatrix}^T P\right\}\bar{x}(t) \leq {} & \bar{x}^T(t)\left\{\Upsilon + \{P^T\begin{bmatrix} 0 \\ A_d \end{bmatrix} - Y^T\}\right. \\ & \left.\times[(1-\mu)S]^{-1}\{P^T\begin{bmatrix} 0 \\ A_d \end{bmatrix} - Y^T\}^T\right\}\bar{x}(t) \\ & + (1-\mu)x^T(t-d(t))Sx(t-d(t)) \end{aligned}$$

where, $\Upsilon = \begin{bmatrix} Y \\ 0 \end{bmatrix} + \begin{bmatrix} Y \\ 0 \end{bmatrix}^T + \begin{bmatrix} -A_d^T P_2 + P_2^T A_d & -A_d^T P_3 \\ \star & 0 \end{bmatrix}$.

Substituting the RHS of the above inequality in the last term of (3.16) one can obtain the following:

$$\dot{V}(t) \leq \bar{x}(t) \begin{bmatrix} \Psi \; P^T \begin{bmatrix} 0 \\ A_d \end{bmatrix} - Y^T \\ \star \quad -(1-\mu)S \end{bmatrix} \bar{x}(t) \tag{3.17}$$

where, $\Psi = P^T \begin{bmatrix} 0 & I \\ A & -I \end{bmatrix} + \begin{bmatrix} 0 & A^T \\ I & -I \end{bmatrix} P + d_u Z + \begin{bmatrix} S & 0 \\ 0 & d_u R \end{bmatrix} + \begin{bmatrix} Y \\ 0 \end{bmatrix} + \begin{bmatrix} Y \\ 0 \end{bmatrix}^T$,

$Z = \begin{bmatrix} Z_{11} & Z_{12} \\ \star & Z_{13} \end{bmatrix}$, and $Y = \begin{bmatrix} Y_{11} & Y_{12} \end{bmatrix}$.

If the LMIs,

$$\begin{bmatrix} \Psi \; P^T \begin{bmatrix} 0 \\ A_d \end{bmatrix} - Y^T \\ \star \quad -(1-\mu)S \end{bmatrix} < 0 \; and \tag{3.18}$$

$$\begin{bmatrix} R & Y \\ \star & Z \end{bmatrix} \geq 0 \tag{3.19}$$

then the system (3.6) with $u(t) = 0$ is asymptotically stable. The LMI (3.18) is due to the use of Moons bounding inequality lemma 2.3 for replacing the quadratic integral term that arises out of derivative of LK functional. Now, replacing the matrix A by $A + BK$ in the LMI (3.18).

Defining $P^{-1} = Q = \begin{bmatrix} Q_1 & 0 \\ Q_2 & Q_3 \end{bmatrix}$ and pre- and post multiply (3.18) by $\Delta = diag\{Q, \; I\}$ and Δ^T respectively, pre- and post multiply (3.19) by $diag\{R^{-1}, \; Q^T\}$ and $diag\{R^{-1}, \; Q\}$ respectively. Choosing following linear changes in variables $Q^T Z Q = \bar{Z}$, $S^{-1} = \bar{S}$, $R^{-1} = \bar{R}$ and $\bar{Y} = \epsilon_1 A_d^T [\bar{P}_2, \quad \bar{P}_3]$ with $\epsilon_1 \; I$ a block diagonal matrix. Now, it is now straight forward to obtain the LMI condition in (3.10) and (3.11), which are the required stabilizing condition for the time-delay systems (3.6). The state feedback gain is computed by the relation $K = \bar{Y} Q_1^{-1}$.

Remark 3.3 The selection of $\bar{Y}$ matrix in the stabilization formulation helps to avoid the NLMI stabilization condition. The stabilization results presented in [12] reveal the fact that, descriptor system formulation of the problem in this case helped to obtain better results than that of [9].

The stability condition (2.94)–(2.95) discussed in Theorem 2.12 (for system σ_2) is extended to obtain the stabilization condition which is presented in the form of theorem below.

Theorem 3.4 *(Theorem 2, [12]) Given the scalars $d_u > 0$, $\mu > 0$, the system (3.6) is asymptotically stabilizable with the state-feedback controller, $u(t) = YX^{-1}x(t)$ for any time-delay satisfying the condition (3.3) if there exist symmetric positive matrices $\bar{P}$, $\bar{Q}$, $\bar{R}$, $\bar{T}$, $\bar{Z}$ and matrices S_i, $(i = 1, ..., 4)$, Y with appropriate dimensions satisfying the following, LMI conditions:*

$$\bar{P} = \begin{bmatrix} X & \bar{P}_{12} \\ \star & \bar{P}_{22} \end{bmatrix}, \quad with \; X > 0 \; and \; \bar{Q} = \begin{bmatrix} \bar{Q}_{11} & \bar{Q}_{12} \\ \star & \bar{Q}_{22} \end{bmatrix} \geq 0 \tag{3.20}$$

and,

$$\begin{bmatrix} \bar{\Omega}_{11} & \bar{\Omega}_{12} & \bar{S}_3 & \bar{\Omega}_{14} & d_u\bar{Q}_{11} & d_u\bar{Q}_{12} & \bar{\Omega}_{17} & 0 & \mu\bar{P}_{12} \\ \star & \bar{\Omega}_{22} & -\bar{S}_3 & \bar{\Omega}_{24} & 0 & 0 & XA_d^T & 0 & 0 \\ \star & \star & -\bar{Q}_{11} & \bar{\Omega}_{34} & 0 & 0 & 0 & \bar{P}_{12}^T & \mu\bar{P}_{22} \\ \star & \star & \star & \bar{\Omega}_{44} & 0 & 0 & 0 & 0 & 0 \\ \star & \star & \star & \star & -\bar{Q}_{11} & -\bar{Q}_{12} & 0 & d_u\bar{Q}_{12} & 0 \\ \star & \star & \star & \star & \star & -\bar{Q}_{22} & 0 & d_u\bar{Q}_{22} & 0 \\ \star & \star & \star & \star & \star & \star & -Z & 0 & 0 \\ \star & \star & \star & \star & \star & \star & \star & -XZ^{-1}X & 0 \\ \star & \star & \star & \star & \star & \star & \star & \star & -\mu\bar{T} \end{bmatrix} < 0 \quad (3.21)$$

where, $\bar{\Omega}_{11} = XA^T + AX + Y^TB^T + BY + \bar{R} + \bar{S}_1^T + \bar{S}_1$, $\bar{\Omega}_{12} = A_dX - \bar{S}_1^T + \bar{S}_2$
$\bar{\Omega}_{14} = \bar{P}_{12} - \bar{S}_1^T + \bar{S}_4$, $\bar{\Omega}_{17} = XA^T + Y^TB^T$, $\bar{\Omega}_{22} = -(1-\mu)\bar{R} + \mu\bar{T} - \bar{S}_2 - \bar{S}_2$
$\bar{\Omega}_{24} = -\bar{S}_2^T - \bar{S}_4$, $\bar{\Omega}_{34} = \bar{P}_{22} - \bar{Q}_{12} - \bar{S}_3^T$ $\bar{\Omega}_{44} = -\bar{Q}_{22} - \bar{S}_4^T - \bar{S}_4$

Proof The stabilization condition (3.21) has been obtained by extending the stability condition (2.95) stated in Theorem 2.12 of Chap. 2. A brief illustration is given as a part of proof for this theorem. Starting with the stability condition (2.95) one can first apply Schur-complement to obtain

$$\Lambda = \begin{bmatrix} \Lambda_{11} & \Lambda_{12} & \Lambda_{13} & \Lambda_{14} & \Lambda_{15} & \Lambda_{16} & \mu P_{12} \\ \star & \Lambda_{22} & \Lambda_{23} & \Lambda_{24} & \Lambda_{25} & \Lambda_{26} & 0 \\ \star & \star & -Q_{11} & \Lambda_{34} & 0 & 0 & \mu P_{22} \\ \star & \star & \star & \Lambda_{44} & 0 & 0 & 0 \\ \star & \star & \star & \star & -Q_{11} & -Q_{12} & 0 \\ \star & \star & \star & \star & \star & -Q_{22} & 0 \\ \star & \star & \star & \star & \star & \star & -\mu T \end{bmatrix} < 0 \quad (3.22)$$

where, $\Lambda_{11} = A^TP_{11} + P_{11}A + R + S_1 + S_1^T$; $\Lambda_{12} = P_{11}A_d - S_1^T + S_2$; $\Lambda_{13} = A^TP_{12} + S_3$
$\Lambda_{14} = P_{12} - S_1^T + S_4$; $\Lambda_{15} = d_u(Q_{11} + A^TQ_{12}^T)$; $\Lambda_{16} = d_u(Q_{12} + A^TQ_{22})$;
$\Lambda_{22} = -(1-\mu)R + \mu T - S_2^T - S_2$; $\Lambda_{23} = A_d^TP_{12} - S_3$; $\Lambda_{24} = -S_2^T - S_4$;
$\Lambda_{25} = d_uA_d^TQ_{12}^T$; $\Lambda_{26} = d_uA_d^TQ_{22}$; $\Lambda_{34} = P_{22} - Q_{12} - S_3^T$; and
$\Lambda_{44} = -Q_{22} - S_4^T - S_4$

Now, Pre- and post-multiplying (3.22) with $diag\{X,\ X,\ X,\ X,\ X,\ X,\ X\}$, where $X = P_{11}^{-1}$ and denoting $X(.)X = \bar{(.)}$, (where (.) indicates any matrix variable) one can get,

$$\bar{\Lambda} = \bar{\Lambda}_0 + \Pi_1^TX^{-1}\Pi_2 + \Pi_2^TX^{-1}\Pi_1 < 0 \quad (3.23)$$

Using Lemma 2.1 for any positive definite matrix Z, the last two terms of (3.23) can be bounded with inequality constraints as

$$\Pi_1^T X^{-1}\Pi_2 + \Pi_2^T X^{-1}\Pi_1 \le \Pi_1^T Z^{-1}\Pi_1 + \Pi_2^T (XZ^{-1}X)^{-1}\Pi_2 \tag{3.24}$$

where, $\Pi_1 = [AX,\ A_dX,\ 0,\ 0,\ 0,\ 0,\ 0]$, $\Pi_2 = [0,\ 0,\ \bar{P}_{12},\ 0,\ d_u\bar{Q}_{12}^T,\ d_u\bar{Q}_{22},\ 0]$, and

$$\bar{\Lambda}_0 = \begin{bmatrix} \bar{\Lambda}_{11} & \bar{\Lambda}_{12} & \bar{S}_3 & \bar{\Lambda}_{14} & d_u\bar{Q}_{11} & d_u\bar{Q}_{11} & \mu\bar{P}_{12} \\ \star & \bar{\Lambda}_{22} & -\bar{S}_3 & \bar{\Lambda}_{24} & 0 & 0 & 0 \\ \star & \star & -\bar{Q}_{11} & \bar{\Lambda}_{34} & 0 & 0 & \mu\bar{P}_{22} \\ \star & \star & \star & \bar{\Lambda}_{44} & 0 & 0 & 0 \\ \star & \star & \star & \star & -\bar{Q}_{11} & -\bar{Q}_{12} & 0 \\ \star & \star & \star & \star & \star & -\bar{Q}_{22} & 0 \\ \star & \star & \star & \star & \star & \star & -\mu\bar{T} \end{bmatrix} < 0$$

The block matrices in $\bar{\Lambda}_0$ are expressed as

$\bar{\Lambda}_{11} = XA^T + AX + \bar{R} + \bar{S}_1 + \bar{S}_1^T$; $\bar{\Lambda}_{12} = A_dX - \bar{S}_1^T + \bar{S}_2$; $\bar{\Lambda}_{14} = \bar{P}_{12} - \bar{S}_1^T + \bar{S}_4$;
$\bar{\Lambda}_{22} = -(1-\mu)\bar{R} + \mu\bar{T} - \bar{S}_2^T - \bar{S}_2$;
$\bar{\Lambda}_{24} = -\bar{S}_2^T - \bar{S}_4$; $\bar{\Lambda}_{34} = \bar{P}_{22} - \bar{Q}_{12} - \bar{S}_3^T$; and $\Lambda_{44} = -\bar{Q}_{22} - \bar{S}_4^T - \bar{S}_4$

Substituting (3.24) into (3.23) and replacing A with $(A + BK)$ and then applying Schur-complement one can easily obtain the stabilizing condition (3.21).

Remark 3.4 One can observe in the condition (3.21) that, the (8, 8) block $(XZ^{-1}X)$ is nonlinear, so standard LMI tools cannot be used to solve this matrix inequalities. Thus cone complementarity algorithm proposed in [16] is used to find the feasible solution of this problem. This linearization iterative algorithm gives suboptimal value of the delay upper bound estimate.

The stability conditions (for system σ_2 with the condition (2.7)) obtained in (2.85) discussed in Theorem 2.11 is extended to obtain the stabilization condition that is presented below in the form of theorem. This stabilization theorem is formulated by the present author in NLMI framework for the purpose of investigating the effect of more free weighting matrices on the convergence of cone-complementarity problem with the use of same number of bounding inequalities.

Theorem 3.5 *Given the scalars $d_u > 0, \mu > 0$, the system (3.6) is asymptotically stabilizable with the state-feedback controller, $u(t) = SY^{-1}x(t)$ for any time-delay satisfying the condition (3.3), if there exist symmetric positive definite matrices Y, X, Q_R, any free matrices T_R, T_S and S with appropriate dimensions satisfying the following, LMI conditions:*

$$\Theta = \begin{bmatrix} \Theta_{11} & \Theta_{12} & \Theta_{13} & T_R \\ \star & \Theta_{22} & d_uYA_d^T & T_S \\ \star & \star & -d_uX & 0 \\ \star & \star & \star & -d_u^{-1}YX^{-1}Y \end{bmatrix} < 0 \tag{3.25}$$

where, $\Theta_{11} = YA^T + AY + BS + S^T B^T + T_R + T_R^T + Q_R$, $\Theta_{12} = A_d Y - T_R + T_S^T$, $\Theta_{13} = d_u(YA^T + S^T B^T)$, $\Theta_{22} = -T_S - T_S^T - (1-\mu)Q_R$.

Proof Considering the stability condition (2.85) of Theorem 2.11, using Schur-complement on it one can write the condition as

$$\Omega = \begin{bmatrix} \Omega_{11} & \Omega_{12} & \Omega_{13} & T_1 \\ \star & \Omega_{22} & d_u A_d^T Q_2 & T_2 \\ \star & \star & -d_u Q_2 & 0 \\ \star & \star & \star & -d_u^{-1} Q_2 \end{bmatrix} < 0 \tag{3.26}$$

where, $\Omega_{11} = A^T P + PA + T_1 + T_1^T + Q_1$, $\Omega_{12} = PA_d - T_1 + T_2^T$, $\Omega_{13} = d_u A^T Q_2$, $\Omega_{22} = -T_2 - T_2^T - (1-\mu)Q_1$.

Using state-feedback control law $u(t) = Kx(t)$ to the system (3.6) and replace A matrix by $(A + BK)$ matrix in (3.26), yields the condition

$$\Xi = \begin{bmatrix} \Xi_{11} & \Xi_{12} & \Xi_{13} & T_1 \\ \star & \Xi_{22} & d_u A_d^T Q_2 & T_2 \\ \star & \star & -d_u Q_2 & 0 \\ \star & \star & \star & -d_u^{-1} Q_2 \end{bmatrix} < 0 \tag{3.27}$$

where, $\Xi_{11} = A^T P + PA + T_1 + T_1^T + Q_1 + PBK + K^T B^T P$, $\Omega_{12} = PA_d - T_1 + T_2^T$, $\Xi_{13} = d_u A^T Q_2 + d_u K^T B^T Q_2$, $\Omega_{22} = -T_2 - T_2^T - (1-\mu)Q_1$.

Pre- and post-multiplying (3.27) by $diag\{P^{-1}, P^{-1}, Q_2^{-1}, P^{-1}\}$, and adopting following changes in variables,

$P^{-1} = Y$, $Q_2^{-1} = X$, $KP^{-1} = KY = S$, $P^{-1}T_1P^{-1} = T_R$, $P^{-1}T_2P^{-1} = T_S$, and $P^{-1}Q_1P^{-1} = Q_R$, where $Y = Y^T > 0$ and $X = X^T > 0$, and substituting this change of variables in (3.27) one obtains the LMI condition in (3.25).

Remark 3.5 One can observe in the condition (3.25) that the (4,4) block is not linear in matrix variable, rather it is a nonlinear, hence the obtained condition is not an LMI and the standard routine of LMI Toolbox of $MATLAB^{®}$ [17] cannot be used to obtain the feasible solution set.

For obtaining feasible solution, one can easily transform this NLMI condition into an LMI by assuming $X = Y$, but the stabilizing results will tend to be conservative [9]. An iterative cone-complementarity algorithm in [16] has been used to solve this NLMI problem which can yield less conservative stabilizing results compared to that of the former assumption ($X = Y$), but the estimate of delay upper bound and the state feedback gains obtained are suboptimal.

The iterative cone-complementarity for solving the NLMI condition (3.25) is illustrated in brief below:

Let us fix,

$$-YX^{-1}Y \leq -L. \tag{3.28}$$

Substituting (3.28) in (3.25), one can write

$$\Theta = \begin{bmatrix} \Theta_{11} & \Theta_{12} & \Theta_{13} & T_R \\ \star & \Theta_{22} & d_u Y A_d^T & T_S \\ \star & \star & -d_u X & 0 \\ \star & \star & \star & -d_u L \end{bmatrix} < 0. \tag{3.29}$$

Using Schur-complement to (3.28), one can rewrite

$$\begin{bmatrix} L^{-1} & Y^{-1} \\ Y^{-1} & X^{-1} \end{bmatrix} \geq 0 \tag{3.30}$$

Now defining, $D = L^{-1},\ J = Y^{-1},\ N = X^{-1}$, one can rewrite (3.30) as

$$\begin{bmatrix} D & J \\ J & N \end{bmatrix} \geq 0 \tag{3.31}$$

Again, one can have the following valid identities valid, $DL = I,\ JY = I,\ NX = I$. Thus, in view of the identities defined, one can write it in the form of matrix inequalities as,

$$\begin{bmatrix} L & I \\ \star & D \end{bmatrix} \geq 0,\ \begin{bmatrix} Y & I \\ \star & J \end{bmatrix} \geq 0,\ and \begin{bmatrix} X & I \\ \star & N \end{bmatrix} \geq 0, \tag{3.32}$$

Now, one can solve (3.25) as a linear minimization problem as:

$$Minimize\ Trace\ (LD + YJ + XN)$$
subject to. (3.29), (3.31), and (3.32)

This routine is iteratively implemented by incrementing the value of the delay bound d_u in small steps and checking the feasible solution of (3.28) at each step, the algorithm stops at a value of d_u where the condition (3.28) is not satisfied.

For convenience of the discussion of the main results of this chapter, some preliminaries including few definitions, basic theorems on stabilization of time-delay systems which are related to the main results are presented in previous sections.

3.4 Main Results on Delay-Dependent Stabilization of Nominal TDS

The stabilization condition is obtained by directly extending the stability condition (2.130)–(2.134) discussed in Theorem 2.18.

Theorem 3.6 *Given a scalar* $0 \leq d(t) \leq d_u$ *(where* $d_u > 0$*), the system (3.6) for* $0 < \mu < 1$ *is asymptotically stabilizable with the state-feedback controller,* $u(t) = Kx(t)$ $(K = YZ^{-1})$ *for any time-delay satisfying the condition (3.3), if there exist symmetric positive definite matrices* $\bar{P}$, $\bar{Q}$, $\bar{R}$, $\bar{T}$, *and any free matrices* $\bar{M}_i$, $\bar{L}_i$ $(i = 1, 2, 3)$ *and* Z *with appropriate dimensions such that the following LMIs hold:*

$$\bar{P} = \begin{bmatrix} \bar{P}_{11} & \bar{P}_{12} \\ \star & \bar{P}_{22} \end{bmatrix} > 0, \quad \begin{bmatrix} \bar{Q}_{11} & 0 \\ 0 & \bar{Q}_{22} \end{bmatrix} > 0 \tag{3.33}$$

$$\begin{bmatrix} \bar{\Omega} & \mu\bar{P} & d_u\bar{M} \\ \star & -\mu\bar{T} & 0 \\ \star & 0 & -\bar{Q}_{22} \end{bmatrix} < 0 \tag{3.34}$$

and,

$$\begin{bmatrix} \bar{\Omega} & \mu\bar{P} & d_u\bar{L} \\ \star & -\mu\bar{T} & 0 \\ \star & 0 & -\bar{Q}_{22} \end{bmatrix} < 0 \tag{3.35}$$

where,

$$\bar{\Omega} = \begin{bmatrix} \bar{\Omega}_{11} & \bar{\Omega}_{12} & \bar{\Omega}_{13} & \bar{\Omega}_{14} & 0 & \bar{\Omega}_{16} \\ \star & \bar{\Omega}_{22} & \bar{\Omega}_{23} & \bar{\Omega}_{24} & 0 & \bar{\Omega}_{26} \\ \star & \star & \bar{\Omega}_{33} & 0 & 0 & 0 \\ \star & \star & 0 & -\bar{Q}_{11} & 0 & \bar{P}_{12}^T \\ 0 & 0 & 0 & 0 & -\bar{Q}_{11} & 0 \\ \star & \star & 0 & \star & 0 & \bar{\Omega}_{66} \end{bmatrix}$$

where, $\bar{\Omega}_{11} = d_u^2\bar{Q}_{11} + AZ + Z^TA^T + BY + Y^TB^T + \bar{P}_{12} + \bar{P}_{12}^T + \bar{R}_1 + \bar{R}_2 + d_u(\bar{L}_1 + \bar{L}_1^T)$
$\bar{\Omega}_{12} = A_dZ - \bar{P}_{12} + d_u(-\bar{L}_1 + \bar{L}_2^T + \bar{M}_1)$
$\bar{\Omega}_{13} = d_u(\bar{L}_3^T - \bar{M}_1)$, $\bar{\Omega}_{14} = \bar{P}_{22}$, $\bar{\Omega}_{16} = \bar{P}_{11} - Z + \alpha Z^TA^T + \alpha Y^TB^T$
$\bar{\Omega}_{22} = -(1-\mu)\bar{R}_1 + \mu\bar{T} + d_u(-\bar{L}_2 - \bar{L}_2^T + \bar{M}_2 + \bar{M}_2^T)$
$\bar{\Omega}_{23} = d_u(-\bar{M}_2 + \bar{M}_3^T - \bar{L}_3^T)$, $\bar{\Omega}_{24} = -\bar{P}_{22}$, $\bar{\Omega}_{26} = \alpha Z^TA_d^T$
$\bar{\Omega}_{33} = d_u(-\bar{M}_3 - \bar{M}_3^T) - \bar{R}_2$, $\bar{\Omega}_{66} = -\alpha(Z + Z^T) + d_u^2\bar{Q}_{22}$
and $Y = KZ$

Proof This is an extension of the stability conditions derived in Theorem 2.18. Replacing matrix A by $A_c = A + BK$ in the matrix Ω (see (2.134)) and set the free variables $G_1 = G$ and $G_2 = \alpha G$. In 2.157 the (6,6) block contains $-G_2 - G_2^T$ which indicates that for negativity of that block the matrix G_2 must be positive definite, thus in this view for guaranteing the negativity of the LMI here, matrix $G_2 = \alpha G$

must also be positive definite which in turns will guarantee the term $-\alpha(G + G^T)$ as negative definite.

Now pre-multiply the matrix Ω in (2.134) by $diag[G^{-T}, G^{-T}, G^{-T}, G^{-T}, G^{-T}, G^{-T}]$ and post-multiply by $diag[G^{-1}, G^{-1}, G^{-1}, G^{-1}, G^{-1}, G^{-1}]$ and subsequently pre-multiply $\begin{bmatrix} P_{11} & P_{12} \\ \star & P_{22} \end{bmatrix}$ and $\begin{bmatrix} Q_{11} & 0 \\ 0 & Q_{22} \end{bmatrix}$ by $diag[G^{-T}, G^{-T}]$ and post-multiply the same matrices by $diag[G^{-1}, G^{-1}]$ one can obtain the LMIs in (3.33)–(3.35) with the following changes in variables $G^{-1} = Z, G^{-1}(.)G^{-1} = \bar{(.)}$ and $KZ = Y$.

Note that, the stabilizing conditions obtained here are convex combination of LMI conditions.

To obtain a realizable solution of gain matrix K for a particular delay bound, one needs to impose constraint on Y and Z matrices that limit the size of the gain matrix K, and it is expressed as

$$K = YZ^{-1} \tag{3.36}$$

Imposing constraint on matrix Y in the following form

$$Y^T Y < \delta I, \ \ \delta > 0 \tag{3.37}$$

or,

$$\begin{bmatrix} -\delta I & Y^T \\ Y & -I \end{bmatrix} < 0 \tag{3.38}$$

Similarly imposing constraint on matrix Z^{-1} in the following form

$$Z^{-1} < \beta I, \ \ \beta > 0 \tag{3.39}$$

or,

$$\begin{bmatrix} \beta I & I \\ I & Z \end{bmatrix} > 0 \tag{3.40}$$

To find the optimal value of the gain K for a particular delay upper bound d_u and α, following minimization problem is considered,

$$\text{Minimize } \delta + \beta$$
$$\text{subject to (3.33)–(3.35), (3.38), (3.40), } \bar{P}_{11} = \bar{P}_{11}^T > 0, \bar{R}_i > 0, \bar{T} > 0.$$

Next the stabilization condition for $\mu = 0$ is obtained from the conditions (3.33)–(3.35) by substituting $\mu = 0$. The stabilization condition is presented below in the form of corollary.

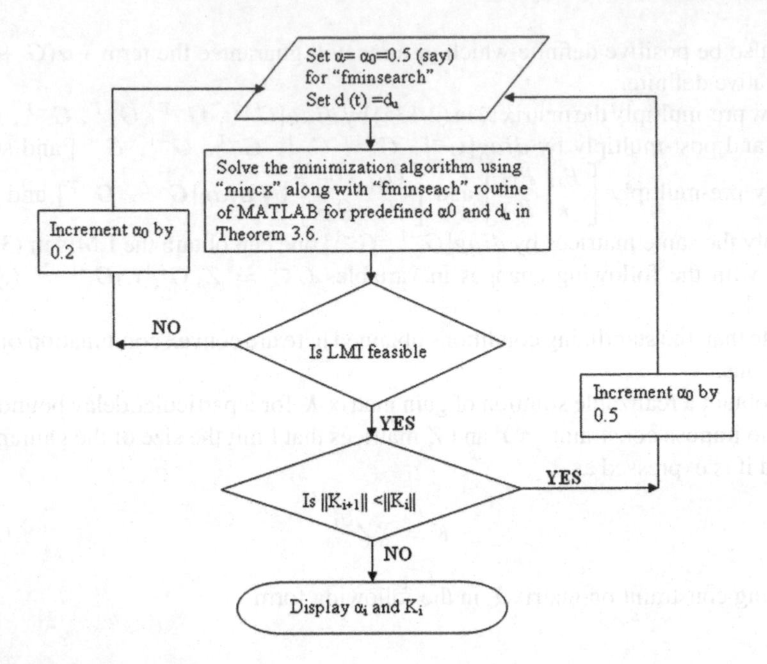

Fig. 3.1 Numerical implementation of Minimization Problem

Numerical Implementation of the Algorithm: The above minimization algorithm is solved using the 'mincx' solver of the LMI toolbox of $MATLAB^{®}$ along with 'fminsearch' routine to tune the value of parameter α for a particular delay value. The numerical implementation of the algorithm is presented in the form of flow chart as shown in Fig. 3.1.

Corollary 3.1 *Given $d_u > 0$, the system (3.6) for $\mu = 0$ is asymptotically stabilizable with the state-feedback controller, $u(t) = Kx(t)$ $(K = YZ^{-1})$ for any time-delay satisfying the condition (3.3), if there exist symmetric positive definite matrices $\bar{P}$, $\bar{Q}$, $\bar{R}$, and any free matrices $\bar{M}_i$, $\bar{L}_i$ $(i = 1, 2, 3)$ and Z with appropriate dimensions satisfying the following LMI constraints:*

$$\bar{P} = \begin{bmatrix} \bar{P}_{11} & \bar{P}_{12} \\ \star & \bar{P}_{22} \end{bmatrix} > 0, \quad \begin{bmatrix} \bar{Q}_{11} & 0 \\ 0 & \bar{Q}_{22} \end{bmatrix} > 0 \tag{3.41}$$

$$\begin{bmatrix} \bar{\Omega} & d_u\bar{M} \\ \star & -\bar{Q}_{22} \end{bmatrix} < 0 \tag{3.42}$$

and,

$$\begin{bmatrix} \bar{\Omega} & d_u\bar{L} \\ \star & -\bar{Q}_{22} \end{bmatrix} < 0 \tag{3.43}$$

Table 3.1 d_u and K for Example 3.1 for $\mu = 0$ using LMI framework

Methods	d_u	$K\ matrix$	Remarks
[6]	0.9999	$[-0.10452, 749058]$	–
[18]	1.28	$[0, -1209100]$	–
[10]	1.51	$[-293.0350, 1]$	–
[19]	3.35	$[-6.0276, -11.03223]$	–
[20]	7	$[-86.92, -98.21]$	–
Cor 3.1	7	$[-47.6658, -54.6150]$	$\alpha = 4.5203$
	8	$[-75.3591, -83.9332]$	$\alpha = 4.6481$
	21	$10^3\times[-1.8832, -1.9405]$	$\alpha = 11.0893$

where, the elements of $\bar{\Omega}$ matrix is same as defined in the Theorem 3.6.

The proof of this corollary is straight forward and can be obtained from Theorem 3.6 by substituting $\mu = 0$. The solution of the state feedback gain matrix is obtained in a similar manner using optimization algorithm presented in Theorem 3.6 subject to the constraints (3.41)–(3.43).

Numerical Example 3.1 *([12]) Consider the system (3.6) with the following constant matrices*

$$A = \begin{bmatrix} 0 & 0 \\ 0 & 1 \end{bmatrix},\ A_d = \begin{bmatrix} -1 & -1 \\ 0 & -0.9 \end{bmatrix},\ and\ B = \begin{bmatrix} 0 \\ 1 \end{bmatrix}$$

The eigenvalues of the matrix $[A + A_d]$ is not Hurwitz and hence the open-loop system is unstable. The proposed stabilization result is compared with the existing LMI based methods and presented in Table 3.1.

Remark 3.6 As pointed out above that the open-loop system considered in Numerical Example 3.1 is unstable, but the eigenvalues of closed-loop system ($A+BK$ matrix) is found to be stable, on the other hand the eigenvalues of $(A+BK-A_d)$ is not Hurwitz thus indicating that the closed-loop system is not delay-independently stable [12].

Remark 3.7 For a particular delay value $d_u = 7secs$ the parameter α is tuned using 'fminsearch' to find the optimal value of the gain K as presented in the flowchart (Fig. 3.1). The variation of parameter α with respect to the control energy (represented as $\| K \|_2$) is shown in the Fig. 3.2.

Remark 3.8 The controller gain K and the delay upper bound d_u for the system described in Example 3.1 using the proposed Corollary 3.1 are presented in Table 3.1. The effectiveness of the designed state feedback controller is obvious from the values of d_u and K compared to the existing methods.

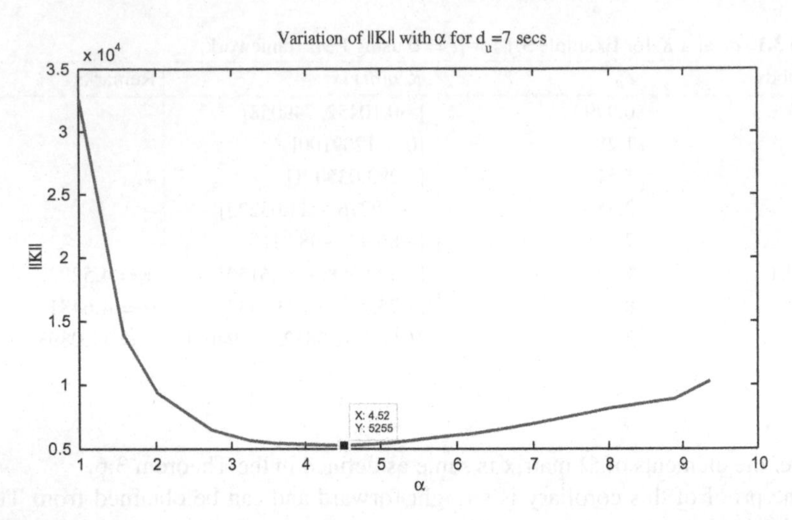

Fig. 3.2 Variation of α with $\| K \|$

Table 3.2 d_u and K for Example 3.1 for $\mu = 0$ using NLMI framework

Methods	d_u	K matrix	Iterations	Decision variables
[12]	1.51	$[-0.7851, -2.0379]$	6	$9 \times n^2 + (3 + m) \times n$
	3.0	$[-2.7835, -5.0543]$	12	
	5.0	$[-8.5157, -11.9412]$	25	
	8.0	$[-65.4058, -76.7778]$	111	
Theorem 3.5	1.5	$[-0.5382, -1.7503]$	9	$3.5 \times n^2 + (1.5 + m) \times n$
	3.0	$[-3.3027, -7.2005]$	61	
	3.5	$[-7.3494, -13.5122]$	189	

n=order of the system, m=no. of inputs

Remark 3.9 Comparison of stabilizing results (d_u *and* K) for Example 3.1 is presented in Table 3.2. One can observe that the number of decision variables used in [12] is more than that of Theorem 3.5, more decision variables indicates use of more free matrices in establishing the stability condition. But, both the methods uses same number of bounding inequalities for obtaining the LMI condition. In view of the above reasons, probably the stabilizing condition in [12] acquired enhanced delay upper bound with sizeable iterations compared to Theorem 3.5. It must be noted here that, the scope of improvement of Theorem 3.5 is still open in terms of enhancing the delay bound by introducing more free matrices and its associated state vectors.

Another significant reason for better stabilizing result for $\mu = 0$ in [12] is due to the use of a matrix variable Q_{12} (expressing relationship between the state vectors $x(s)$ and $\dot{x}(s)$). This parameter is not used in our stability formulation. The reason for not using this variable is that it can enhance the result of delay bound only for

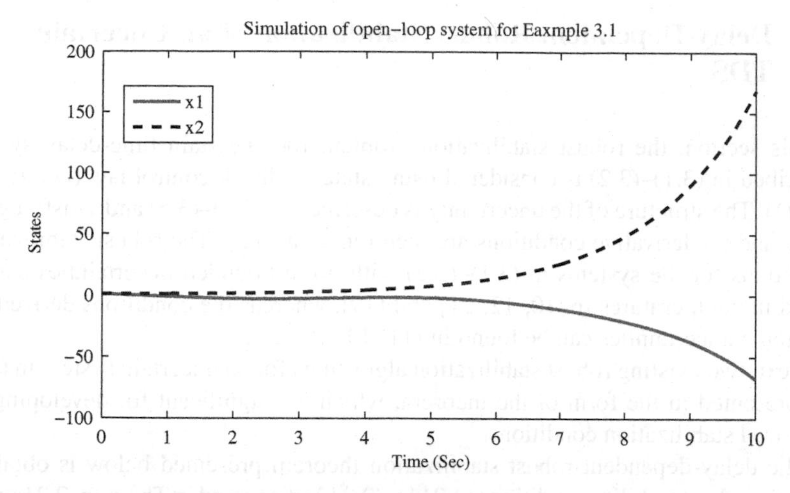

Fig. 3.3 Open-loop simulation of system in Example 3.1

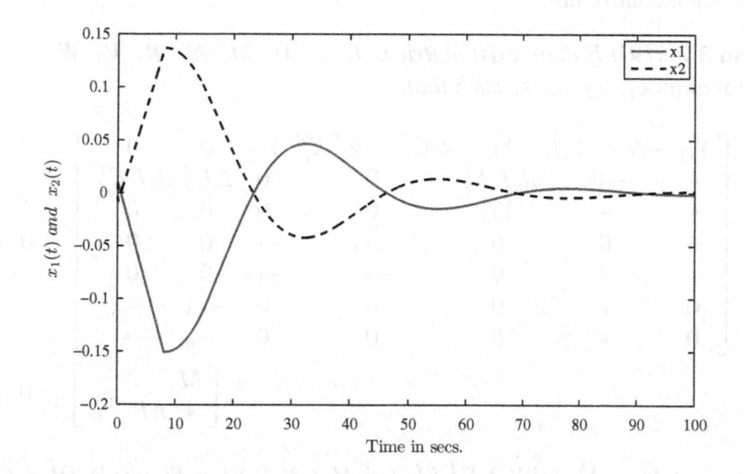

Fig. 3.4 Closed-loop simulation of system in Example 3.1

the case $\mu = 0$ but for other μ values it fails to improve the results and subsequently incorporates more number of decision variables in the formulation.

The simulation results of the system in Numerical Example 3.1 for constant delay (i.e, $\mu = 0$), considering $d_u = 8.0\, secs$. with and without controller are presented in Fig. 3.3 and Fig. 3.4 respectively. It may be observed that the open-loop time delay system response is unstable whereas the closed-loop system response with the state feedback gain $K = [-75.3591, \ -83.9332]$ and $d_u = 8secs$ stabilizes the unstable system.

3.5 Delay-Dependent Robust Stabilization of an Uncertain TDS

In this section, the robust stabilization problem for uncertain time-delay system described in (3.1)–(3.2) is considered using state feedback control law (i.e, $u(t) = Kx(t)$). The structure of the uncertainty is described in (3.4)–(3.5) and satisfying the delay and its derivative conditions are mentioned in (3.3). The robust stabilization conditions for the systems in (3.1)–(3.2) with norm bounded uncertainties can be found in the literatures [6–10, 12, 21] and [19], whereas the conditions derived for polytopic uncertainties can be found in [11, 14, 18, 22].

Next, two existing robust stabilization algorithms for an uncertain system in (3.1) are presented in the form of the theorem, which are significant for developing an improved stabilization conditions.

The delay-dependent robust stabilization theorem presented below is obtained from the robust stability conditions (2.211)–(2.212) discussed in Theorem 2.21 under state feedback control law.

Theorem 3.7 *([9]) If there exist matrices $L > 0, M, N, R, V, W > 0$ and positive scalars $\epsilon_1, \epsilon_2, ..., \epsilon_6$ such that,*

$$\begin{bmatrix} Y_{11} & -N + A_d L & Y_{13} & LE_a^T + V^T E_b^T & Y_{15} & 0 & 0 \\ \star & -W & d_u L A_d^T & 0 & 0 & LE_d^T & d_u LE_d^T \\ \star & \star & Y_{33} & 0 & 0 & 0 & 0 \\ \star & 0 & 0 & -\epsilon_1 & -\epsilon_3 & 0 & 0 \\ \star & 0 & 0 & -\epsilon_3 & -\epsilon_2 & 0 & 0 \\ 0 & \star & 0 & 0 & 0 & -\epsilon_4 & -\epsilon_6 \\ 0 & \star & 0 & 0 & 0 & -\epsilon_6 & -\epsilon_5 \end{bmatrix} < 0 \quad (3.44)$$

$$\begin{bmatrix} M & N \\ \star & LR^{-1}L \end{bmatrix} \geq 0 \quad (3.45)$$

where, $Y_{11} = LA^T + AL + BV + V^T B^T + d_u M + N + N^T + W + \epsilon_1 D_a D_a^T + \epsilon_4 D_d D_d^T$

$Y_{13} = d_u(LA^T + V^T B^T) + \epsilon_3 D_a D_a^T + \epsilon_6 D_d D_d^T$

$Y_{15} = d_u(LE_a^T + V^T E_b^T)$; $Y_{33} = -d_u R + \epsilon_2 D_a D_a^T + \epsilon_5 D_d D_d^T$

then the system in (3.1)–(3.2) with the control law $u(t) = VL^{-1}x(t)$ is asymptotically stable for any constant time-delay d satisfying the condition $0 \leq d \leq d_u$ and all admissible uncertainties defined in (3.4)–(3.5).

Remark 3.10 The proof of this theorem is straightforward as it is an extension of the robust stability condition stated in Theorem 2.21 and hence omitted. The derived condition is a NLMI, it is solved using iterative cone-complementarity algorithm as discussed in [9]. The nature of time-delay is assumed to be constant.

The delay-dependent robust stability condition (2.219)–(2.221) discussed in Theorem 2.23 has been extended for solution of robust stabilization problem using state feedback control law and is presented below in the form of theorem.

Theorem 3.8 *([12]) Given the scalars* $d_u > 0$, $\mu > 0$ *, system (3.1)–(3.2) is robustly asymptotically stabilizable with the memoryless state-feedback controller,* $u(t) = YX^{-1}x(t)$ *for any time-delay satisfying (3.3) and for the admissible uncertainties (3.4) satisfying (3.5) if there exist symmetric positive definite matrices* $\bar{P}$, $\bar{Q}$, $\bar{R}$, $\bar{T}$, $\bar{Z}$ *matrices* $\bar{S}_i$, $(i = 1, 2, ..., 4)$, Y *and scalars* $\epsilon_i^{'}s(i = 1, .., 4)$ *satisfying the following LMIs:*

$$\bar{P} = \begin{bmatrix} X & \bar{P}_{12} \\ \star & \bar{P}_{22} \end{bmatrix} \geq 0, \; with\; X = X^T > 0, \; \bar{P}_{12} = \bar{P}_{12}^T, \; \begin{bmatrix} \bar{Q}_{11} & \bar{Q}_{12} \\ \star & \bar{Q}_{22} \end{bmatrix} \geq 0 \tag{3.46}$$

$$0 > \left[\begin{array}{ccccccccccccc} \bar{\Sigma}_{11} & \bar{\Sigma}_{12} & \bar{S}_3 & \bar{\Sigma}_{14} & d_u\bar{Q}_{11} & d_u\bar{Q}_{12} & \bar{\Sigma}_{17} & 0 & XE_a^T & 0 & Y^TE_b^T & \mu\bar{P}_{12} \\ \star & \bar{\Sigma}_{22} & \bar{S}_3 & \bar{\Sigma}_{24} & 0 & 0 & XA_d^T & 0 & 0 & XE_d^T & 0 & 0 \\ \star & \star & -\bar{Q}_{11} & \bar{\Sigma}_{34} & 0 & 0 & 0 & \bar{P}_{12} & 0 & 0 & 0 & \mu\bar{P}_{22} \\ \star & \star & \star & \bar{\Sigma}_{44} & 0 & 0 & 0 & 0 & 0 & 0 & 0 & 0 \\ \star & \star & \star & \star & -\bar{Q}_{11} & -\bar{Q}_{12} & 0 & d_u\bar{Q}_{12} & 0 & 0 & 0 & 0 \\ \star & \star & \star & \star & \star & -\bar{Q}_{22} & 0 & {}_u\bar{Q}_{22} & 0 & 0 & 0 & 0 \\ \star & \star & \star & \star & \star & \star & \bar{\Sigma}_{77} & 0 & 0 & 0 & 0 & 0 \\ \star & \star & \star & \star & \star & \star & \star & -XZ^{-1}X & 0 & 0 & 0 & 0 \\ \star & \star & \star & \star & \star & \star & \star & \star & -\epsilon_1 & 0 & 0 & 0 \\ \star & \star & \star & \star & \star & \star & \star & \star & \star & -\epsilon_2 & 0 & 0 \\ \star & \star & \star & \star & \star & \star & \star & \star & \star & \star & \epsilon_3 & 0 \\ \star & \star & \star & \star & \star & \star & \star & \star & \star & \star & \star & -\mu\bar{T} \end{array}\right] \tag{3.47}$$

where, $\bar{\Sigma}_{11} = XA^T + AX + Y^TB^T + BY + \bar{R} + \bar{S}_1^T + \bar{S}_1 + \Delta$

$\bar{\Sigma}_{12} = A_dX - S_1^T + \bar{S}_2^T$; $\bar{\Sigma}_{14} = \bar{P}_{12} - \bar{S}_1^T + \bar{S}_4$; $\bar{\Sigma}_{17} = XA^T + Y^TB^T + \Delta$

$\bar{\Sigma}_{22} = -(1-\mu)\bar{R} + \mu\bar{T} - S_2^T - \bar{S}_2$; $\bar{\Sigma}_{24} = -\bar{S}_2^T - \bar{S}_4$; $\bar{\Sigma}_{34} = \bar{P}_{22} - \bar{Q}_{12} - S_3^T$

$\bar{\Sigma}_{44} = -\bar{Q}_{22} - S_4^T - \bar{S}_4$; $\bar{\Sigma}_{77} = -\bar{Z} + \Delta$; $\Delta = \epsilon_1 D_a D_a^T + \epsilon_2 D_d D_d^T + \epsilon_3 D_b D_b^T$

Proof Replacing A, A_d and B with $A(t)$, $A_d(t)$ and $B(t)$ as defined in (3.4) respectively in the stabilization condition (3.21) of (Theorem 3.4) and then decomposing the resulting matrix inequality into nominal and uncertain parts which will take the form:

$$\bar{\Sigma}_{nom} + \bar{\Sigma}_{unc}^T + \bar{\Sigma}_{unc} < 0 \tag{3.48}$$

where, $\bar{\Sigma}_{unc} = D_3 F_a(t) E_3 + D_4 F_d(t) E_4 + D_5 F_b(t) E_5$
$D_3 = [D_a^T\ 0\ 0\ 0\ 0\ 0\ D_a^T\ 0\ 0]^T$, $D_4 = [D_d^T\ 0\ 0\ 0\ 0\ 0\ D_d^T\ 0\ 0]^T$,
$D_5 = [D_b^T\ 0\ 0\ 0\ 0\ 0\ D_b^T\ 0\ 0]^T$, $E_3 = [E_a X\ 0\ 0\ 0\ 0\ 0\ D_a^T\ 0\ 0]$,
$E_4 = [0\ E_d X\ 0\ 0\ 0\ 0\ D_a^T\ 0\ 0]$, $E_5 = [E_b Y\ 0\ 0\ 0\ 0\ 0\ D_a^T\ 0\ 0]$.

Using Lemma 2.6 on the last two terms of (3.48) and then using Schur-complement one can get (3.47).

Remark 3.11 If uncertainties are described as $D_a = D_d = D$ and $F_a(t) = F_d(t) = F(t)$ and $\Delta B(t) = 0$, then the robust stabilizability condition is reduced to following corollary.

Corollary 3.2 *Given the scalars $d_u > 0$, $\mu > 0$ $\epsilon > 0$, system (3.1)–(3.2) is robustly asymptotically stabilizable with the memoryless state-feedback controller, $u(t) = YX^{-1}x(t)$ for any time-delay satisfying (3.3) and for the admissible uncertainties defined above (in Remark 3.11) if there exist symmetric positive definite matrices $\bar{P}$, $\bar{Q}$, $\bar{R}$, $\bar{T}$, $\bar{Z}$ and any matrices $\bar{S}_i$, $(i = 1, 2, ..., 4)$, Y such that the condition (3.46) as well as the LMI holds:*

$$\begin{bmatrix} (1,1) & \bar{\Sigma}_{12} & \bar{S}_3 & \bar{\Sigma}_{14} & d_u\bar{Q}_{11} & d_u\bar{Q}_{12} & (1,7) & 0 & XE_a^T & \mu\bar{P}_{12} \\ \star & \bar{\Sigma}_{22} & \bar{S}_3 & \bar{\Sigma}_{24} & 0 & 0 & XA_d^T & 0 & XE_d^T & 0 \\ \star & \star & -\bar{Q}_{11} & \bar{\Sigma}_{34} & 0 & 0 & 0 & \bar{P}_{12} & 0 & \mu\bar{P}_{22} \\ \star & \star & \star & \bar{\Sigma}_{44} & 0 & 0 & 0 & 0 & 0 & 0 \\ \star & \star & \star & \star & -\bar{Q}_{11} & -\bar{Q}_{12} & 0 & d_u\bar{Q}_{12} & 0 & 0 \\ \star & \star & \star & \star & \star & -\bar{Q}_{22} & 0 & d_u\bar{Q}_{22} & 0 & 0 \\ \star & \star & \star & \star & \star & \star & (7,7) & 0 & 0 & 0 \\ \star & \star & \star & \star & \star & \star & \star & -XZ^{-1}X & 0 & 0 \\ \star & \star & \star & \star & \star & \star & \star & \star & -\epsilon & 0 \\ \star & \star & \star & \star & \star & \star & \star & \star & \star & -\mu\bar{T} \end{bmatrix} < 0 \tag{3.49}$$

where, $(1,1) = \bar{\Sigma}_{11}|_{\Delta=0} + \epsilon DD^T$, $(1,7) = \bar{\Sigma}_{17}|_{\Delta=0} + \epsilon DD^T$, $(7,7) = -Z + \epsilon DD^T$.

Note: Delay-dependent robust stabilization condition for $\mu = 0$ can be obtained from corollary 3.2 by substituting the value of $\mu = 0$ and $T = 0$ in (3.49).

For better understanding of the main results of this chapter, some basic theorems on robust stabilization of time-delay systems relevant to the main results are presented in preceding section.

3.6 Main Results on Delay-dependent Robust Stabilization of an Uncertain TDS

In this section, two different robust stabilization conditions for an uncertain TDS (3.1)are derived (i) in a nonlinear matrix inequality (NLMI) framework and (ii) in a linear matrix inequality (LMI) framework, which are presented in the form of theorems below. The effectiveness of the proposed stabilization criteria is validated by comparing the results with existing methods.

Theorem 3.9 *([23]) System (3.1) with the state feedback control law* $u(t) = Kx(t)$ *is stabilizable if there exist symmetric positive-definite matrices* $Y,\ X,\ Q_r$ *and any free matrices* $T_r,\ T_s$ *and* S*, positive scalars* $\epsilon_1,\ \epsilon_2,\ \epsilon_3$ *and* d_u*, such that the following LMI conditions holds:*

$$\begin{bmatrix} M_{11} & M_{12} & M_{13} & T_r & YE_a^T & 0 & S^T E_b^T \\ \star & M_{22} & M_{23} & T_s & 0 & YE_d^T & 0 \\ \star & \star & M_{33} & 0 & 0 & 0 & 0 \\ \star & \star & \star & -d_u Y X^{-1} Y & 0 & 0 & 0 \\ \star & \star & \star & \star & -\epsilon_1 I & 0 & 0 \\ \star & \star & \star & \star & \star & -\epsilon_2 I & 0 \\ \star & \star & \star & \star & \star & \star & -\epsilon_3 I \end{bmatrix} < 0 \qquad (3.50)$$

where, $M_{11} = YA^T + AY + BS + S^T B^T + T_r + T_r^T + Q_r + \Delta$; $M_{12} = A_d Y + T_s^T - T_r$
$M_{13} = d_u Y A^T + d_u S^T B^T + d_u \Delta$; $M_{22} = -T_s - T_s^T - (1-\mu) Q_r$; $M_{23} = d_u Y A_d^T$
$M_{33} = -d_u X + d_u^2 \Delta$; $\Delta = \epsilon_1 D_a D_a^T + \epsilon_2 D_d D_d^T + \epsilon_3 D_b D_b^T$

Proof Consider an uncertain time-delay system (3.1) satisfying the delay and its derivative conditions (3.3), to prove the above robust stabilizability condition we consider following LK functional candidate is considered.

$$V(t) = V_1 + V_2 + V_3 \qquad (3.51)$$

$$V_1(t) = x^T(t) P x(t),\ P = P^T > 0 \qquad (3.52)$$

$$V_2(t) = \int_{t-d(t)}^{t} x^T(s) Q_1 x(s) d,\ Q_1 = Q_1^T > 0 \qquad (3.53)$$

$$V_3(t) = \int_{-d_u}^{0} \int_{\beta}^{0} \dot{x}(t+\alpha) Q_2 \dot{x}(t+\alpha) d\alpha d\beta,\ Q_2 = Q_2^T > 0 \qquad (3.54)$$

Finding the time-derivative of (3.53) and substituting the value of $\dot{x}(t)$ from (3.1) with $u(t) = Kx(t)$, one can get

$$\dot{V}_1(t) = \xi^T(t) \begin{bmatrix} \Theta_{11} & \Theta_{12} \\ \star & \Theta_{22} \end{bmatrix} \xi(t) \tag{3.55}$$

where, $\xi(t) = [x^T(t) \;\; x^T(t - d(t)]^T$; $\Theta_{11} = (A(t) + B(t)K)^T P + P(A(t) + B(t)K)$ and Θ_{12} , $= PA_d(t)$

The time-derivative of (3.54) and (3.54) are

$$\dot{V}_2(t) \le \xi^T(t) \begin{bmatrix} Q_1 & 0 \\ 0 & -(1-\mu)Q_1 \end{bmatrix} \xi(t) \tag{3.56}$$

and,

$$\dot{V}_3(t) = d_u \dot{x}^T(t) Q_2 \dot{x}(t) - \int_{t-d_u}^{t} \dot{x}^T(\alpha) Q_2 \dot{x}(\alpha) d\alpha \tag{3.57}$$

The last integral term of (3.57) is approximated as

$$-\int_{t-d_u}^{t} \dot{x}^T(\alpha) Q_2 \dot{x}(\alpha) d\alpha \le -\int_{t-d(t)}^{t} \dot{x}^T(\alpha) Q_2 \dot{x}(\alpha) d\alpha \tag{3.58}$$

Using Lemma 2 of [24], (3.58) may be written as

$$-\int_{t-d(t)}^{t} \dot{x}^T(\alpha) Q_2 \dot{x}(\alpha) d\alpha \le \xi^T(t) \left\{ \begin{bmatrix} T_1 + T_1^T & -T_1 + T_2^T \\ \star & -T_2 - T_2^T \end{bmatrix} + d_u \begin{bmatrix} T_1 \\ T_2 \end{bmatrix} Q_2^{-1} \begin{bmatrix} T_1 \\ T_2 \end{bmatrix}^T \right\} \xi(t) \tag{3.59}$$

Substituting the value of $\dot{x}(t)$ from (3.1) and applying state feedback control law (i.e. $u(t) = Kx(t)$) in the first term of (3.57) and subsequently approximating the integral term by (3.59), one can obtain after simple algebraic manipulations the following expression for $\dot{V}_3(t)$ as

$$\dot{V}_3(t) = \xi^T(t) \left\{ \begin{bmatrix} \Upsilon_{11} & \Upsilon_{12} \\ \star & \Upsilon_{22} \end{bmatrix} + d_u \begin{bmatrix} T_1 \\ T_2 \end{bmatrix} Q_2^{-1} \begin{bmatrix} T_1 \\ T_2 \end{bmatrix}^T \right\} \xi(t) \tag{3.60}$$

where, $\Upsilon_{11} = d_u(A(t) + B(t)K)^T Q_2(A(t) + B(t)K) + T_1 + T_1^T$;
$\Upsilon_{22} = d_u A_d(t)^T Q_2 A_d(t) - T_2 - T_2^T$;
$\Upsilon_{12} = d_u(A(t) + B(t)K)^T Q_2 A_d(t) - T_1 + T_2^T$.

Now, in view of (3.57) and invoking (3.55), (3.56) and (3.60) one obtains

$$\dot{V}(t) = \dot{V}_1 + \dot{V}_2 + \dot{V}_3$$
$$\dot{V}(t) = \xi^T(t)\left\{\begin{bmatrix}\Lambda_{11} & \Lambda_{12}\\ \star & \Lambda_{22}\end{bmatrix} + d_u\begin{bmatrix}T_1\\T_2\end{bmatrix}Q_2^{-1}\begin{bmatrix}T_1\\T_2\end{bmatrix}^T\right\}\xi(t) \tag{3.61}$$

where, $\Lambda_{11} = (A(t)+B(t)K)^T P + P(A(t)+B(t)K) + d_u(A(t)+B(t)K)^T Q_2(A(t)+B(t)K) + T_1 + T_1^T + Q_1$;
$\Lambda_{22} = d_u A_d(t)^T Q_2 A_d(t) - T_2 - T_2^T - (1-\mu)Q_1$;
$\Lambda_{12} = PA_d(t) + d_u(A(t)+B(t)K)^T Q_2 A_d(t) - T_1 + T_2^T$.

Now, for stability $\dot{V}(t)$ must be less than zero, i.e, the following conditions must be satisfied

$$\begin{bmatrix}\Lambda_{11} & \Lambda_{12}\\ \star & \Lambda_{22}\end{bmatrix} + d_u\begin{bmatrix}T_1\\T_2\end{bmatrix}Q_2^{-1}\begin{bmatrix}T_1\\T_2\end{bmatrix}^T < 0 \tag{3.62}$$

Using Schur-complement equation (3.62) can be rewritten as

$$\begin{bmatrix}\Pi_{11} & \Pi_{12} & \Pi_{13} & T_1\\ \star & \Pi_{22} & \Pi_{23} & T_2\\ \star & \star & -d_u Q_2 & 0\\ \star & \star & 0 & -d_u^{-1}Q_2\end{bmatrix} < 0 \tag{3.63}$$

where, $\Pi_{11} = (A(t)+B(t)K)^T P + P(A(t)+B(t)K) + T_1 + T_1^T + Q_1$; $\Pi_{12} = PA_d(t) - T_1 + T_2^T$;
$\Pi_{13} = d_u(A(t)+B(t)K)^T Q_2$; $\Pi_{22} = -T_2 - T_2^T - (1-\mu)Q_1$; $\Pi_{23} = d_u A_d^T Q_2$.

Pre- and post-multiplying (3.63) by $diag\{P^{-1},\ P^{-1},\ Q_2^{-1},\ P^{-1}\}$ and defining the linear changes in variables as $Y = P^{-1}$, $X = Q_2^{-1}$, $KY = S$, $YT_1Y = T_r$, $YT_2Y = T_s$ and $YQ_1Y = Q_r$, one obtain,

$$\begin{bmatrix}W_{11} & W_{12} & W_{13} & T_r\\ \star & W_{22} & d_u Y A_d^T(t) & T_s\\ \star & \star & -d_u X & 0\\ \star & \star & 0 & -d_u^{-1} Y X^{-1} Y\end{bmatrix} < 0 \tag{3.64}$$

where, $W_{11} = YA^T(t) + A(t)Y + B(t)S + S^T B^T(t) + T_r + T_r^T + Q_r$; $W_{12} = A_d(t)Y - T_r + T_s^T$; $W_{13} = d_u Y A^T(t) + d_u S^T B^T(t)$; $W_{22} = -T_s - T_s^T - (1-\mu)Q_r$.

The matrices $A(t)$ and $A_d(t)$ in (3.64) are replaced by $(A + \Delta A(t)$ and $(A_d + \Delta A_d(t)$ with (3.4) and then decomposing the (3.64) as nominal and uncertain parts as

$$U + V < 0 \tag{3.65}$$

where,

$$U = \begin{bmatrix} U_{11} & U_{12} & U_{13} & T_r \\ \star & U_{22} & d_u Y A_d^T & T_s \\ \star & \star & -d_u X & 0 \\ \star & \star & 0 & -d_u Y X^{-1} Y \end{bmatrix},$$

$$V = \begin{bmatrix} V_{11} & V_{12} & V_{13} & 0 \\ \star & 0 & V_{23} & 0 \\ \star & \star & 0 & 0 \\ 0 & 0 & 0 & 0 \end{bmatrix}$$

U_{11}=Y$A^T + AY + BS + S^T B^T + T_r + T_r^T + Q_r$; $U_{12} = A_d Y - T_r + T_s^T$;
$U_{13} = d_u Y A^T + d_u S^T B^T$; $U_{22} = -T_s - T_s^T - (1-\mu) Q_r$;
$V_{11} = Y E_a^T F_a^T D_a^T + D_a F_a E_a Y + S^T E_b^T F_b^T D_b^T + D_b F_b E_b S$;
$V_{12} = D_d F_d E_d Y$; $V_{13} = d_u (Y E_a^T F_a^T D_a^T + S^T E_b^T F_b^T D_b^T)$; $V_{23} = d_u Y E_d^T F_d^T D_d^T$

Rearranging (3.65) one may write

$$U + J_u + J_u^T < 0 \tag{3.66}$$

where,

$$J_u = \begin{bmatrix} D_a \\ 0 \\ d_u D_a \\ 0 \end{bmatrix} F_a \begin{bmatrix} E_a Y & 0 & 0 & 0 \end{bmatrix} + \begin{bmatrix} D_d \\ 0 \\ d_u D_d \\ 0 \end{bmatrix} F_d \begin{bmatrix} 0 & E_d Y & 0 & 0 \end{bmatrix} + \begin{bmatrix} D_b \\ 0 \\ d_u D_b \\ 0 \end{bmatrix} F_b \begin{bmatrix} E_b S & 0 & 0 & 0 \end{bmatrix}$$

Applying Lemma 2.6 thrice on $(J_u + J_u^T)$ term of (3.66), eliminates the uncertain matrices $F_a(t)$, $F_b(t)$ and $F_d(t)$, and one obtains the LMI condition (3.50).

Remark 3.12 If uncertainties are described as $D_a = D_d = D$ and $F_a(t) = F_d(t) = F(t)$ and $\Delta B(t) = 0$, then the robust stabilizability condition is reduced to following corollary.

Corollary 3.3 *System (3.1) with state feedback control law $u(t) = Kx(t)$ is stabilizable if there exist (a) symmetric positive-definite matrices Y, X, Q_r and (b) any free matrices T_r, T_s and S, positive scalars ϵ and d_u, such that the following LMI conditions holds:*

$$\begin{bmatrix} \bar{M}_{11} & \bar{M}_{12} & \bar{M}_{13} & YE_a^T & T_r \\ \star & \bar{M}_{22} & d_u Y A_d^T & YE_d^T & T_s \\ \star & \star & \bar{M}_{33} & 0 & 0 \\ \star & \star & 0 & -\epsilon I & 0 \\ \star & \star & 0 & 0 & -d_u Y X^{-1} Y \end{bmatrix} < 0 \qquad (3.67)$$

where, $\bar{M}_{11} = YA^T + AY + BS + S^T B^T + T_r + T_r^T + Q_r + \epsilon DD^T$; $\bar{M}_{12} = A_d Y + T_s^T - T_r$
$\bar{M}_{13} = d_u Y A^T + d_u S^T B^T + d_u \epsilon DD^T$; $\bar{M}_{22} = -T_s - T_s^T - (1-\mu) Q_r$;
$\bar{M}_{33} = -d_u X + d_u^2 \epsilon DD^T$

Remark 3.13 The Stabilizability conditions in (3.50) and (3.67) are NLMIs due to the presence of nonlinear term $YX^{-1}Y$. In order to solve numerical such NLMIs cone-complementarity algorithm [16] has been used here.

Next, the step-by-step numerical implementation of cone complementarity algorithm is used to solve this NLMI problem.

Algorithmic Computation:
Let us fix

$$-YX^{-1}Y \le -L \qquad (3.68)$$

substituting (3.68) in (3.67), one can rewrite

$$\begin{bmatrix} \bar{M}_{11} & \bar{M}_{12} & \bar{M}_{13} & YE_a^T & T_r \\ \star & \bar{M}_{22} & d_u Y A_d^T & YE_d^T & T_s \\ \star & \star & \bar{M}_{33} & 0 & 0 \\ \star & \star & 0 & -\epsilon & 0 \\ \star & \star & 0 & 0 & -d_u L \end{bmatrix} < 0 \qquad (3.69)$$

using Schur-complement to (3.68), one can write

$$\begin{bmatrix} L^{-1} & Y^{-1} \\ Y^{-1} & X^{-1} \end{bmatrix} \ge 0 \qquad (3.70)$$

Now defining, $D = L^{-1}$, $J = Y^{-1}$, $N = X^{-1}$, one can rewrite (3.70) as

$$\begin{bmatrix} D & J \\ J & N \end{bmatrix} \ge 0 \qquad (3.71)$$

Again, one can have the following valid identities, $DL = I$, $JY = I$, $NX = I$. Thus, in view of the identities defined, one can write it in the form of matrix inequalities as

$$\begin{bmatrix} L & I \\ \star & D \end{bmatrix} \ge 0, \quad \begin{bmatrix} Y & I \\ \star & J \end{bmatrix} \ge 0, \quad and \quad \begin{bmatrix} X & I \\ \star & N \end{bmatrix} \ge 0, \qquad (3.72)$$

As the nonlinear LMI condition in 3.67 cannot be solved as a feasibility problem by standard routine of LMI toolbox of MATLAB, so the NLMI in (3.67) can be solved as a cone complementarity problem suggested in [16] which is recast as

$$\text{Minimize Trace}(LD + YJ + XN)$$
$$\text{subject to (3.67), (3.71) and (3.71)}$$

Such problems are solved by considering the linear approximation of $(Trace\ LD + YJ + XN)$ in the form $Trace\ (D_0L + L_0D + J_0Y + Y_0J + N_0X + X_0N)$ at a given point $(D_0, L_0, J_0, Y_0, N_0, X_0)$ [16]. Note that, (3.72) confronts the exact solution when these are at the boundary, i.e., the inequalities are rank-deficient. Now we are ready to present algorithmic steps of the linearization algorithm.

Algorithmic Steps. ***Step 1****: Select initially a small value of delay bound d_u and set $j = 0$.*

Step 2: *Find a feasible set of $(D_0,\ J_0,\ L_0,\ Y_0,\ N_0,\ S_0,\ T_{r0},\ T_{s0},\ X_0,\ Q_{r0},\ \epsilon_0)$ satisfying (3.68), (3.71) and (3.72) with $Y > 0$ and $X > 0$.*

Step 3*: Solve the following LMI optimization problem for the variables $(D,\ J,\ L,\ Y,\ N,\ S,\ T,\ T,\ X,\ Q, \epsilon)$*

Minimize Trace $(LD_j + DL_j + YJ_j + JY_j + XN_j + NX_j)$ subject to (3.68), (3.71) and (3.72) with $Y > 0$ and $X > 0$. The LMIs are solved using the standard routines available with LMI toolbox of MATLAB [6].

Set $(D_{j+1} = D,\ L_{j+1} = L,\ J_{j+1} = J,\ Y_{j+1} = Y,\ N_{j+1} = N,\ X_{j+1} = X)$.

Step 4: *If $L_{j-1} \le Y_j X_j^{-1} Y_j$ is satisfied then increase d_u by small value and go to Step 2. If this is condition is not satisfied within a prespecified number of iterations then stop. Otherwise set j=j+1 and go to step 3.*

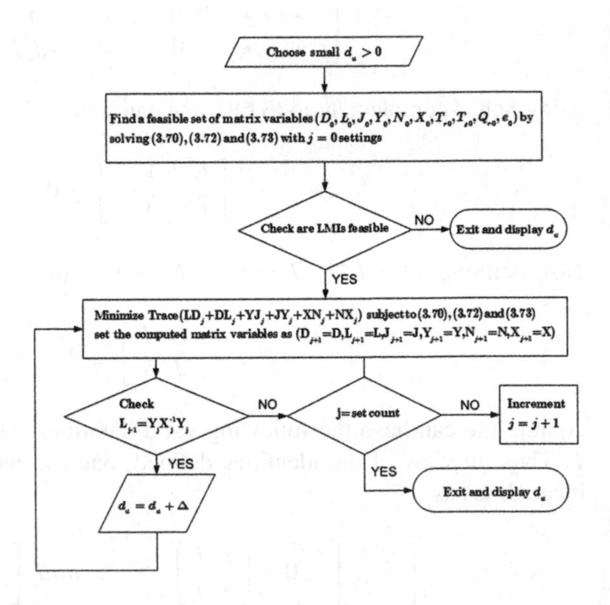

Fig. 3.5 Flow-chart for Cone Complementarity Algorithm

The above algorithmic steps are presented in the form of flow-chart in Fig. 3.5 for better understanding of the numerical implementation of the algorithm for stabilization problem of time-delay system.

An LMI based robust stabilizing conditions are derived next by extending the stabilizing conditions obtained in Theorem 3.6 and its associated corollaries.

Theorem 3.10 *Given the scalar* $d_u > 0$*, the system (3.1) for* $0 < \mu < 1$ *is asymptotically robustly stabilizable with the state-feedback controller,* $u(t) = YZ^{-1}x(t)$ *for any time-delay satisfying the condition (3.3) with admissible uncertainties, if there exist symmetric positive definite matrices* $\bar{P}$, $\bar{Q}$, $\bar{R}, \bar{T}$, *and any free matrices* $\bar{M}_i$, $\bar{L}_i$ $(i = 1, 2, 3)$, Z *and the scalars* $\epsilon_i > 0$, $(i = 1, 2)$ *with appropriate dimensions satisfying the following LMI constraints:*

$$\bar{P} = \begin{bmatrix} \bar{P}_{11} & \bar{P}_{12} \\ \star & \bar{P}_{22} \end{bmatrix} > 0, \quad \begin{bmatrix} \bar{Q}_{11} & 0 \\ 0 & \bar{Q}_{22} \end{bmatrix} > 0 \tag{3.73}$$

$$\begin{bmatrix} \bar{\Omega}_{per} & \mu\tilde{P}_{cl} & E_1^T & E_2^T & d_u\bar{M} \\ \star & -\mu\bar{T} & 0 & 0 & 0 \\ \star & 0 & -\epsilon_1 I & 0 & 0 \\ \star & 0 & 0 & -\epsilon_2 I & 0 \\ \star & 0 & 0 & 0 & -\bar{Q}_{22} \end{bmatrix} < 0 \tag{3.74}$$

and,

$$\begin{bmatrix} \bar{\Omega}_{per} & \mu\tilde{P}_{cl} & E_1^T & E_2^T & d_u\bar{L} \\ \star & -\mu\bar{T} & 0 & 0 & 0 \\ \star & 0 & -\epsilon_1 I & 0 & 0 \\ \star & 0 & 0 & -\epsilon_2 I & 0 \\ \star & 0 & 0 & 0 & -\bar{Q}_{22} \end{bmatrix} < 0 \tag{3.75}$$

where,

$$E_1 = \begin{bmatrix} E_a Z \; 0 \; 0 \; 0 \; 0 \; 0 \end{bmatrix}$$

$$E_2 = \begin{bmatrix} 0 \; E_a Z \; 0 \; 0 \; 0 \; 0 \end{bmatrix}$$

$$\tilde{P}_{cl} = \begin{bmatrix} \bar{P}_{12}^T \; 0 \; 0 \; \bar{P}_{22} \; 0 \; 0 \end{bmatrix}^T$$

$$\bar{\Omega}_{per} = \begin{bmatrix} \bar{\Omega}_{11} & \bar{\Omega}_{12} & \bar{\Omega}_{13} & \bar{\Omega}_{14} & 0 & \bar{\Omega}_{16} \\ \star & \bar{\Omega}_{22} & \bar{\Omega}_{23} & \bar{\Omega}_{24} & 0 & \bar{\Omega}_{26} \\ \star & \star & \bar{\Omega}_{33} & 0 & 0 & 0 \\ \star & \star & 0 & -\bar{Q}_{11} & 0 & \bar{P}_{12}^T \\ 0 & 0 & 0 & 0 & -\bar{Q}_{11} & 0 \\ \star & \star & 0 & \star & 0 & \bar{\Omega}_{66} \end{bmatrix},$$

where, $\bar{\Omega}_{11}=d_u^2\bar{Q}_{11} + AZ + Z^TA^T + BY + Y^TB^T + \bar{P}_{12} + \bar{P}_{12}^T + \bar{R}_1 + \bar{R}_2 + d_u(\bar{L}_1 + \bar{L}_1^T) + \Delta$
$\bar{\Omega}_{12} = A_dZ - \bar{P}_{12} + d_u(-\bar{L}_1 + \bar{L}_2^T + \bar{M}_1)$
$\bar{\Omega}_{13} = d_u(\bar{L}_3^T - \bar{M}_1)$, $\bar{\Omega}_{14} = \bar{P}_{22}$, $\bar{\Omega}_{16} = \bar{P}_{11} - Z + \alpha Z^TA^T + \alpha Y^TB^T + \alpha\Delta$
$\bar{\Omega}_{22} = -(1-\mu)\bar{R}_1 + \mu\bar{T} + d_u(-\bar{L}_2 - \bar{L}_2^T + \bar{M}_2 + \bar{M}_2^T)$
$\bar{\Omega}_{23} = d_u(-\bar{M}_2 + \bar{M}_3^T - \bar{L}_3^T)$, $\bar{\Omega}_{24} = -\bar{P}_{22}$, $\bar{\Omega}_{26} = \alpha Z^TA_d^T$
$\bar{\Omega}_{33} = d_u(-\bar{M}_3 - \bar{M}_3^T) - \bar{R}_2$, $\bar{\Omega}_{66} = -\alpha(Z + Z^T) + d_u^2\bar{Q}_{22} + \alpha^2\Delta$
$\Delta = \epsilon_1 D_aD_a^T + \epsilon_2 D_dD_d^T$

Proof Replace A and A_d matrices in the $\bar{\Omega}$ block matrix of the stabilizing condition (3.34)–(3.35) by $A + D_aF_aE_a$ and $A_d + D_dF_dE_d$ respectively, this replacement will give rise to a new matrix of the form,

$$\tilde{\Omega} = \bar{\Omega}_{nom} + \bar{\Omega}_{unc} \tag{3.76}$$

where, $\bar{\Omega}_{nom} = \bar{\Omega}$ and the $\bar{\Omega}_{unc}$ is defined below,

$$\bar{\Omega}_{unc} = \begin{bmatrix} (1,1) & (1,2) & 0\,0\,0 & (1,6) \\ \star & 0 & 0\,0\,0 & (2,6) \\ 0 & 0 & 0\,0\,0 & 0 \\ 0 & 0 & 0\,0\,0 & 0 \\ 0 & 0 & 0\,0\,0 & 0 \\ \star & \star & 0\,0\,0 & 0 \end{bmatrix} \tag{3.77}$$

where, $(1,1) = D_aF_a(t)E_aZ + (D_aF_a(t)E_aZ)^T$, $(1,2) = D_dF_d(t)E_dZ$,
$(1,6) = \alpha(D_aF_a(t)E_aZ)^T$, $(2,6) = \alpha(D_dF_d(t)E_dZ)^T$

Further one can rewrite (3.77) in the form,

$$\bar{\Omega}_{unc} = \Sigma + \Sigma^T = D_1F_aE_1 + D_2F_dE_2 + (D_1F_aE_1)^T + (D_2F_dE_2)^T \tag{3.78}$$

where,

$$D_1 = \left[\, D_a^T\ 0\,0\,0\,0\ \alpha D_a^T \,\right]^T,\quad D_2 = \left[\, D_d^T\ 0\,0\,0\,0\ \alpha D_d^T \,\right]^T,$$

and

$$E_1 = \left[\, E_aZ\ 0\,0\,0\,0\,0 \,\right],\quad E_2 = \left[\, 0\ E_dZ\ 0\,0\,0\,0 \,\right],$$

Now using Lemma 3.2 one can write (3.78) as,

$$\Sigma + \Sigma^T \leq \epsilon_1 D_1 D_1^T + E_1^T \epsilon_1^{-1} E_1 + \epsilon_2 D_2 D_2^T + E_2^T \epsilon_2^{-1} E_2 \tag{3.79}$$

Substituting (3.79) in (3.76) one can get following block matrices,

$$\begin{bmatrix} \bar{\Omega}_{per} & \mu \tilde{P}_{cl} & d_u \bar{M} \\ \star & -\mu \bar{T} & 0 \\ \star & 0 & -\bar{Q}_{22} \end{bmatrix} + \epsilon_1^{-1} E_1^T E_1 + \epsilon_2^{-1} E_2^T E_2 \tag{3.80}$$

$$\begin{bmatrix} \bar{\Omega}_{per} & \mu \tilde{P}_{cl} & d_u \bar{L} \\ \star & -\mu \bar{T} & 0 \\ \star & 0 & -\bar{Q}_{22} \end{bmatrix} + \epsilon_1^{-1} E_1^T E_1 + \epsilon_2^{-1} E_2^T E_2 \tag{3.81}$$

now using Schur-complement twice on the block matrices in (3.80) provides the robust-stabilizing condition (3.74)–(3.75), rest of the LMIs are same as in stabilization theorem as they do not contain any system matrices in it.

To obtain a realizable solution of gain matrix $K = YZ^{-1}$ for a particular delay bound, one can impose constraint on the size of matrix K elements as,

$$K = YZ^{-1} \tag{3.82}$$

with

$$Y^T Y < \delta I, \ \ \delta > 0 \tag{3.83}$$

and

$$Z^{-1} < \beta I, \ \ \beta > 0 \tag{3.84}$$

the above two constraints (3.83) and (3.84) further can be rewritten as

$$\begin{bmatrix} -\delta I & Y^T \\ Y & -I \end{bmatrix} < 0 \tag{3.85}$$

and

$$\begin{bmatrix} \beta I & I \\ I & Z \end{bmatrix} > 0 \tag{3.86}$$

To find the value of the gain K for a particular delay upper bound d_u and α following minimization problem is proposed:

Minimize $\delta + \beta$
s.t. (3.73)–(3.75), $P_{11} = P_{11} > 0$, $R_i > 0$, $T > 0$ and $\epsilon_i > 0$, $i = 1, 2$

If $D_a = D_d = D$ and $F_a(t) = F_d(t) = F(t)$ then the resulting norm bounded uncertainties will take the form $\Delta A = DFE_a$ and $\Delta A_d = DFE_d$, the robust stabilizing condition with norm bounded uncertainties for $0 < \mu < 1$ can be stated in the form of corollary presented below:

Corollary 3.4 *Given the scalars $d_u > 0$, the system (3.1) with $\Delta B = 0$ for $0 < \mu < 1$ is asymptotically robustly stabilizable with the state-feedback controller $u(t) = YZ^{-1}x(t)$ for any time-delay satisfying the condition (3.3) with admissible uncertainties, if there exist symmetric positive definite matrices $\bar{P}$, $\bar{Q}$, $\bar{R}, \bar{T}$, and any free matrices $\bar{M}_i$, $\bar{L}_i$ $(i = 1, 2, 3), Z$ and a scalar $\epsilon > 0$ with appropriate dimensions satisfying the following LMI constraints:*

$$\bar{P} = \begin{bmatrix} \bar{P}_{11} & \bar{P}_{12} \\ \star & \bar{P}_{22} \end{bmatrix} > 0, \quad \begin{bmatrix} \bar{Q}_{11} & 0 \\ 0 & \bar{Q}_{22} \end{bmatrix} > 0 \tag{3.87}$$

$$\begin{bmatrix} \bar{\Omega}_{per} & \mu\tilde{P}_{cl} & E^T & d_u\bar{M} \\ \star & -\mu\bar{T} & 0 & 0 \\ \star & 0 & -\epsilon I & 0 \\ \star & 0 & 0 & -\bar{Q}_{22} \end{bmatrix} < 0 \tag{3.88}$$

and,

$$\begin{bmatrix} \bar{\Omega}_{per} & \mu\tilde{P}_{cl} & E^T & d_u\bar{L} \\ \star & -\mu\bar{T} & 0 & 0 \\ \star & 0 & -\epsilon I & 0 \\ \star & 0 & 0 & -\bar{Q}_{22} \end{bmatrix} < 0 \tag{3.89}$$

where, the elements of $\bar{\Omega}_{per}$ are all same as defined in Theorem 3.10 except the terms,

$$\begin{aligned} \bar{\Omega}_{11} =& d_u^2\bar{Q}_{11} + AZ + Z^TA^T + BY + Y^TB^T + \bar{P}_{12} + \bar{P}_{12}^T + \bar{R}_1 + \bar{R}_2 \\ &+ d_u(\bar{L}_1 + \bar{L}_1^T) + \epsilon DD^T \\ \bar{\Omega}_{16} =& \bar{P}_{11} - Z + \alpha Z^TA^T + \alpha Y^TB^T + \alpha\epsilon DD^T \\ \bar{\Omega}_{66} =& -\alpha(Z + Z^T) + d_u^2\bar{Q}_{22} + \alpha^2\alpha\epsilon DD^T \end{aligned}$$

The proof this corollary is straightforward and can be obtained in a similar manner as in Theorem 3.10, by replacing A and A_d matrices with $A + DFE_a$ and $A_d + DFE_d$ respectively and along with the choices of the matrix E defined above and the matrix D as

$$D = \begin{bmatrix} D\,0\,0\,0\,0\,\alpha D \end{bmatrix}.$$

$$E = \begin{bmatrix} E_aZ\ E_dZ\,0\,0\,0\,0 \end{bmatrix}$$

If $\mu = 0$ (delay-derivative) and the uncertainties are as defined in the Corollary 3.4 then one can get robust stabilizing condition directly from Corollary 3.4 by substituting $\mu = 0$.

Corollary 3.5 *Given the scalars $d_u > 0$, the system (3.1) with $\Delta B = 0$ for $0 < \mu < 1$ is asymptotically robustly stabilizable with the state-feedback controller, $u(t) = YZ^{-1}x(t)$ for any time-delay satisfying the condition (3.3) with admissible uncertainties, if there exist symmetric positive definite matrices $\bar{P}$, $\bar{Q}$, $\bar{R}, \bar{T}$, and any free matrices $\bar{M}_i$, $\bar{L}_i$ $(i = 1, 2, 3), Z$ and a positive scalar ϵ with appropriate dimensions satisfying the following LMI constraints,*

$$\bar{P} = \begin{bmatrix} \bar{P}_{11} & \bar{P}_{12} \\ \star & \bar{P}_{22} \end{bmatrix} > 0, \quad \begin{bmatrix} \bar{Q}_{11} & 0 \\ 0 & \bar{Q}_{22} \end{bmatrix} > 0 \tag{3.90}$$

$$\begin{bmatrix} \bar{\Omega}_{per} & E^T & d_u\bar{M} \\ \star & -\epsilon I & 0 \\ \star & 0 & -\bar{Q}_{22} \end{bmatrix} < 0 \tag{3.91}$$

and,

$$\begin{bmatrix} \bar{\Omega}_{per} & E^T & d_u\bar{L} \\ \star & -\epsilon I & 0 \\ \star & 0 & -\bar{Q}_{22} \end{bmatrix} < 0 \tag{3.92}$$

where, the elements of $\bar{\Omega}_{per}$ are all same as defined in Theorem 3.10, whereas following term gets modified in view of the structure of uncertainty assumed for obtaining this corollary

where, $\bar{\Omega}_{11} = d_u^2\bar{Q}_{11} + AZ + Z^TA^T + BY + Y^TB^T + \bar{P}_{12} + \bar{P}_{12}^T + \bar{R}_1 + \bar{R}_2 + d_u(\bar{L}_1 + \bar{L}_1^T) + \epsilon DD^T$

$\bar{\Omega}_{16} = \bar{P}_{11} - Z + \alpha Z^TA^T + \alpha Y^TB^T + \alpha\epsilon DD^T$

$\bar{\Omega}_{66} = -\alpha(Z + Z^T) + d_u^2\bar{Q}_{22} + \alpha^2\alpha\epsilon DD^T$

Remark 3.14 It may be noted that both the corollaries 3.4 and 3.5 are equivalently same and can be applied for robust stabilization of time-delay systems with admissible uncertainties (with $\Delta B = 0$) for $0 \leq \mu \leq 1$. The LMIs (3.88) and (3.89) involved in corollary 3.4 are replaced by the lower dimensional LMIs (3.91) and (3.92). While a sequence Schur-complement is employed on them, this indeed requires less number of LMI variables and in turn improves the upper bound estimate.

Numerical Example 3.2 *[6, 12] Consider an uncertain time-delay system,*

$$\dot{x}(t) = [A + \Delta A(t)]x(t) + [A_d + \Delta A_d(t)]x(t - d(t) + Bu(t)$$

$$where,\ A = \begin{bmatrix} 0 & 0 \\ 0 & 1 \end{bmatrix},\ A_d = \begin{bmatrix} -1 & -1 \\ 0 & -0.9 \end{bmatrix},\ B = \begin{bmatrix} 0 \\ 1 \end{bmatrix},$$

$$D_a = D_d = D = 0.2I,\ and\ E_a = E_d = I$$

Note: Minimization algorithm as discussed in Theorem 3.10 is considered along with the delay-dependent robust stabilization conditions obtained in Corollary 3.4 and 3.5 for computing the delay bound d_u and the stabilizing gain matrix K.

(n= Order of the system and m= number of inputs).

The simulation results of the system considered in Example 3.2 (for a constant delay, i.e, $\mu = 0$) are obtained by considering the uncertainty matrix as in [12],

$$F(t) = \begin{bmatrix} \cos t & 0 \\ 0 & \sin t \end{bmatrix}$$

The considered uncertain TDS is found to be unstable under open-loop with time-delay set to $d_u = 1.3\ seconds$. Now applying the stabilizing control law with state feedback gain $K = [-1.1923, \ -4.1754]$ and the corresponding $d_u = 1.3$ seconds (see Table 3.4), the system responses are shown in the Fig. 3.6 and its corresponding control input plot is shown in Fig. 3.7. It may be mentioned that, proposed corollary 3.3 provides less control effort as well as less number of iterations

Table 3.3 d_u and K for Example 3.2 with $\mu = 0$ via LMI framework

Methods	d_u	$K\ matrix$	Remark
[6]	0.2250	–	'α' is not used as parameter
[7]	0.3346	–	'α' is not used as parameter
Cor. 3.5	1.3	$[-2.4249, -6.2278]$	$\alpha = 1.4119$
	1.35	$[-19.5454, -43.1380]$	$\alpha = 1.3094$
	1.3544	$[-180.4840, -391.5516]$	$\alpha = 1.2432$

'–' means result is not available in the reference

Table 3.4 d_u and K for Example 3.1 for $\mu = 0$ via NLMI framework

Methods	d_u	$K\ matrix$	Iterations	Decision Variables
Cor. 6 [12]	1.2	$[-1.1110, -3.6432]$	19	$9 \times n^2 + (4 + m) \times n$
	1.3	$[-2.1485, -5.6948]$	41	
Cor. 3.3	1.2	$[-0.7413, -3.0261]$	16	
	1.3	$[-1.1923, -4.1754]$	24	$3.5 \times n^2 + (2.5 + m) \times n$
	1.4	$[-2.7088, -7.5578]$	50	

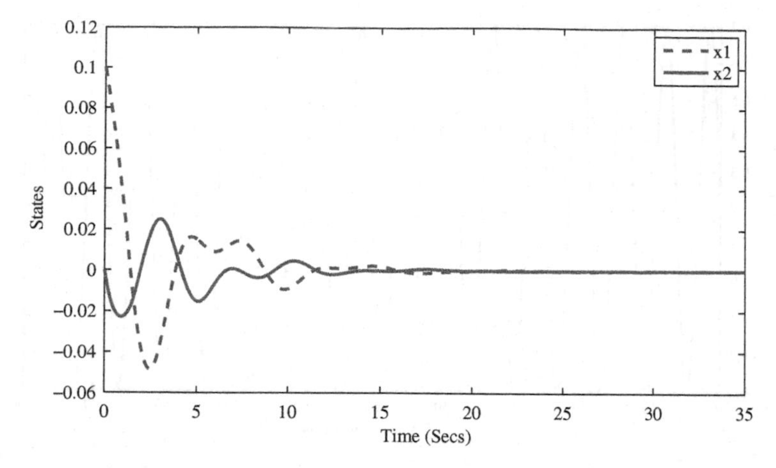

Fig. 3.6 Closed-loop simulation of system in Example 3.2

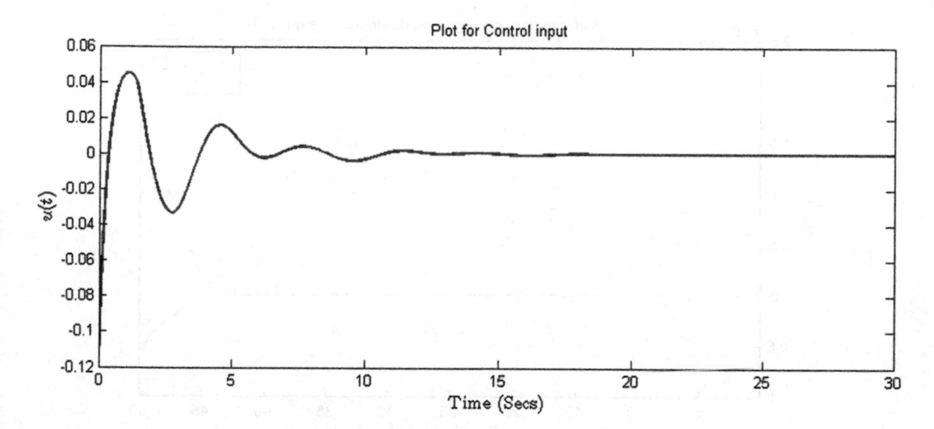

Fig. 3.7 Control input of system in Example 3.2

are required compared to existing method [12] for the same value of delay upper bound $d_u = 1.3\ secs.$

Numerical Example 3.3 *Consider an uncertain time-delay system, [6, 12]*

$$\dot{x}(t) = [A + \Delta A(t)]x(t) + [A_d + \Delta A_d(t)]x(t - d(t) + Bu(t)$$

$$where,\ \ A = \begin{bmatrix} 0 & 0 \\ 0 & 1 \end{bmatrix},\ A_d = \begin{bmatrix} -2 & -0.5 \\ 0 & -1 \end{bmatrix},\ B = \begin{bmatrix} 0 \\ 1 \end{bmatrix},$$

$$D_a = D_d = D = 0.2I,\ and\ E_a = E_d = I$$

Responses of an uncertain TDS described in Numerical Example 3.3 are obtained with the following data; $F_a(t) = F_d(t) = F(t) = I$ (as $\| F(t) \| \leq 1$) and the time-varying delay of value $d(t) = 0.2 + 0.5sin(t)$ (Fig. 3.8). Under open-loop the

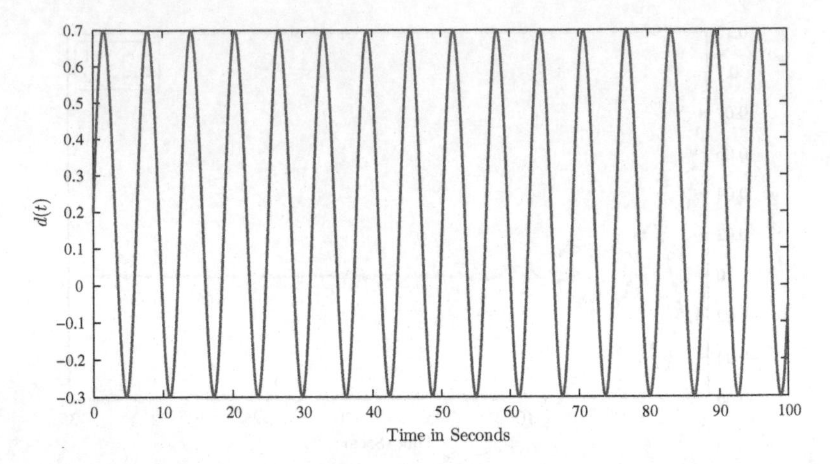

Fig. 3.8 Time-varying delay considered for Example 3.3

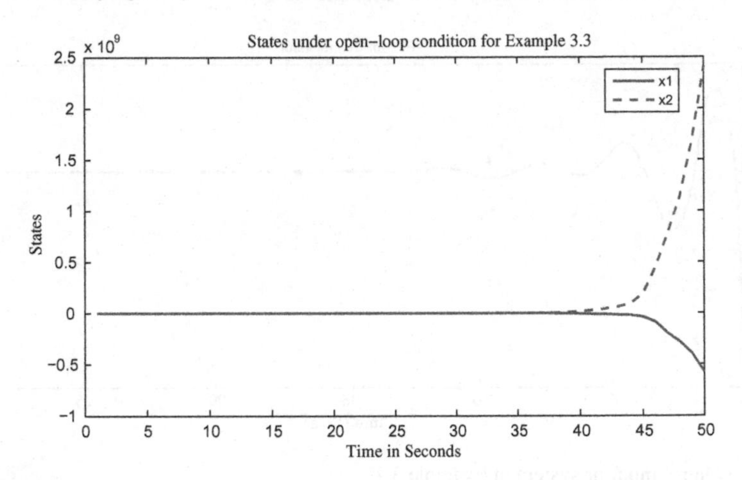

Fig. 3.9 Open-loop simulation of system in Example 3.3

system response is found to be unstable as shown in the Fig. 3.9, it is stabilized using $K = [-105.4272, \ -69.6643]$ for the delay $d_u = 0.7sec$, $\mu = 0s.5$ (see Table 3.5). The closed-loop system response is shown in Fig. 3.10.

Remark 3.15 The robust stabilization results obtained using Corollary 3.3 (in a NLMI framework) for the systems described in Examples 3.2 and 3.3 are presented in Tables 3.4 and 3.6 respectively. One can observe from Table 3.5 that proposed stabilizing controllers provide less conservative delay bound d_u and less control effort $\| K \|_\infty$ than those obtained using delay-dependent stabilization criteria with or without system uncertainties. Furthermore, it can be noted from Table 3.6 that the proposed stabilizing controller while solved via NLMI framework provides significant reduction in number of iterations to achieve the same delay bound estimate

Table 3.5 d_u and K for Example 3.3 for different μ using LMI framework

Methods	d_u	$K\ matrix$	Remarks
		$\mu = 0$	
[25]	0.3015	–	'α' is not used as parameter
[6]	0.2716	$[-8.701 \times 10^{-6}, -1.009]$	'α' is not used as parameter
[10]	0.5865	$[-0.3155, -4.4417]$	'α' is not used as parameter
[14]	0.5500	$[-0.0229, -52.8656]$	'α' is not used as parameter
[19]	0.671	$[-8.3397, -11.3527]$	'α' is not used as parameter
[20]	0.84†	$[-34.72, -18.41]$	'α' is not used as parameter
Cor. 3.5	0.7226	$[-66.1514, -45.1055]$	$\alpha = 0.7191$
	0.7229	$[-93.6270, 63.0454]$	$\alpha = 0.7138$

† The result appears to be erroneous owing to fact that, the reduced order LMI of (30) in [20] obtained by following the steps of derivation and Remark 3 for $d_l = 0$ and hence may not be treated as a basis for comparing the results.

Methods	d_u	$K\ matrix$	Remark
		$\mu = 0.5$	
[10]	0.4960	$[-0.34, -5.168]$	'α' is not used as parameter
[14]	0.489	$[-0.2884, -13.8558]$	'α' is not used as parameter
Cor. 3.4	0.7	$[-105.4271, -69.6643]$	$\alpha = 0.6225$
	0.703	$[-182.8925, -117.3064]$	$\alpha = 0.5920$

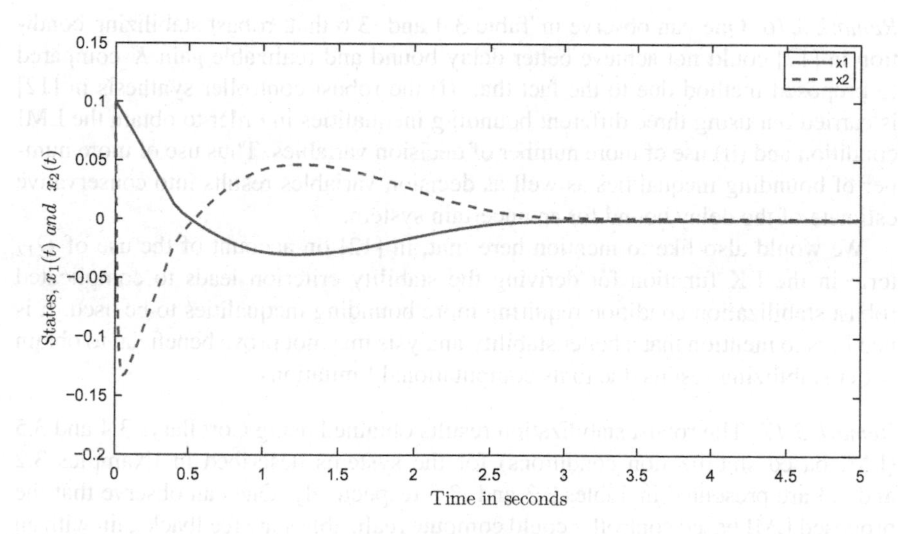

Fig. 3.10 Closed-loop simulation of system in Example 3.3

Table 3.6 d_u and K for Example 3.3 for different μ using NLMI framework

Methods	d_u	$K\ matrix$	Iterations	Decision variables
		$\mu = 0$		
[9]	0.4500	$[-4.8122, -7.7129]$	99	
Cor. 6 [12]	0.6300	$[-1.5829, -4.1376]$	19	$9 \times n^2 + (4+m) \times n$
	0.6900	$[-23.2572, -26.1488]$	192	
[26]	0.7226	$[-1850, 1256]$	–	
Cor. 3.3	0.7	$[-0.4606, -1.6763]$	15	$3.5 \times n^2 + (2.5+m) \times n$
	0.8	$[-35.312, -36.7487]$	208	
Methods	d_u	$K\ matrix$	Iterations	Decision variables
		$\mu = 0.5$		
[26]	0.694	$[-137, -99.4]$	–	
Cor. 5 [12]	0.5500	$[-1.1095, -3.1773]$	54	$9.5 \times n^2 + (4.5+m) \times n$
	0.6000	$[-9.5735, -2.9742]$	106	
Cor. 3.3	0.6	$[-0.7192, -2.1450]$	20	$3.5 \times n^2 + (2.5+m) \times n$
	0.7	$[-8.8698, -11.8340]$	80	

and thus it is computationally more attractive than the existing methods due to the number of decision variables involved in the derivation is less.

Remark 3.16 One can observe in Table 3.4 and 3.6 that, robust stabilizing condition in [12] could not achieve better delay bound and realizable gain K compared to proposed method due to the fact that, (i) the robust controller synthesis in [12] is carried out using three different bounding inequalities in order to obtain the LMI condition and (ii) use of more number of decision variables. Thus use of more number of bounding inequalities as well as decision variables results into conservative estimate of the delay bound for an uncertain system.

We would also like to mention here that, in [12] on account of the use of Q_{12} term in the LK function for deriving the stability criterion leads to complicated robust stabilization condition requiring more bounding inequalities to be used. It is needless to mention that a better stability analysis may not prove beneficial to obtain better stabilizing results due to its computational limitations.

Remark 3.17 The robust stabilization results obtained using Corollarys 3.4 and 3.5 (LMI based stabilization conditions) for the systems described in Examples 3.2 and 3.3 are presented in Tables 3.3 and 3.5 respectively. One can observe that the proposed LMI based controller could compute realizable state feedback gain with an enhanced delay upper bound compared to the existing results which are mainly due to (i) solving the proposed LMI conditions along with gain minimization routine [27], and (ii) use of more free matrices to express the relationship between various state vectors.

3.7 State Feedback H_∞ Control of TDS

While designing controller, the primary objective is to construct systems with guaranteed cost performance measure. A popular performance measure of a stable linear time-invariant system is the H_∞ norm of its transfer function. Reliable control problems for time-delay systems using LMI technique. The H_∞ control theory has gained significant advances over past few decades [28, 29] and references cited therein.

The problem of H_∞ control of linear delay system with delayed state feedback has become the focus of research just over a decade and has been investigated in delay-independent framework [30–32] and [33].

Some of the delay-dependent H_∞ control of time-delay system with or without parametric uncertainties can be found in [14, 18, 21, 34–36], and [22]. The work in [14, 18, 36], and [22] and references cited therein are all based on descriptor method, while the methods in [21, 34], and [35] are all based on first model transformation and expected to give conservative results due to the presence of additional dynamics in the model transformation discussed earlier in Chap. 2.

In this chapter, a state feedback H_∞ control of a nominal time-delay system is dealt in presence of disturbance input.

3.7.1 Problem Statement

Consider a general class of linear time-delay systems described by the following state equations

$$\dot{x}(t) = Ax(t) + A_d x(t - d(t)) + Dw(t) + Bu(t)$$

$$x(t) = \phi(t), \quad \forall t \in [-d_u, 0], \tag{3.93}$$

$$z(t) = Cx(t) + Fu(t) \tag{3.94}$$

where, $x(t) \in \mathcal{R}^n$ is the state vector, $u(t) \in \mathcal{R}^m$ is the control input, $w(t) \in \mathcal{R}^p$ is the exogenous disturbance which belongs to $\mathcal{L}_2\,[0, \infty]$, $z(t) \in \mathcal{R}^q$ is the regulated output, and $\phi(.)$ is the initial condition. The matrices A, A_d, B, D, C and F are known real constant matrices of appropriate dimensions. The delay $d(t)$ is time-varying and satisfies the condition (3.3).

The state feedback H_∞ control problem consists of two parts, (i) firstly, one develops the condition of H_∞ performance analysis for an unforced system (3.93) with $u(t) = 0$, such that the said system is stable with disturbance attenuation γ (where $\gamma > 0$ is a scalar) subject to zero initial condition and $\| z \|_2 < \gamma \| w \|_2$ for any non zero exogenous input $w(t)$, and (ii) secondly, the obtained condition in (i) is further extended to design of a H_∞ state feedback control law $u(t) = Kx(t)$ such that the resultant close-loop system is asymptotically stable and transfer function from w to z satisfy $\| T_{wz} \|_\infty < \gamma$. Specifically, the problem of H_∞ control can be defined as given below.

State Feedback H-Infinity Control *[21] For given scalars $d_u > 0$ and $\gamma > 0$, find a state feedback control law $u(t) = Kx(t)$ for the system (3.93)–(3.94) such that the resulting closed loop system is asymptotically stable and satisfying the disturbance attenuation γ for any time-delay $d(t)$ with $0 \leq d(t) \leq d_u$. In this case the system (3.93)–(3.94) is said to be stabilizable with the disturbance attenuation γ.*

In [18] and [14] it has been demonstrated that the choice of appropriate LK functional and selection of suitable bounding technique for approximating the cross terms arising out of the LK functional derivatives are the two important factors for deriving less conservative delay bound as well as for achieving the H_∞ performance conditions.

3.8 Stabilization of LFC Problem for Time-Delay Power System Based on H_∞ Approach

Load frequency control (LFC) is of importance in electric power system operation to damp frequency and voltage oscillations originated from load variations of real and reactive powers. Many control strategies, e.g., Proportional-Integral (PI) control [37–39], control using state feedback [33, 40–43], variable structure control [44], adaptive control [45], have been investigated to obtain a suitable LFC strategy. In view of the structure of existing power system model used for LFC [33, 38, 40–42, 44, 45] and [43], the area control error (ACE) is used as a control input to suppress the frequency deviation automatically. In general, the ACE signals are sought through high speed communication channel and may involve negligible communication delay. In [38, 42, 46–48], the need for open communication network has been highlighted that may cause a significant amount of communication delay present in the ACE signal. However, to the best of our knowledge, there are only few literature which investigate the LFC design problem considering communication delay in the ACE signal since such a design has only been considered in [48]. In this section, the design of LFC problem based on H_∞ controller with communication delays in the ACEs is considered.

The control structures suggested for stability analysis of interconnected power system are, (i) Decentralized control (ii) Quasi-Decentralized control (iii) Centralized Control and (iv) Hierarchial or Multilevel control [41]. Literature reveals that, the load-frequency control (LFC) of an inter-connected power system with or without time-delays are based on decentralized control structures [38, 40, 43, 48–50] and references cited therein, which is a network of local controller receiving local signals at each sub-system and sends control signals to the same subsystem, whereas a centralized control consists of one controller that uses all systems outputs to generate each system input, example of one such control structure in power system is wide-area-measurement-system (WAMS) centralized control [41] and references cited therein. So, far no works have been reported on application of delay-dependent H_∞ state feedback control using centralized LFC control structure for an interconnected time-delay power system in an LMI frame work, but the delay-independent

H_∞ state feedback control formulation for a linear interconnected time-delay power system can be found in [48] for LFC problem and in [42] for power system stabilizer (PSS) control problem. In this thesis an attempt has been made to design a delay-dependent H_∞ state feedback controller using centralized LFC control structure such that the interconnected power system asymptotically stable and satisfying norm from exogenous input $w(t)$ to regulated output $z(t)$.

The state-feedback controllers used for LFC may be classified, on the basis of whether delayed state (in addition to the present state) information have been used or not in implementing two types of controllers—(a) one-term controller (no delayed state) and (b) two-term controller (with delayed state) [33, 42, 48]. Note that, the latter one may yields better performance due to the use of past state information. However, existing designs of such two-term controllers [33, 42, 48] consider only delay-independent design technique. Such designs consider the delay at infinity as a special case of it and yields conservative results. Clearly, the controller performance may be improved if one considers these delayed states belong to only the recent past times. This can be reflected in the design if the delay is considered to be limited and correspondingly delay-dependent design approach is made. Design of a two-term state feedback LFC for an interconnected power system having two areas is considered in this thesis. The system model under consideration takes care of the time-delays in the ACE signals as state delays. With the proper selection of Lyapunov-Krasovskii functional and use of tighter bounding inequality constraints, the system under consideration is asymptotically stable while two-term controller is designed with and without delayed state information via LMI framework with a view to achieve closed-loop system performance requirements γ.

Let $\gamma > 0$ be a given constant, then the system (3.1)–(3.2) is said to be with H_∞ performance index no larger than γ if, the system (3.1) is asymptotically stable subject to $x(0) = 0$, then the transfer function matrix satisfies,

$$\| T_{wz}(j\omega) \|_\infty \leq \gamma, \ \forall\omega \tag{3.95}$$

$$\| T_{wz}(j\omega) \|_\infty = \frac{\| z \|_2}{\| w \|_2} = \frac{\sqrt{\int_0^\infty z^T(t)z(t)dt}}{\sqrt{\int_0^\infty w^T(t)w(t)dt}} \leq \gamma \tag{3.96}$$

Equation (3.96) is equivalent to $\int_0^\infty z^T(t)z(t)dt \ \leq \ \gamma^2 \ \int_0^\infty w^T(t)w(t)dt, \ \ \forall\omega$.

Remark 3.18 Note that, γ is a load disturbance rejection measure of the controller. Clearly, the system performance is better as γ is smaller and this indicates better the disturbance rejection. Therefore, for obtaining an optimal H_∞ controller one attempts to minimize the γ in order to have minimal effect of the load variation in the system performance.

Table 3.7 Notations used in LFC model with their meanings

Sl. No.	Notations	Meanings
1	ΔP_{vi}	Generator Valve position
2	ΔP_{mi}	Mechanical power output of the generator
3	Δf_i	Frequency deviations
4	ΔE_i	ACE Signals
5	ΔP_{12}	Tie-line power flow from area 1 to area 2
6	B_i	Proportional gain of local PI controller
7	ΔP_{di}	Load disturbances
8	M_i	Moment of inertia of the generator
9	D_i	Generator Damping coefficient
10	T_{gi}	Generator time-constant
11	T_{chi}	Turbine time-constant
12	R_i	Speed droop
13	T_i	Stiffness coefficient
14	k_i	Integral gain of local PI controller
15	T_{pi}	Power system time-constant

In next following two sections, mathematical model of an interconnected time delayed LFC system is considered. Next, a two-term controller design criterion is proposed for the solution of LFC problem with the inequality 3.95 describes restraint disturbance ability.

3.8.1 Load-Frequency Control (LFC) of Power Systems with Communication Delay

A two-area interconnected power system model with communication delays is shown in Fig. 3.11, both the areas are identical in structure but having different generation capacities. The notations used for the ith area, $(i = 1, 2)$, are given in the following.

Further, d_i represents communication delays present in the ith-area that arises in the ACEs due to the time taken in measuring frequency and tie line power flow from remote terminal units (RTUs) to local control center. Note that, the local controller is a PI controller that is embedded in the system as an integral part of the model.

The dynamics of the two-area interconnected LFC model with communication delay is shown in Fig. 3.11 and it may be described in state space form (for $i,\ j = 1, 2$ and $i \neq j$)

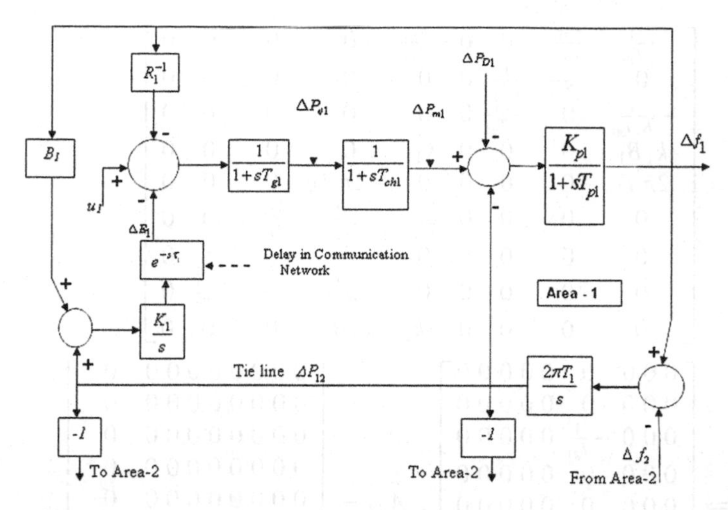

Fig. 3.11 Two-area LFC system

$$\Delta\dot{P}_{mi}(t) = \frac{\Delta P_{vi}(t)}{T_chi} - \frac{\Delta P_{mi}(t)}{T_chi} \tag{3.97}$$

$$\Delta\dot{E}_i(t) = +k_i\Delta P_{ij}(t) + k_iB_i\Delta f_i(t) \tag{3.98}$$

$$\Delta\dot{P}_{ij}(t) = 2\pi T_i\Delta f_i(t) - 2\pi T_i\Delta f_j(t) \tag{3.99}$$

$$\Delta\dot{P}_{vi}(t) = -\frac{\Delta f_i(t)}{R_iT_{gi}} - \frac{\Delta P_{vi}(t)}{T_{gi}} - \frac{\Delta E_i(t-d_i)}{T_{gi}} + \frac{u_i(t)}{T_{gi}} \tag{3.100}$$

$$\Delta\dot{f}_i(t) = -\frac{K_{pi}\Delta P_{di}(t)}{T_{pi}} - \frac{K_{pi}\Delta P_{ij}(t)}{T_{pi}} + \frac{K_{pi}\Delta P_{mi}(t)}{T_{pi}} - \frac{\Delta f_i(t)}{T_{pi}} \tag{3.101}$$

where, $\Delta P_{ij} = -\Delta Pji$. Now, defining a state vector as,

$$\left[\Delta f_1\ \Delta P_{m1}\ \Delta P_{v1}\ \Delta E_1\ \Delta P_{12}\ \Delta f_2\ \Delta P_{m2}\ \Delta P_{v2}\ \Delta E_2\right]^T$$

The above equations (3.97)–(3.101) may be represented in a compact form as:

$$\dot{x}(t) = Ax(t) + \sum_{i=1}^{2} A_{di}x(t-d_i(t)) + Bu(t) + Dw(t) \tag{3.102}$$

$$z(t) = Cx(t) \tag{3.103}$$

where, $d_i(t)$ is a time-varying delay in the model, but in the LFC model it is assumed to be of constant nature and $w(t) = \Delta P_d(t) = [\Delta P_{d1},\ \Delta P_{d2}]^T$ is a load disturbance vector and the constant matrices associated with the (3.102) and (3.103) are given below:

$$A = \begin{bmatrix} -\frac{1}{T_{p1}} & \frac{k_{p1}}{T_{p1}} & 0 & 0 & -\frac{k_{p1}}{T_{p1}} & 0 & 0 & 0 & 0 \\ 0 & -\frac{1}{T_{ch1}} & \frac{1}{T_{ch1}} & 0 & 0 & 0 & 0 & 0 & 0 \\ -\frac{1}{R_1 T_{g1}} & 0 & -\frac{1}{T_{g1}} & 0 & 0 & 0 & 0 & 0 & 0 \\ k_1 B_1 & 0 & 0 & 0 & k_1 & 0 & 0 & 0 & 0 \\ 2\pi T_1 & 0 & 0 & 0 & 0 & -2\pi T_1 & 0 & 0 & 0 \\ 0 & 0 & 0 & 0 & \frac{k_{p2}}{T_{p2}} & -\frac{1}{T_{p2}} & \frac{k_{p2}}{T_{p2}} & 0 & 0 \\ 0 & 0 & 0 & 0 & 0 & 0 & -\frac{1}{T_{ch2}} & \frac{1}{T_{ch2}} & 0 \\ 0 & 0 & 0 & 0 & 0 & -\frac{1}{R_2 T_{g2}} & 0 & -\frac{1}{T_{g2}} & 0 \\ 0 & 0 & 0 & 0 & -k_2 & k_2 B_2 & 0 & 0 & 0 \end{bmatrix}$$

$$A_{d1} = \begin{bmatrix} 0 & 0 & 0 & 0 & 0 & 0 & 0 & 0 & 0 \\ 0 & 0 & 0 & 0 & 0 & 0 & 0 & 0 & 0 \\ 0 & 0 & 0 & -\frac{1}{T_{g1}} & 0 & 0 & 0 & 0 & 0 \\ 0 & 0 & 0 & 0 & 0 & 0 & 0 & 0 & 0 \\ 0 & 0 & 0 & 0 & 0 & 0 & 0 & 0 & 0 \\ 0 & 0 & 0 & 0 & 0 & 0 & 0 & 0 & 0 \\ 0 & 0 & 0 & 0 & 0 & 0 & 0 & 0 & 0 \\ 0 & 0 & 0 & 0 & 0 & 0 & 0 & 0 & 0 \\ 0 & 0 & 0 & 0 & 0 & 0 & 0 & 0 & 0 \end{bmatrix}, \quad A_{d2} = \begin{bmatrix} 0 & 0 & 0 & 0 & 0 & 0 & 0 & 0 & 0 \\ 0 & 0 & 0 & 0 & 0 & 0 & 0 & 0 & 0 \\ 0 & 0 & 0 & 0 & 0 & 0 & 0 & 0 & 0 \\ 0 & 0 & 0 & 0 & 0 & 0 & 0 & 0 & 0 \\ 0 & 0 & 0 & 0 & 0 & 0 & 0 & 0 & 0 \\ 0 & 0 & 0 & 0 & 0 & 0 & 0 & 0 & 0 \\ 0 & 0 & 0 & 0 & 0 & 0 & 0 & 0 & 0 \\ 0 & 0 & 0 & 0 & 0 & 0 & 0 & 0 & -\frac{1}{T_{g2}} \\ 0 & 0 & 0 & 0 & 0 & 0 & 0 & 0 & 0 \end{bmatrix}$$

$$B = \begin{bmatrix} 0 & 0 & \frac{1}{T_{g1}} & 0 & 0 & 0 & 0 & 0 & 0 \\ 0 & 0 & 0 & 0 & 0 & 0 & 0 & \frac{1}{T_{g2}} & 0 \end{bmatrix}^T, \quad D = \begin{bmatrix} -\frac{k_{p1}}{T_{p1}} & 0 & 0 & 0 & 0 & 0 & 0 & 0 & 0 \\ 0 & 0 & 0 & 0 & 0 & -\frac{k_{p2}}{T_{p2}} & 0 & 0 & 0 \end{bmatrix}^T$$

The objective of the control problem for system (3.102) and (3.103) is to design a suitable control law $u(t)$ such that, the closed-loop system exhibits good load-disturbance rejection property in the sense that it attains certain H_∞ performance, which will be discussed in succeeding sections on load-frequency controller synthesis.

The detailed discussion on need for evolution of LFC model under dergulated power market scenario for interconnected power system involving communication delay can be found in [51].

3.8.2 Existing H_∞ Control Design For LFC Model

In this section, the H_∞ control design technique (one-term and two-term controller) of [33] for the solution of LFC problem (subsection (3.8.1)) is presented briefly. It must be mentioned here that, the H_∞ control strategy of [33] has been applied in [48] and [42] that are designed based on delay-independent stability analysis approach. The system model in [33] as well as in [42] and [48] considers a single state time-delay, and it is presented as:

$$\dot{x}(t) = Ax(t) + A_d x(t-d) + Bu(t) + Dw(t) \tag{3.104}$$
$$y(t) = Cx(t) \tag{3.105}$$

Two types of control strategies are discussed in [42] for the solutions of PSS and LFC problems. The structure of both one-term and two-term controllers are described as
One-term:

$$u(t) = Kx(t) \tag{3.106}$$

Two-term:

$$u(t) = Kx(t) + K_d x(t-d) \tag{3.107}$$

for the solution of H_∞ LFC problem satisfying the H_∞ performance index no larger than 'γ'.

3.8.2.1 One-Term H_∞ Control

Theorem 3.11 *[33, 42] and [48] The system described in (3.104) with the control law (3.106) is asymptotically stable and* $\| T_{wy} \| \leq \gamma, \gamma > 0$ *for any time-delay d, if there exist matrices,* $Y = Y^T > 0,\ \bar{Q}_1 = \bar{Q}_1^T > 0$*, such that the following LMI is satisfied provided* (A, B) *is stabilizable,*

$$W = \begin{bmatrix} W_{11} & YC^T & A_d Y & D \\ \star & -I & 0 & 0 \\ \star & 0 & -\bar{Q}_1 & 0 \\ \star & 0 & 0 & -\gamma^2 I \end{bmatrix} < 0 \tag{3.108}$$

where, $W_{11} = AY + Y^T A^T + BS + S^T B^T + \bar{Q}_1$. The state feedback controller law is given by $u(t) = Kx(t)$ with $K = SY^{-1}$.

3.8.2.2 Two-Term H_∞ Control

Theorem 3.12 *[33, 42] and [48] The system described in (3.104) with the control law (3.107) is asymptotically stable and* $\| T_{wy} \| \leq \gamma, \gamma > 0$ *for any time-delay d, if there exist matrices,* $Y = Y^T > 0,\ \bar{Q}_1 = \bar{Q}_1^T > 0$*, positive scalars* σ *and* κ*, such that the following LMI is satisfied provided* (A, B) *is stabilizable,*

$$W = \begin{bmatrix} W_{11} & \sigma A_d Y & \kappa B V & Y C^T & D \\ \star & -\bar{Q}_1 & 0 & 0 & 0 \\ \star & 0 & -\bar{Q}_1 & 0 & 0 \\ \star & 0 & 0 & -I & 0 \\ \star & 0 & 0 & 0 & -\gamma^2 I \end{bmatrix} < 0 \qquad (3.109)$$

where, $W_{11} = AY + Y^T A^T + BS + S^T B^T + \bar{Q}_1$. The state feedback control law is given by $u(t) = Kx(t) + K_d x(t-d)$ with $K = SY^{-1}$, and $K_d = VY^{-1}$.

3.9 Main Results on H_∞ Based LFC of an Interconnected Time-Delay Power System

In this section, the existing delay-independent one-term as well as two-term control design techniques discussed above are extended for an interconnected power system LFC problem having multiple delays (see equations (3.102)–(3.103)). An improved feedback delay-dependent H_∞ two-term controller is proposed for LFC of an interconnected power systems with the constraint on H_∞ performance index 'γ'.

3.9.1 *One-Term H_∞ Control*

As pointed out earlier that, the design of one-term controller in [33, 42] and [48] can be applied to a single time-delay systems (or equivalently to one-area LFC model). An extension of the result of [42] to LFC problem of an interconnected two-area power system model (3.102)–(3.103) is presented in the form of following lemma.

Lemma 3.1 *The system described in (3.102)–(3.103) with the control law (3.106) is asymptotically stable and* $\| T_{wy} \| \leq \gamma, \gamma > 0$ *for any time-delay d, if there exist matrices,* $Y = Y^T > 0,\ \bar{Q}_i = \bar{Q}_i^T > 0, (i = 1, 2)$, *such that the following LMI is satisfied provided* (A, B) *is stabilizable*

$$W = \begin{bmatrix} W_{11} & YC^T & A_{d1}Y & A_{d2}Y & D \\ \star & -I & 0 & 0 & 0 \\ \star & 0 & -\bar{Q}_1 & 0 & 0 \\ \star & 0 & 0 & -\bar{Q}_2 & 0 \\ \star & 0 & 0 & 0 & -\gamma^2 I \end{bmatrix} < 0 \qquad (3.110)$$

where, $W_{11} = AY + Y^T A^T + BS + S^T B^T + \sum_{i=1}^{2} \bar{Q}_i$.

The state feedback control law is given by $u(t) = Kx(t)$ with $K = SY^{-1}$.

Proof The proof of Lemma 3.1 is straightforward following [42] and hence omitted.

The result of LMI condition for one-term control in (3.110) can be further extended for an n-interconnected power systems as,

$$M = \begin{bmatrix} M_{11} & YC^T & A_{d1}Y & A_{d2}Y & \dots & A_{dn}Y & D \\ \star & -I & 0 & 0 & \dots & 0 & 0 \\ \star & 0 & -\bar{Q}_1 & 0 & \dots & 0 & 0 \\ \star & 0 & 0 & -\bar{Q}_2 & \dots & 0 & 0 \\ \vdots & \vdots & \vdots & \vdots & \vdots & \vdots & \vdots \\ \star & 0 & 0 & 0 & \dots & -\bar{Q}_n & 0 \\ \star & 0 & 0 & 0 & \dots & 0 & -\gamma^2 I \end{bmatrix} < 0 \qquad (3.111)$$

where, $M_{11} = AY + Y^T A^T + BS + S^T B^T + \sum_{i=1}^{n} \bar{Q}_i$

3.9.2 *Two-Term H_∞ Control*

A two-term H_∞ controller design using delay-independent analysis for power system stabilizer (P.S.S) model has been considered in [42]. In this section, the result of [42] is adopted for solution of time-delay LFC control problem (3.102) and its delay-independent stabilization is presented below in the form of lemma. However in the design stage, similar to [42] one can consider the d in (3.107) as $d \in max[d_1, d_2]$. In the present synthesis, assume $d = d_2$ so the closed loop system (3.102) with the control law (3.107) becomes

$$\dot{x}(t) = A_c x(t) + A_{d1} x(t - d_1) + A_{d2c} x(t - d_2) + Dw(t) \qquad (3.112)$$

$$y(t) = Cx(t) \qquad (3.113)$$

where, $A_c = A + BK$ and $A_{d2c} = A_{d2} + BK_d$.

Lemma 3.2 *System (3.112) with the controller (3.107) and assumption $d = d_2$ satisfies the H_∞ performance defined in (3.95), if there exist positive definite symmetric matrices Y, $\bar{Q}_1$, $\bar{Q}_2$ and any matrices S, V, positive scalars σ and κ, such that the following LMI holds:*

$$\Theta = \begin{bmatrix} \Theta_{11} & \sigma A_{d1}Y & \sigma A_{d2}Y & \kappa BV & YC^T & D \\ \star & -\bar{Q}_1 & 0 & 0 & 0 & 0 \\ \star & 0 & -\bar{Q}_2 & 0 & 0 & 0 \\ \star & 0 & 0 & -I & 0 & 0 \\ \star & 0 & 0 & 0 & -I & 0 \\ \star & 0 & 0 & 0 & 0 & \gamma^2 I \end{bmatrix} < 0 \qquad (3.114)$$

where, $\Theta_{11} = AY + Y^T A^T + BS + S^T B^T + \bar{Q}_1 + \bar{Q}_2$. The corresponding H_∞ two-term controller gains may then be obtained as $K = SY^{-1}$ and $K_d = VY^{-1}$.

Remark 3.19 Note that, if $d_2 \to \infty$ then feedback delay 'd' also tends to infinity and this situation is equivalent to delay-independent one-term controller design. Hence, at this limiting situation the use of delayed states in feedback term is insignificant. This fact has been observed by solving two-area LFC problem using LMI (3.114).

The proposed feedback delay-dependent H_∞ two-term control strategy in the form of following theorem is presented, where the two-term control law (3.107) in the present situation is modified to

$$u(t) = Kx(t) + K_\tau x(t-\tau) \tag{3.115}$$

where, τ is the delay in feedback signal and its upper bound is unknown.

Remark 3.20 Here, τ is a feedback delay involved in the control law, which is not equal to that of the state delay d_i $(i = 1, 2)$ available in the system model. In all the existing H_∞ control formulations irrespective of delay-independent analysis [33, 42, 48] or delay-dependent [14, 18, 22] and [21], the delay information used in the LK functional is the state delay of the system model, whereas in the present synthesis, the designed controller uses a delayed state information 'τ' in the LK functional corresponding to the delay-dependent term. The modification of the control law (3.115) leads to the choice of a new delay-dependent LK functional that avoids the demerit of the limiting situation (as mentioned in Remark (3.19), thus in practice, it is suitable for the solution of LFC problem. To the best of the present author knowledge, the stabilization of two-area LFC problem satisfying $H-\infty$ performance bound based on proposed delay-dependent control strategy in LMI framework has not been reported so far in literature.

Theorem 3.13 *The system (3.102)–(3.103) with the controller (3.115) is asymptotically stable and satisfies the H_∞ performance (3.95), if there exist positive definite matrices X, $\bar{Q}_i$ $(i = 1, 2..4)$ and matrices S, Y, $\bar{T}_i (i = 1, 2)$, V and positive scalars γ, τ such that the following LMI holds:*

$$\Lambda = \begin{bmatrix} \Lambda_{11} & A_{d1}Y^T & A_{d2}Y^T & \Lambda_{14} & \Lambda_{15} & \bar{T}_1 & D & YC^T \\ \star & -\bar{Q}_1 & 0 & YA_{d1}^T & YA_{d1}^T & 0 & 0 & 0 \\ \star & \star & -\bar{Q}_2 & YA_{d2}^T & YA_{d2}^T & 0 & 0 & 0 \\ \star & \star & \star & \Lambda_{44} & -Y^T + V^T B^T & \bar{T}_2 & D & 0 \\ \star & \star & \star & \star & -Y^T - Y + d\bar{Q}_4 & 0 & D & 0 \\ \star & \star & \star & \star & \star & -d^{-1}\bar{Q}_4 & 0 & 0 \\ \star & \star & \star & \star & \star & 0 & -\gamma^2 I & 0 \\ \star & \star & \star & \star & \star & 0 & 0 & -I \end{bmatrix} < 0 \tag{3.116}$$

where, $\Lambda_{11} = YA^T + AY + BS + S^T B^T + \bar{Q}_1 + \bar{Q}_2 + \bar{Q}_3 + \bar{T}_1 + \bar{T}_1^T$,

$\Lambda_{14} = BV + YA^T + BS + S^T B^T - \bar{T}_1 + \bar{T}_2^T$, $\Lambda_{15} = -Y^T + X + YA^T + S^T B^T$, and

$\Lambda_{44} = BV + V^T B^T - \bar{T}_2 - \bar{T}_2^T - \bar{Q}_3$.

Proof The closed-loop system with the implementable control law (3.115) is expressed as:

$$\dot{x}(t) = A_c x(t) + B_\tau x(t-\tau) + A_{d1}x(t-d_1) + A_{d2}x(t-d_2) + Dw(t)$$
$$y(t) = Cx(t) \tag{3.117}$$

where, $A_c = A + BK$ and $B_\tau = BK_\tau$.

It is assumed that the pair $(A,\ B)$ is stabilizable. As the design of H_∞ controller is delay-dependent with respect to feedback delay 'τ' one needs to choose appropriately LK functional of the form:

$$V(t) = x^T Px(t) + \int_{t-d_1}^{t} x^T(s)Q_1x(s)ds + \int_{t-d_2}^{t} x^T(s)Q_2x(s)ds + \int_{t-\tau}^{t} x^T(s)Q_3x(s)ds$$
$$+ \int_{-\tau}^{0}\int_{t+\alpha}^{t} \dot{x}(s)Q_4\dot{x}(s)dsd\alpha \tag{3.118}$$

Finding the time-derivative of (3.118) one can get,

$$\dot{V}(t) = x^T(t)(Q_1+Q_2+Q_3)x(t) - x^T(t-d_1)Q_1x(t-d_1) + x^T(t-d_2)Q_2x(t-d_2)$$
$$-x^T(t-\tau)Q_3x(t-\tau) + \tau\dot{x}^T(t)Q_4\dot{x}(t) - \int_{t-\tau}^{t}\dot{x}^T(s)Q_4\dot{x}(s)ds + 2x^T(t)Px(t) \tag{3.119}$$

Now, to approximate the quadratic integral term $-\int_{t-\tau}^{t}\dot{x}^T(s)Q_4\dot{x}(s)ds$ in (3.119) one can use Lemma 2 of [23] which yields

$$-\int_{t-\tau}^{t}\dot{x}^T(s)Q_4\dot{x}(s)ds \leq \begin{bmatrix} x(t) \\ x(t-\tau)\end{bmatrix}^T \left\{\begin{bmatrix} T_1+T_1^T & -T_1+T_2^T \\ \star & -T_2-T_2^T\end{bmatrix} + \right.$$
$$\left. \tau\begin{bmatrix} T_1 \\ T_2\end{bmatrix} Q_4^{-1}\begin{bmatrix} T_1 \\ T_2\end{bmatrix}^T\right\}\begin{bmatrix} x(t) \\ x(t-\tau)\end{bmatrix} \tag{3.120}$$

Now, for a free matrix G of appropriate dimension following equality is valid,

$$0 = 2[x^T(t)G + x^T(t-\tau)G + \dot{x}^T G]$$
$$\times[-\dot{x}(t) + A_c x(t) + B_\tau x(t-\tau) + A_{d1}x(t-d_1) + A_{d2}x(t-d_2) + Dw(t)] \tag{3.121}$$

On expansion of (3.121), one can get

$$\xi^T(t)\begin{bmatrix}\phi_{11} & GA_{d1} & GA_{d2} & \phi_{14} & \phi_{15}\\ \star & 0 & 0 & \phi_{24} & \phi_{24}\\ \star & 0 & 0 & \phi_{34} & \phi_{35}\\ \star & \star & \star & \phi_{44} & \phi_{45}\\ \star & \star & \star & \star & \phi_{55}\end{bmatrix}\xi(t)+2\xi^T(t)\begin{bmatrix}G\\0\\0\\G\\G\end{bmatrix}Dw(t)=0 \quad (3.122)$$

where, $\xi(t)=[x^T(t),\ x^T(t-d_1),\ x^T(t-d_2),\ x^T(t-\tau),\ \dot{x}^T(t)]^T$ and

$\phi_{11}=GA_c+A_c^TG^T, \phi_{14}=GB_\tau+A_c^TG^T, \phi_{15}=-G+A_c^TG^T,$

$\phi_{24}=A_{d1}^TG^T, \phi_{25}=A_{d1}^TG^T, \phi_{34}=A_{d2}^TG^T, \phi_{35}=A_{d2}^TG^T,$

$\phi_{44}=GB_\tau+B_\tau^TG^T, \phi_{45}=-G+B_\tau^TG^T, \phi_{55}=-G-G^T$

Using the bounding Lemma 2.1, one can treat the cross term $2\xi^T(t)\left[\,G^T\ 0\ 0\ G^T\ G^T\,\right]^T Dw(t)$ in (3.122) and rewrite it as

$$\xi^T(t)\left\{\begin{bmatrix}\phi_{11} & GA_{d1} & GA_{d2} & \phi_{14} & \phi_{15}\\ \star & 0 & 0 & \phi_{24} & \phi_{24}\\ \star & 0 & 0 & \phi_{34} & \phi_{35}\\ \star & \star & \star & \phi_{44} & \phi_{45}\\ \star & \star & \star & \star & \phi_{55}\end{bmatrix}+\begin{bmatrix}GD\\0\\0\\GD\\GD\end{bmatrix}\gamma^{-2}\begin{bmatrix}GD\\0\\0\\GD\\GD\end{bmatrix}^T\right\}\xi(t)+\gamma^2w^T(t)w(t)=0 \quad (3.123)$$

where, γ is any positive scalar quantity. Invoking (3.120) and (3.123) in (3.119), one can obtain

$$\dot{V}(t)\le\xi^T(t)\left\{\begin{bmatrix}\psi_{11} & GA_{d1} & GA_{d2} & \psi_{14} & \psi_{15}\\ \star & 0 & 0 & \psi_{24} & \psi_{24}\\ \star & 0 & 0 & \psi_{34} & \psi_{35}\\ \star & \star & \star & \psi_{44} & \psi_{45}\\ \star & \star & \star & \star & \psi_{55}\end{bmatrix}+\begin{bmatrix}T_1\\0\\0\\T_2\\0\end{bmatrix}\tau Q_4^{-1}\begin{bmatrix}T_1\\0\\0\\T_2\\0\end{bmatrix}^T\right.$$

$$\left.+\begin{bmatrix}GD\\0\\0\\GD\\GD\end{bmatrix}\gamma^{-2}\begin{bmatrix}GD\\0\\0\\GD\\GD\end{bmatrix}^T\right\}\xi(t)+\gamma^2w^T(t)w(t) \quad (3.124)$$

where, $\psi_{11}=\phi_{11}+Q_1+Q_2+Q_3+T_1+T_1^T, \psi_{14}=\phi_{14}-T_1+T_2^T,$
$\psi_{15}=\phi_{15}+P,$
$\psi_{24}=\phi_{24}, \psi_{25}=\phi_{25}, \psi_{34}=\phi_{34}, \psi_{35}=\phi_{35},$
$\psi_{44}=\phi_{44}-Q_3-T_2-T_2^T, \psi_{45}=\phi_{45}, \psi_{55}=\phi_{55}+\tau Q_4$

Using Schur-complement one can easily rewrite (3.124) as,

$$\dot{V}(t) \leq \xi^T(t)\Psi\xi(t) + \gamma^2 w^T(t)w(t) \tag{3.125}$$

$$\text{where, } \Psi = \begin{bmatrix} \psi_{11} & GA_{d1} & GA_{d2} & \psi_{14} & \psi_{15} & T_1 & GD \\ \star & 0 & 0 & \psi_{24} & \psi_{24} & 0 & 0 \\ \star & 0 & 0 & \psi_{34} & \psi_{35} & 0 & 0 \\ \star & \star & \star & \psi_{44} & \psi_{45} & T_2 & GD \\ \star & \star & \star & \star & \psi_{55} & 0 & GD \\ \star & \star & \star & \star & \star & -\tau^{-1}Q_4 & 0 \\ \star & \star & \star & \star & \star & 0 & -\gamma^2 I \end{bmatrix}$$

In (3.125), if $\dot{V}(t)$ is negative definite then it is guaranteed that the system considered in (3.102) is stabilizable with the control law (3.115). In case of H_∞ state feedback control with $x(0) = 0$, and additional constraint $\| z(t) \|_2 \leq \gamma \| w(t) \|_2$, $\gamma > 0$ that describes the restraint disturbance ability must be included in the delay-dependent stability condition (3.125). One can rewrite (3.125) as,

$$\dot{V}(t) + z^T(t)z(t) - \gamma^2 w^T(t)w(t) \leq \xi^T(t)\Psi\xi(t) + z^T(t)z(t) - \gamma^2 w^T(t)w(t) \tag{3.126}$$

Integrate (3.126) from 0 to ∞ on both the sides, it follows then

$$\int_0^\infty \{z^T(t)z(t) - \gamma^2 w^T(t)w(t) + \dot{V}(t)\}dt \leq \int_0^\infty \{\xi^T(t)\Psi\xi(t) + z^T(t)z(t)\}dt \tag{3.127}$$

Substituting $z(t) = C\,x(t)$ in (3.127), one can rewrite

$$\int_0^\infty \{z^T(t)z(t) - \gamma^2 w^T(t)w(t) + \dot{V}(t)\}dt \leq \int_0^\infty \{x^T(t)C^T Cx(t) + \xi^T(t)\Psi\xi(t)\}dt \tag{3.128}$$

After simple algebraic manipulation of (3.128) one can rewrite it as

$$\int_0^\infty \{z^T(t)z(t) - \gamma^2 w^T(t)w(t) + \dot{V}(t)\}dt \leq \int_0^\infty \{\xi^T(t)\bar{\Omega}\xi(t)\}dt \tag{3.129}$$

If we define the H_∞ performance index as

$$J_{wz} = \int_0^\infty \{z^T(t)z(t) - \gamma^2 w^T(t)w(t)\}dt \tag{3.130}$$

thus in view of (3.130) one can write

$$J_{wz} \leq \int_0^{\infty} \{z^T(t)z(t) - \gamma^2 w^T(t)w(t) + \dot{V}(t)\}dt \tag{3.131}$$

Now, in view of (3.131) and (3.129) it is obvious that following is true,

$$J_{wz} \leq \int_0^{\infty} \{\xi^T(t)\bar{\Omega}\xi(t)\}dt \tag{3.132}$$

$$\bar{\Omega} = \begin{bmatrix} \psi_{11} + C^T C & GA_{d1} & GA_{d2} & \psi_{14} & \psi_{15} & T_1 & GD \\ \star & 0 & 0 & \psi_{24} & \psi_{24} & 0 & 0 \\ \star & 0 & 0 & \psi_{34} & \psi_{35} & 0 & 0 \\ \star & \star & \star & \psi_{44} & \psi_{45} & T_2 & GD \\ \star & \star & \star & \star & \psi_{55} & 0 & GD \\ \star & \star & \star & \star & \star & -\tau^{-1}Q_4 & 0 \\ \star & \star & \star & \star & \star & 0 & -\gamma^2 I \end{bmatrix} \tag{3.133}$$

Using Schur-complement one can rewrite (3.133) equivalently as,

$$\Omega = \begin{bmatrix} \psi_{11} & GA_{d1} & GA_{d2} & \psi_{14} & \psi_{15} & T_1 & GD & C^T \\ \star & 0 & 0 & \psi_{24} & \psi_{24} & 0 & 0 & 0 \\ \star & 0 & 0 & \psi_{34} & \psi_{35} & 0 & 0 & 0 \\ \star & \star & \star & \psi_{44} & \psi_{45} & T_2 & GD & 0 \\ \star & \star & \star & \star & \psi_{55} & 0 & GD & 0 \\ \star & \star & \star & \star & \star & -\tau^{-1}Q_4 & 0 & 0 \\ \star & \star & \star & \star & \star & 0 & -\gamma^2 I & 0 \\ \star & \star & \star & \star & \star & 0 & 0 & -I \end{bmatrix} \tag{3.134}$$

If $\Omega < 0$ in (3.134), then it guarantees both $\dot{V}(t) < 0$ as well as $J_{wz} < 0$ thus satisfying the condition of (3.96). Now, substitute $A_c = A + BK$ and $B_\tau = BK_\tau$ in (3.133) first, then pre- and post-multiplying (3.133) by $\{G^{-1}, G^{-1}, G^{-1}, G^{-1}, G^{-1}, G^{-1}, I, I\}$ and its transpose respectively and adopting following linear changes of variables,

$$G^{-1} = Y,\ G^{-T} = Y^T,\ G^{-1}K^T = YK^T = S^T,\ KG^{-T} = KY^T = S,$$
$$G^{-1}K_\tau^T = YK_\tau^T = V,$$

$$G^{-1}Q_1G^{-T} = \bar{Q}_1,\ G^{-1}Q_2G^{-T} = \bar{Q}_2,\ G^{-1}Q_3G^{-T} = \bar{Q}_3,\ G^{-1}Q_4G^{-T} = \bar{Q}_4,$$

$$G^{-1}PG_{-T} = X,\ G^{-1}T_1G^{-T} = \bar{T}_1,\ G^{-1}T_2G^{-T} = \bar{T}_2$$

After carrying out above linear changes of matrix variables in Ω matrix, one can obtain $\Lambda < 0$ in (3.116) which is the required stabilizing condition for the LFC system. This completes the proof. ■

Remark 3.21 Note that, one may obtain controller gains using $K = SY^{-1}$ and $K_\tau = VY^{-1}$ from feasible solution of (3.116) for a specified γ. However, by defining $\bar{\gamma} = \gamma^2$ and then obtaining a solution of (3.116) by minimizing $\bar{\gamma}$ yields an optimal controller in the sense that γ gets optimized. But such optimal controllers generally have high gains and significantly amplify noises causing performance degradation. However, these high control gains may be reduced if one attempts to obtain a sub-optimal controller by exploiting the trade-off between the control gains and the H_∞ performance index γ. An attempt is made to design such suboptimal controllers by minimizing the γ as well as restricting the size of the control gains K and K_τ simultaneously. Such an attempt is not new in literature, for example see [27], where suboptimal controllers have been obtained to avoid the problem that arises due to high control gains. For this purpose, note that, computing the control gains K and K_τ involves the LMI variables S, V and Y. In view of this, one can define the following multi-objective optimization algorithm for computing the controller gains and simultaneously the H_∞ performance index γ.

Multi-objective Optimization Algorithm:

$$\text{Minimize } \bar{\gamma} + p + s + v$$

$$subject\ to\ (3.117), \begin{bmatrix} sI & S \\ \star & I \end{bmatrix}, \begin{bmatrix} vI & V \\ \star & I \end{bmatrix}, \begin{bmatrix} Y & I \\ \star & pI \end{bmatrix} \geq 0,$$

$$\bar{\gamma} > 0,\ p > 0,\ s > 0,\ and\ v > 0$$

3.9.3 Simulation Results

To illustrate the effectiveness of the proposed LFC H_∞ control problem satisfying performance index 'γ', the following two-area power system model has been considered. The area-1 is equivalent to a single generator and area-2 is equivalent to 4-interconnected generator units as in ([48]). The plant parameters are given as follows,

Area -1 (Parameters are in p.u):

$T_{ch1} = 0.3\,sec,\ T_{g1} = 0.1\,sec,\ R_1 = 0.05,\ D_1 = 1,\ M_1 = 10,\ k_1 = 0.5$, and $\frac{1}{T_{p1}} = \frac{D_1}{M_1},\ T_{p1} = \frac{M_1}{D_1},\ \frac{k_{p1}}{T_{p1}} = \frac{1}{M_1},\ k_{p1} = \frac{1}{D_1},\ B_1 = \frac{1}{R_1} + D_1$

Area -2 (Parameters are in p.u):

$T_{ch2} = 0.17\,sec,\ T_{g2} = 0.4\,sec,\ R_2 = 0.05,\ D_2 = 1.5,\ M_2 = 12,\ k_2 = 0.5$, and

$T_{p2} = \frac{M_2}{D_2},\ \frac{k_{p2}}{T_{p2}} = \frac{1}{M_2},\ k_{p2} = \frac{1}{D_2},\ B_2 = \frac{4}{R_2} + D_2$

Open-loop simulation: Without control input (i.e, $u(t) = 0$) the system in (3.102) with $d_1 = 0.1\ sec$ and $d_2 = 0.6\ sec$ is simulated with constant load disturbances of 1 p.u in both the areas. It can be observed that the frequency deviations $\Delta f_1(t)$ and $\Delta f_2(t)$ of the system are unstable as shown in the Fig. (3.12). It must be mentioned here that the PI controllers are inherently involved in the respective areas of the system model (3.102), (see Fig. 3.11).

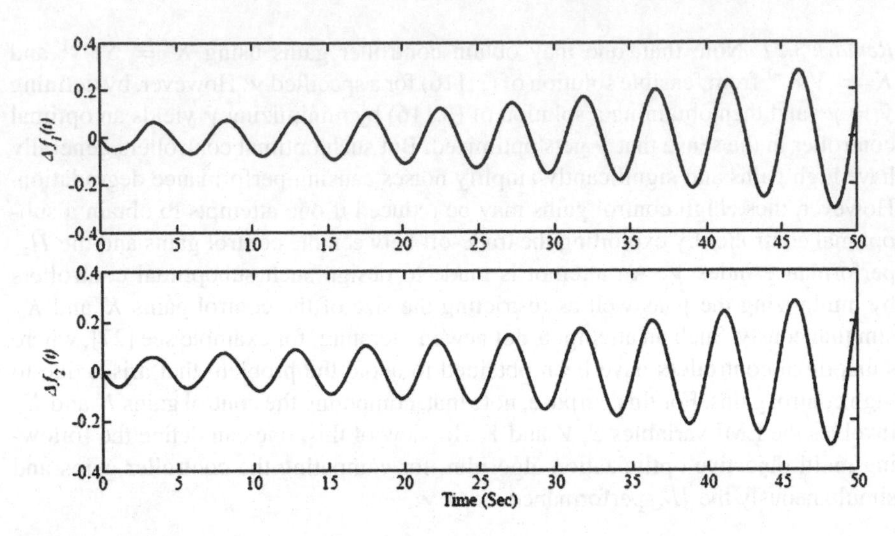

Fig. 3.12 Deviation in frequency for open-loop system

Closed-loop simulation: The designed H_∞ controller gains for the LFC problem are computed by solving the LMI conditions using 'mincx' optimization solver of LMI control toolbox ([17]).

One-term control: Solving the LMI in (3.110), one obtains the γ as 0.4493 and the corresponding gain matrix K as:

$$K = \begin{bmatrix} -16.3022 & -0.3319 & -0.0917 & -1.5967 & -0.1824 & 0.4176 & -0.0224 & -0.1252 \\ -0.4749 & -0.0208 & -0.0099 & 0.4723 & -0.3316 & -22.6162 & -0.2747 & -0.4027 \end{bmatrix}$$
$$\begin{bmatrix} -0.1959 \\ -1.1464 \end{bmatrix}$$

Delay-independent two-term control [33, 42]: Solving the LMI in (3.114) with the choice $\sigma = 1$ and $\kappa = 1$, the control gains are obtained as:

$$K = 1 \times 10^5 \times$$
$$\begin{bmatrix} -4.0871 & -0.0004 & 0.0004 & -0.9696 & -0.3088 & 3.7540 & 0.0005 & -0.0007 & -0.9837 \\ 4.9949 & 0.0005 & -0.0005 & -2.4130 & -0.7636 & -5.7216 & -0.0008 & 0.0011 & -2.3906 \end{bmatrix}$$

$$K_d = [0]_{2\times 9},\ \textit{corresponding}\ \gamma = 8.4336 \times 10^{-4}$$

Remark 3.22 It is mentioned in the Remark 4.4 of [33] that, the direct implementation of the LMI (59) of Theorem 4.2 for the system considered in (50a) will yield smisleading result for computing the controller gain associated with the delayed term due to the fact that (1,1) entry of the LMI (59) does not contain a symmetric term associated with the variable V which in turn yields $V = [0]$ and consequently

$K_d = VY^{-1} = [0]$, this fact can be observed in the result presented above for delay-independent two-term control.

To overcome this difficulty an iterative optimization procedure has been suggested in [33] by introducing some additional terms in the (1,1) entry of the LMI condition (3.114) to minimize the γ. The drawbacks of this iterative algorithm are:

(i) selecting the initial conditions for several scalar tuning parameters involved in the LMI (like κ and σ)
(ii) selecting the arbitrary initial Y matrix
(ii) selecting predetermined tolerance for $\| Y^* - Y^j \| < \delta$.

As these selections are arbitrary and has no specific guidelines, so one can conclude that the accuracy of the solution is not guaranteed immediately from the solution of this algorithm.

Also the result presented for PSS problem in [42] returned a K_d matrix whose elements are very small whereas the elements of the K matrix are relatively very large (of the order of 10^5), the same trend of the gain matrices are observed in the results presented above (delay-independent two-term controller) for the LFC problem.

The above drawbacks of the delay-independent two-term controller design have been eliminated in the proposed delay-dependent two-term control algorithm (i) introducing an arbitrary finite delay 'τ' in the feedback-loop that consequently avoids the limiting situation of delay-independent design (i.e, when state delay tends to infinity the feedback loop is still closed as 'τ' is finite) and (ii) use of modified LMI conditions are established along with the solution of multi-objective optimization algorithm.

Proposed delay-dependent two-term control: Delay-dependent two-term H_∞ controller gains are obtained by solving the multi-objective optimization algorithm presented in Theorem 3.13. This yields controller gains as,

$$K = \begin{bmatrix} -57.9577 & -1.6131 & 0.4879 & -4.8317 & 1.1070 & 33.1766 & 0.4794 & 0.3582 & 0.3828 \\ -17.2892 & 0.0140 & 0.0355 & -0.1347 & -2.8920 & -106.9613 & -1.5252 & -2.0749 & -3.7771 \end{bmatrix}$$

$$K_\tau = \begin{bmatrix} 2.0425 & -0.0350 & -0.0205 & 0.2533 & 0.1331 & -5.4343 & -0.0359 & -0.0234 & -0.2004 \\ -2.8702 & -0.0311 & -0.0049 & -0.0126 & -0.1211 & 7.1215 & 0.0383 & -0.0666 & 0.3837 \end{bmatrix}$$

The corresponding γ is 4.0121.

Now, simulation results for one-term and delay-dependent two-term controllers with different load disturbances are compared in terms of performances. First, considering the unit-step load disturbance, the variations in frequency deviations in both the areas are shown in Fig. 3.13 whereas the control inputs are presented in Fig. 3.14.

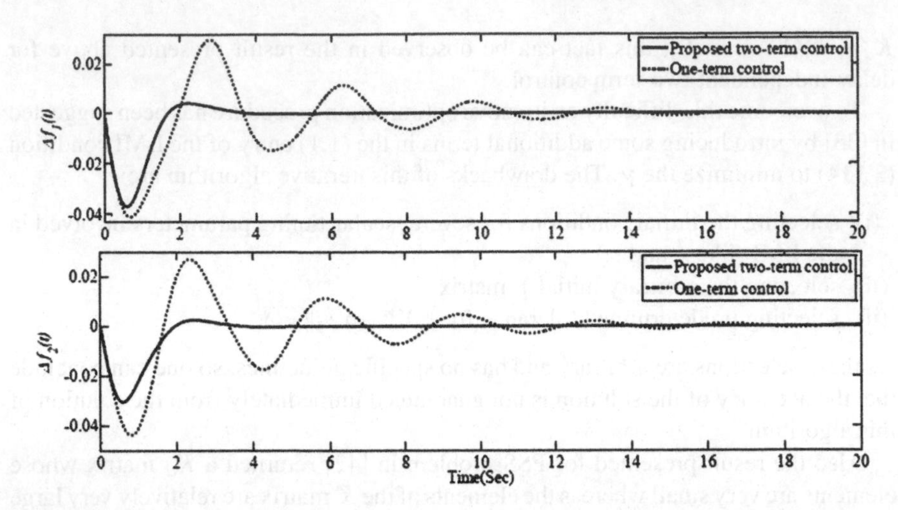

Fig. 3.13 Deviation in frequency for the closed-loop system for unit-step load disturbance for feedback delay $(\tau) = 0.7$ Sec

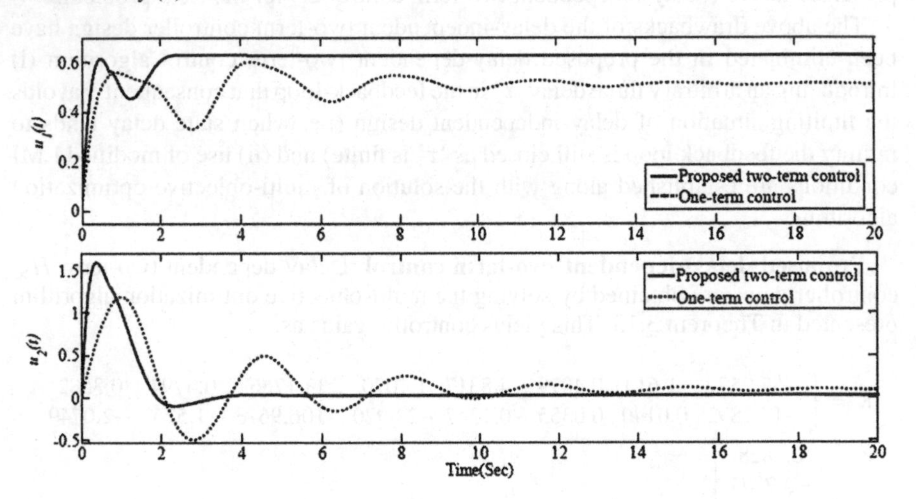

Fig. 3.14 Control inputs for unit-step load disturbance for feedback delay $(\tau) = 0.7$ Sec

Next, the simulation results of the closed-loop system for time-varying disturbance are presented in Figs. 3.15 and 3.16. From these results, it is clear that the transient response of the proposed delay-dependent two-term controller is superior than the one-term controller and in both the cases the disturbance rejection capability appears to be nearly same at the steady-state condition.

Remark 3.23 A linear model of LFC problem for an interconnected power system with communication delay is considered with zero initial condition for the stabilization and disturbance rejection problem. A closed loop simulation study of proposed two-term controller for the same system under non-zero initial conditions is carried

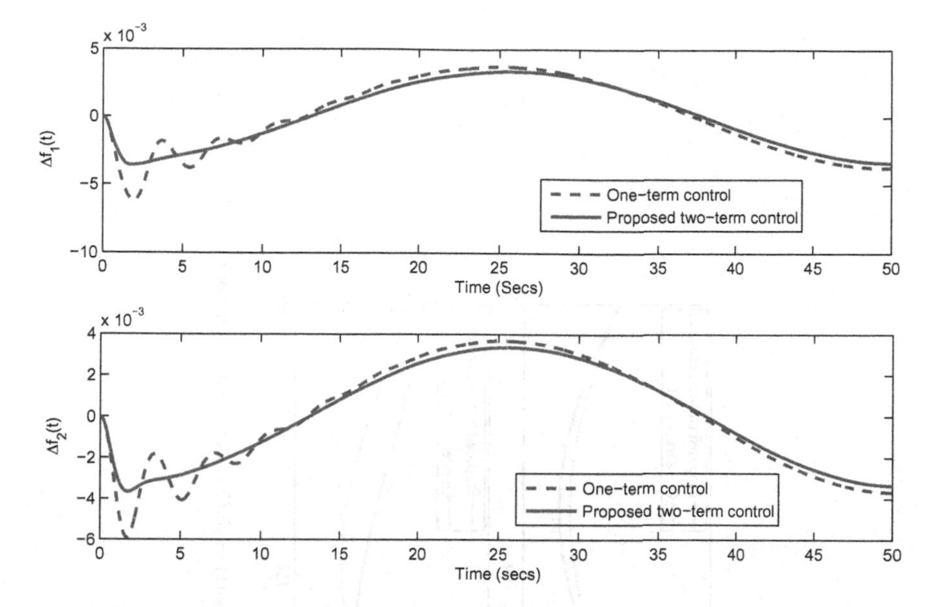

Fig. 3.15 Deviation in frequency for the closed-loop system for time-varying load disturbance of $w(t) = \sin(2\pi t)$ and feedback delay $\tau = 0.7$ Sec

out, the result reveals that there is a tendency for the system to deteriorate the transient response little bit, but the disturbance rejection capability will not be lost i.e, the steady state response is similar to that under zero initial condition.

3.10 Conclusions

The first part of this chapter discusses stabilization and robust stabilization of a linear time-varying delay system with state delay in the feedback control law. Improved delay-dependent stabilization as well as robust stabilization conditions in an LMI and NLMI frameworks have been derived for the linear time-delay system. The proposed delay-dependent LMI based stabilization as well as robust stabilization conditions are formulated using both convex combination of LMIs and improved bounding technique along with the multi-objective optimization algorithm to compute the controller gains for a given delay upper bound. On the other hand, the NLMI based proposed delay-dependent stabilization condition (Theorem 3.5) is formulated with much lesser decision variables and could not yield delay upper bound estimate comparable to that of [12] while solving through cone-complementarity algorithm. It may be emphasized here that the extension of the NLMI stabilization condition (Theorem 3.5) for an uncertain TDS requires lesser bounding inequalities thus yielding improved robust stabilization results than that of [12]. Several numerical examples

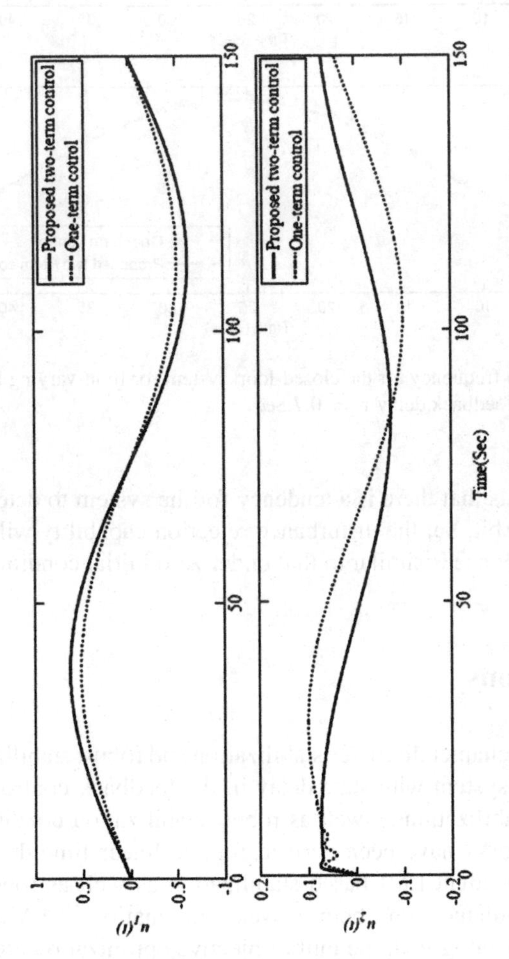

Fig. 3.16 Control inputs for time-varying load disturbance of $w(t) = \sin(2\pi t)$ and feedback delay $\tau = 0.7$ Sec

are considered to illustrate the effectiveness of the proposed delay-dependent stabilizing conditions to achieve improved delay upper bound and lesser control effort.

The last part of this chapter dealt with H_∞ state feedback controller for the solution of LFC problem of an interconnected power system with communication delays in an LMI framework. It must be mentioned at this stage that, the existing results of H_∞ state feedback controller design are all based on delay-dependent formulation with delayed states in feedback signals and however, these results have not been utilized to solve LFC problem yet. The H_∞ controller that has been applied so far for an LFC problem of a multi-area inter-connected power system is delay-independent one with decentralized control structures having (i) one-term control and (ii) two-term control. The proposed two-term H_∞ controller based on performance index 'γ' is delay-dependent formulation with respect to the feedback delays and not the state delays as opposed to other existing delay-dependent H_∞ controller designs. Simulation results of an LFC of a two-area inter-connected power system are presented to show the effectiveness of (i) the proposed two-term delay-dependent controller over delay-independent one in terms of the control effort, (ii) implementation of LMI conditions are less complex computationally compared to the results presented in [33, 42] and [48] and (iii) the superiority of the proposed two-term controller over one-term controller under different types of load-disturbances.

References

1. J.C. Shen, B.S. Chen, F.C. Kung, Memoryless stabilization of uncertain dynamic delay system: Ricatti equation approach. IEEE Trans. Autom. Control **36**, 638–640 (1991)
2. S. Phoojaruenchanachai, K. Furuta, Memoryless stabilization of uncertain linear system including time-varying sate delay. IEEE Trans. Autom. Control **37**, 1022–1026 (1992)
3. M.S. Mahmoud, N.F. Al-muthairi, Quadratic stabilization of continuous time system with the state delay and norm bounded uncertainties. IEEE Trans. Autom. Control **39**, 2135–2139 (1994)
4. A. Trinh, M. Aldeen, Stabilization of uncertain dynamic delay system by memoryless state feedback controllers. Int. J. Control **56**, 1525–1542 (1994)
5. A. Trinh, M. Aldeen, "Robust stabilization and disturbance attenuation for uncertain time-delay systems," in *European Control conference*, 1993, pp. 0 – 0
6. X. Li, C.E. de Souza, Criteria for robust stability and stabilization of uncertain linear systems with state delays. Automatica **33**, 1657–1662 (1997)
7. X. Li, C.E. de Souza, Delay-dependent robust stability and stabilization of uncertain linear delay system:a linear matrix inequality approach. IEEE Trans. Autom. Control **42**, 1144–1148 (1997)
8. Y.Y. Cao, Y.X. Sun, C. Cheng, Delay-dependent robust stabilizion of uncertain systems with multiple state delays. IEEE Trans. Autom. Control **43**, 1608–1612 (1998)
9. Y.S. Moon, P. Park, W.H. Kwon, Y.S. Lee, Delay-dependent robust stabilization of uncertain state delayed system. Int. J. control **74**, 1447–1455 (2001)
10. E. Fridman, U. Shaked, An improved stabilization method for linear time-delay system. IEEE Trans. Autom. Control **47**, 1931–1937 (2002)
11. E. Fridman, U. Shaked, Parameter dependent stability and stabilization of uncertain time-delay systems. IEEE Trans. Autom. Control **48**, 861–866 (2003)
12. M.N.A. Parlakci, Improved robust stability criteria and design of robust stabilizing controller for uncertain linear time-delay system. Int. J. Robust Nonlinear Control **16**, 599–636 (2006)

13. M.N.A. Parlakci, Delay-dependent stability and state feedback stabilization criterion for linear time delay system, in *International Conference on Modeling and Simulation, Coimbator-Vol.2*, 2007, pp. 963 – 968
14. M.N.A. Parlakci, Delay-dependent stability and H_∞ control: Constant and time-varying delays. Int. J. Control **76**, 48–60 (2003)
15. K. Gu, S.I. Niculescu, Additional dynamics in transformed time-dealy systems. IEEE Trans. Autom. Control **45**, 00–00 (2000)
16. E. Ghaoui, F. Oustry, M.A. Rami, A cone-complementary linearization algorithm for static output feedback and related problems. IEEE Trans. Autom. Control **42**, 1171–1176 (1997)
17. P. Gahinet, A. Nemirovski, A.J. Laub, M. Chilali, *LMI control toolbox users guide* (Mathworks, Cambridge, 1995)
18. E. Fridamn, U. Shaked, A descriptor system approach to H_∞ control of linear time-delay systems. IEEE Trans. Autom. Control **47**, 253–270 (2002)
19. S. Xu, J. Lam, Y. Zou, Delay-dependent guaranteed cost control for uncertain system with state and input delays. IEE Proc. Control Theory Appl. **153**, 307–313 (2006)
20. T. Li, L. Guo, Y. Zhang, Delay-range-dependent robust stability and stabilization for uncertain systems with time-varying delay. Int. J. Robust Nonlinear Control **18**, 1372–1387 (2008)
21. C.E. de Souza, X. Li, Delay-dependent robust H_∞ control of uncertain linear state-delayed systems. Automatica **35**, 1313–1321 (1999)
22. V. Suplin, E. Fridman, U. Shaked, H_∞ control of linear uncertain time-delay systems-a projection approach. IEEE Trans. Autom. Control **31**, 680–685 (2006)
23. R. Dey, S. Ghosh, G. Ray, A. Rakshit, State feedback stabilization of uncertain linear time-delay systems: A nonlinear matrix inequality appraoch. Numer. Linear Algebra with Appl. **18**(3), 351–361 (2011)
24. R. Dey, G. Ray, S. Ghosh, A. Rakshit, Stability analysis for contious system with additive time-varying delays: A less conservative result. Appl. Mathematics Comput. **215**, 3740–3745 (2010)
25. S. Niculescu, A.T. Neto, J.M. Dion, L. Dugard, Roust stability and stabilization of uncertain linear systems with state delay: Multiple delay case (I), in *IFAC symposium on Robust control Design*, 1994
26. J.W. Ko, P.G. Park, Delay-dependent robust stabilization for systems with time-varying delays. Int. J. Control Autom. Sys. **7**, 711–722 (2009)
27. D.D. Siljak, D.M. Stipanovic, A.I. Zecevic, Robust decentralized turbine/governor control using linear matrix inequalities. IEEE Trans. Power Sys. **17**, 715–722 (2002)
28. P.P.K.J.C. Doyle, K. Glover, B.A. Francis, State-space solutions to the standard H_2 and H_∞ control problems. IEEE Trans. Autom. Control **34**, 831–927 (1989)
29. B.A. Francis, *A course in H_∞ Control Theory* (Springer Verlag, New York, 1987)
30. H.H. Choi, M.J. Chung, Observer-based H_∞ controller design for state delayed linear system. Automatica **32**, 1073–1075 (1996)
31. S.W.K.J.H. Lee, W.H. Kwon, Memoryless H_∞ controller for state delayed systems. IEEE Trans. Autom. Control **39**, 159–162 (1994)
32. S. Niclescu, C. E. D. Souza, J. M. Dion, L. Dugard, Robust H_∞ memoryless control of uncertain linera systems with time-varying delay, in *Proceedings of 1995 Euoropean Control Conference, Rome*, 1995, pp. 1814 – 1819
33. M.S. Mahmoud, M. Zribi, H_∞ controllers for linearized time-delay systems using LMI. J. Optim. Theory and Appl. **100**, 89–112 (1999)
34. M.S. Mahmoud, M. Zribi, "Robust stabilization and H_∞ controlfor uncertain linear time-delay system: An LMI approach," in 13th *IFAC World Congress, San Francisco*, 1996, pp. 113 – 118
35. Y.Y. Cao, Y.X. Sun, J. Lam, Delay-dependent robust H_∞ control for uncertain system with time-varying delays. IEE Proc. of CTA **145**, 338–344 (1998)
36. X. Li, E. Fridman, U. Shaked, Robust H_∞ control of distributed delay systems with application to combustion control. IEEE Trans. Autom. control **46**, 1930–1933 (2001)
37. G. Ray, A.N. Prasad, G.D. Prasad, Design of robust lfc for interconnected power system based on singular value decomposition method. Electric Power Syst. Res. **37**, 209–219 (1996)

38. H. Bevrani, T. Hiyama, Robust decentralized PI based LFC design for time-delay system. Energy Conversion Manage. **49**, 193–204 (2008)
39. T.C. Yang, H. Cimen, Q.M. Zhu, Decentralized LFC design based on structured singular value. IEE Proc. Generation, Transm. Distribution **145**, 7–14 (1998)
40. D. Rerkpreedapong, A. Feliachi, "Decentralized H_∞ load frequency control using lmi control toolbox," in *IEEE proceedings of the circuit and systems (ISCAS'03)*, 2003, pp. 411 – 414
41. D. Dotta, A.S. Silva, I.C. Decker, Wide-area measurements-based two-level control design considering signal transmission delay. IEEE Trans. Power Sys. **24**, 208–216 (2009)
42. M. Zribi, M.S. Mahmoud, M. Karkoub, T.T. Lie, H_∞ controller for linearized time-delay power system. IEE proc. Generation Transm. Distribution **147**, 401–403 (2000)
43. T. Ishii, G. Shirai, G. Fujita, Decentralized load frequency based on H_∞ control. Electr. Eng. Japan **136**, 28–38 (2001)
44. A.Y. Sivaramakrishnan, M.V. Hariharan, M.C. Srisailam, Design of variable structure load frequency controller using pole placement technique. IEEE Trans. Autom. Control **14**, 487–498 (1984)
45. C.T. Pan, C.M. Liaw, An adaptive controller for power system LFC. IEEE Trans. Power Sys. **14**, 122–128 (1989)
46. E.A.A. Nedzad, N ERC compliant decentralized LFC design using MPC. IEEE Power Eng. Soc. Gen. Meeting **2**, 13–17 (2003)
47. S. Bhowmick, K. Tomsovic, A. Bose, Communication models for third party LFC. IEEE Trans. Power Sys. **19**, 543–548 (2004)
48. X. Yu, K. Tomsovic, Application of linear matrix inequalities for load frequency control with communication delay. IEEE Trans. Power Sys. **19**, 1508–1515 (2004)
49. H. Jiang, H. Cai, J.F. Dorsey, Z. Qu, Towards a globally robust decentralized control for large-scale power systems. IEEE Trans. Control Sys. Technol. **5**, 309–319 (1997)
50. G.J. Li, T.T. Lie, C.B. Soh, G.H. Yang, Decentralized H_∞ control for power system stability enhancement. Electr. Power and Energy Sys. **20**, 453–464 (1998)
51. R. Dey, S. Ghosh, G. Ray, A. Rakshit, h_∞ load frequency control of interconnected power system with communication delays. Electr. Power Energy Sys. **42**, 672–684 (2012)

Chapter 4
Control of Time-Delay Systems with Actuator Saturation

4.1 Introduction

The time-delay in a system model is inherent to any physical system and it induces infinite roots in its characteristic equation forcing the exact analysis to be computationally difficult tasks. This motivated several researchers to develop approximate but computationally efficient methods for analysis and synthesis of time-delay systems via Lyapunov-Krasovskii or Razumikhin method in an LMI framework. The presence of the time-delays in physical systems usually degrade system performance and even frequently a source of instability and these has been discussed in the previous chapters. In last few decades, large number of LMI based delay-dependent and/or generalized delay-dependent stability criteria have been reported to achieve improved delay upper bound estimate as cited in the references of previous chapters. These delay-dependent stability conditions are then extended to stabilization and robust stabilization problems in LMI framework using state feedback control law that have been discussed in Chap. 3. The state feedback stabilization of linear parameter varying (LPV) systems with time-delay in the states can be found in [1].

In practice, both time-delay and actuator saturation controls are commonly encountered in various engineering problems [2]. The presence of actuator saturation alone in the linear system may cause the following, (i) degradation of the system performances (e.g., large overshoot and large settling time) (ii) cannot track large set points and (iii) source of system instability if saturation effect is not taken into account properly. The research article on stability, stabilization and H_∞ control (in presence of $\mathcal{L}_2$ disturbances) of linear systems with actuator saturation (or bounded control) can be found in [3–6] and references therein. The problem of stabilizing linear-parameter varying (LPV) systems with saturating actuators using gain scheduling controllers can be found in [7, 8], this problem is solved by considering parameter-dependent LK functional approach.

In last decade, a considerable research attention has been paid by many authors on stability and stabilization issues of linear system and some mature methods have been widely used to deal with the above problems [2, 9–20] and reference therein. Very

R. Dey et al., *Stability and Stabilization of Linear and Fuzzy Time-Delay Systems*, Intelligent Systems Reference Library 141, https://doi.org/10.1007/978-3-319-70149-3_4

few literature are available for robust stabilization of linear TDS with norm bounded parametric uncertainties subjected to actuator saturation, [21, 22] and references therein. The work in [22] is carried out for an open-loop stable system in presence of uncertainty assuming the nonlinear function inside a sector.

The problem of stability analysis and control of linear systems with actuator saturation with or without time-delay in the system states are generally dealt through two main different strategies initiated by two different school of thoughts [4], (i) anti-wind up scheme, which introduces additional feedbacks in such a way that the control effort never attains the saturation limits, this scheme neglects the saturation nonlinearities in the first stage of control design and (ii) it analyzes the closed-loop system under actuator saturation systematically taking into consideration saturation nonlinearities as *a priori* to meet either the performance or stability requirements. The literature on the former method can be found in [23] and reference therein, this method is beyond the scope of the present work, rather latter method is adopted for stabilization of linear time-delay systems with actuator saturation in this thesis. While adopting the latter method, it needs to include the saturation nonlinearities in the controller design or stability analysis problems as *a priori* and that is accomplished by approximating the saturation function with two different approaches: (i) polytopic differential inclusions [3, 4, 7, 14, 15, 24] and (ii) sector nonlinearity [18, 23]. The demerits polytopic representation are two folds (i) it provides local stability (ii) 2^m number of LMIs are required to be solved where m is the number of inputs. These difficulties are overcome using the latter representation as in this case both local and global stability can be ensured as well as the number of LMIs to be solved are reduced to m only.

In this chapter the problem of designing state feedback control for a linear time-delay system with actuator saturation is considered with a view to achieve improved delay upper bound and simultaneously obtain the estimate of domain of attraction using LK functional approach in an LMI framework. The estimation of the domain of attraction directly refers to the local stabilization problem.

4.2 Characterization of Actuator Saturation

In order to carry out the stability analysis or control design of a linear dynamical systems subjected to actuator saturation using Lyapunov's second method, one can express adequately the actuator saturation function mathematically in two different approaches as discussed below (refer Sect. 1.7 of Chap. 1):

- A Polytopic representation, based on the use of differential inclusions [4].
- A sector nonlinearity representation, which includes the saturation function in a sector [23].

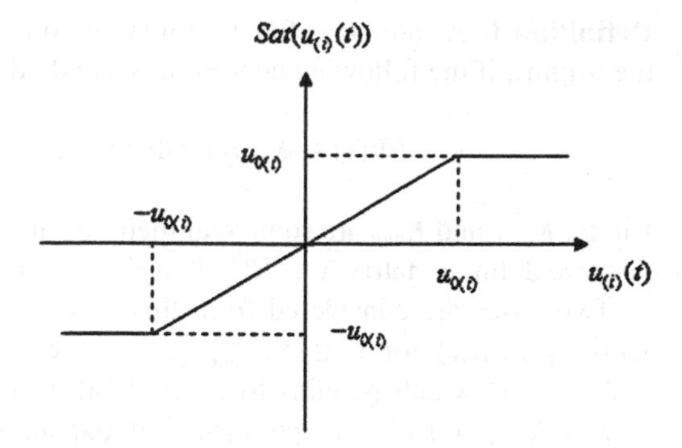

Fig. 4.1 Saturation function

A general mathematical representation of an actuator saturation is given by,

$$Sat(u_{(i)}) = \begin{cases} u_{max(i)}, & \text{if } u_{(i)} > u_{max(i)}; \\ u_{(i)}, & \text{if } -u_{min(i)} \leq u_{(i)} \leq u_{max(i)}; \\ -u_{min(i)}, & \text{if } u_{(i)} < -u_{min(i)} \end{cases}$$

For simplicity in treatment one can consider symmetrical saturation function, where it is assumed that, $u_{max(i)} = u_{min(i)} = u_{0(i)}$, which is shown in Fig. 4.1. The actuator saturation function considered above has two distinct features with the function value being different for different sets of inputs and memoryless owing to the fact that function values do not depend on the past inputs.

4.2.1 *Polytopic Representation [3, 4]*

In this method, the saturation nonlinearity is replaced by the convex combinations of two linear feedbacks such that the saturation function lies in the convex hull[1] of these linear feedbacks. The lemma proposed in [3, 4] is employed in forming the convex hull for the saturation nonlinearity and it is described in Lemma 1.2, for ready reference refer (Sect. 1.7 of Chap. 1).

4.2.2 *Sector Nonlinearities Representation [23]*

First a general definition of sector condition is presented for better understanding of the saturation function as a sector nonlinearity,

---

[1]The convex hull of a set $X = \{x_1, x_2, \ldots, x_k\}$ is the minimal convex set that contains X. For a group $\{x_1, x_2, \ldots, x_k\} \in \mathcal{R}^n$, the convex hull of these points is $co\{x_1, x_2, \ldots, x_k\} = \{\sum_i^k \alpha_i x_i : \sum_i^k \alpha_i = 1, \alpha_i \geq 0\}$ [4].

Definition 1 A saturation function $\Phi(y)$ belongs to sector $\mathbf{S}$, (a set which contains the origin), if the following condition is satisfied,

$$(\Phi(y) - K_{min}y)^T(\Phi(y) - K_{max}y) \leq 0 \,\forall\, y \in \mathcal{S}$$

where, K_{max} and K_{min} are some real matrices and also $(K_{max} - K_{min})$ is a symmetric positive definite matrix, $y \in \mathcal{R}^m$, $\Phi_{y0}(y) \in \mathcal{R}^m$ and $K_{max}, K_{min} \in \mathcal{R}^{m\times m}$.

Two cases are considered from this definition, (i) if $K_{min} > 0$, then $|\Phi(y)| \in sect(K_{min}, K_{max})$ for $y \in \mathbf{S}(K_{min}, y_0) = \{y \in \mathcal{R}^m; -y_{0(i)} \leq y_{(i)} \leq y_{0(i)}, i = 1, 2, \ldots, m\}$ which pertains to local stability and (ii) if $K_{min} = 0$, then $\Phi(y) \in sect\,(0, K_{max})$ for $y \in \mathcal{R}^m$, pertains to global stability, the graphical interpretation of the above sector conditions are given in Figs. 4.2 and 4.3 respectively.

The saturation nonlinearity can equivalently be replaced with a dead-zone nonlinearity in combination with a linear element that is shown in Fig. 4.4 and expressed mathematically as,

$$Sat(u(t)) = u(t) - \Psi(u(t))$$

Now one can express dead-zone nonlinearity equivalently in terms of saturation function as,

$$\Psi(u(t)) = u(t) - Sat(u(t))$$

where, $u(t) = Kx(t)$. The ith component of dead-zone nonlinearity function, output of the controller and output of the saturating actuator are expressed as $\Psi(u_i(t))$, $u_i(t)$ and $Sat(u_i(t))$ respectively. Where $u(t) \in \mathcal{R}^m$, $x(t) \in \mathcal{R}^n$ and k_i is the ith row of the

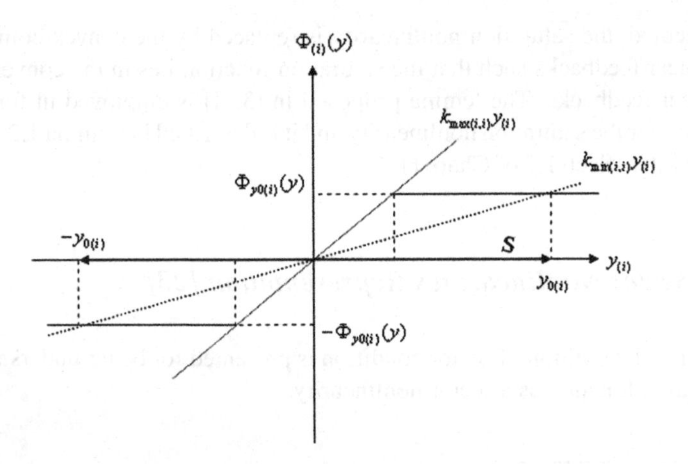

Fig. 4.2 Sector interpretation for local stability

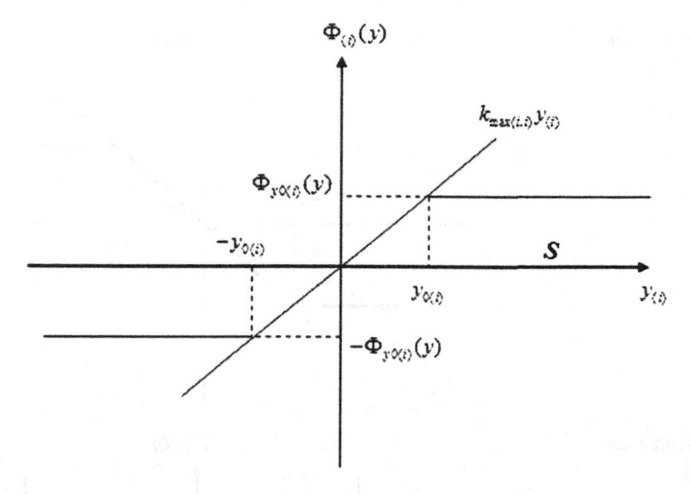

Fig. 4.3 Sector interpretation for global stability

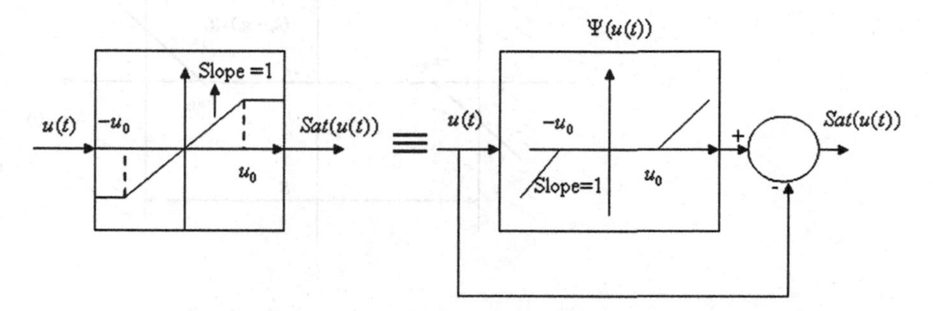

Fig. 4.4 Equivalent representation of saturation nonlinearity

feedback gain matrix K having proper dimension $K \in \mathcal{R}^{m\times n}$, then one can write the ith element of dead-zone nonlinearity function as,

$$\Psi_i(k_i x(t)) = k_i x(t) - Sat(k_i x(t)), i = 1, 2, \dots m$$

From Figs. 4.5 and 4.6 one can express decentralized dead-zone nonlinear element as,

$$\Psi_i(k_{(i)}x(t)) = \begin{cases} k_i x(t) - u_{0(i)} > 0, & \text{if } k_i x(t) > u_{0(i)}; \\ k_i x(t) - u_{(i)} = 0, & \text{if } -u_{0(i)} \le k_i x(t) \le u_{0(i)}; \\ k_i x(t) + u_{0(i)} < 0, & \text{if } k_i x(t) < -u_{0(i)} \end{cases}$$

where, $i = 1, 2, \dots m$.

As the saturation function is replaced by the dead-zone nonlinearity, one needs to satisfy the modified sector condition as stated in Lemma 1.2 (refer Sect. 1.7 of Chap. 1),

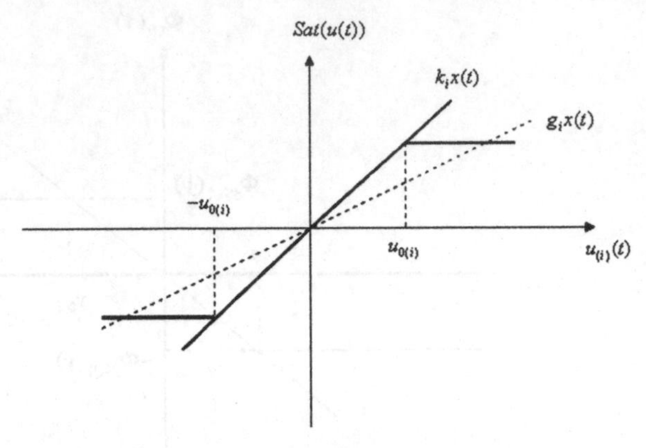

Fig. 4.5 Saturation nonlinearity

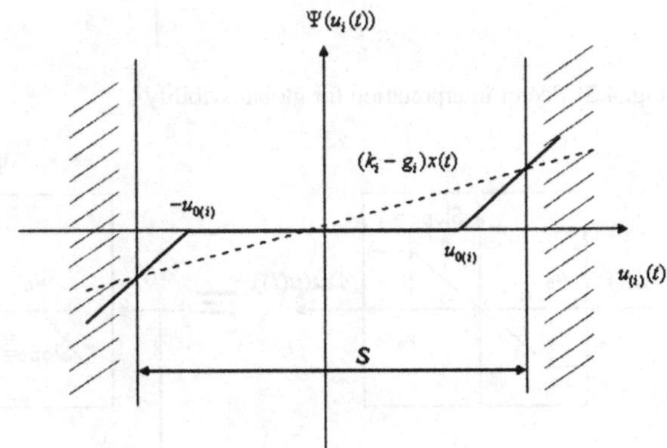

Fig. 4.6 Dead zone nonlinearity

Lemma 4.1 ([18]) *Consider a dead-zone nonlinear function $\Psi_i(k_i x(t))$, with $x(t) \in \mathcal{S}$ then the relation*

$$\Psi_i(k_i x(t)) D_{i,i} [\ \Psi_i(k_i x(t)) - g_i x(t)] \leq 0 \tag{4.1}$$

is valid for any scalar $D_{i,i} > 0$. g_i is the ith component of the auxiliary feedback gain matrix. The set $\mathcal{S}$ is indicated in Fig. 4.6.

The result of Lemma 4.1 reveals the following statements:

Case I: If $\Psi_i(k_i x(t)) > 0$, then it follows from the Lemma 4.1 that,

$$\Psi_i(k_i x(t)) D_{(i,i)}(\Psi_i(k_i x(t)) - g_i x(t)) = \Psi_i(k_i x(t)) D_{(i,i)}(k_i x(t) - u_{0(i)} - g_i x(t)) \leq 0$$

provided, $D_{(i,i)} > 0$ and $k_i x(t) - u_{0(i)} - g_i x(t) \leq 0$, which means,

$$k_i x(t) - g_i x(t) \leq u_{0(i)}$$

Case II: If $\Psi_i(k_i x(t)) = 0$ with $k_i x(t) \le u_{0(i)}$ then it follows from the Lemma 4.1 that,

$$\Psi_i(k_i x(t)) D_{(i,i)} (\Psi_i(k_i x(t)) - g_i x(t)) = 0$$

Case III: If $\Psi_i(k_i x(t)) < 0$, then it follows from the Lemma 4.1 that,

$$\Psi_i(k_i x(t)) D_{(i,i)} (\Psi_i(k_i x(t)) - g_i x(t)) = \Psi_i(k_i x(t)) D_{(i,i)} (k_i x(t) + u_{0(i)} - g_i x(t)) \le 0$$

provided, $D_{(i,i)} > 0$ and $k_i x(t) + u_{0(i)} - g_i x(t) \ge 0$, which means,

$$k_i x(t) - g_i x(t) \ge -u_{0(i)}$$

So, from the above three cases one can conclude that, in order to satisfy the sector condition for the local stability of the dead-zone nonlinearity one has to impose the constraint,

$$|k_i x(t) - g_i x(t)| \le u_{0(i)}$$

Remark 4.1 The actuator saturation has been approximated using dead-zone nonlinearity (or referred as sector nonlinearity) for stability analysis and controller synthesis of a linear time-delay systems with actuator saturation in succeeding sections. This approximate representation of saturation nonlinearity is adopted due to the fact that, the number of LMI conditions to be solved are lesser compared to the polytopic representation.

In this thesis the discussion is limited to the local stability and stabilization of time-delay systems in presence of actuator saturation.

4.3 Problem Formulation and Preliminaries

Let us consider a time-delay system with input (or actuator) saturation

$$\dot{x}(t) = Ax(t) + A_d x(t - d(t)) + BSat(u(t)) \tag{4.2}$$

$$x(\theta) = \phi(\theta), \theta \in [-d_u, 0] \tag{4.3}$$

Considering state feedback control law

$$u(t) = Kx(t) \tag{4.4}$$

The system described in (4.2), can be rewritten as

$$\dot{x}(t) = Ax(t) + A_d x(t - d(t)) + BSat(Kx(t)) \tag{4.5}$$

where $x(t) \in \mathcal{R}^n$ is the state vector, A and A_d are constant matrices with appropriate dimensions, $K \in \mathcal{R}^{m\times n}$ is the state feedback gain matrix. The actuator saturation is mathematically described as,

$$Sat(u(t)) = [Sat(k_1x(t)), \ldots Sat(k_mx(t))]^T \tag{4.6}$$

where,

$$Sat(k_ix(t)) = \begin{cases} u_{0(i)}, & \text{if } k_ix(t) > u_{0(i)}; \\ k_ix(t), & \text{if } -u_{0(i)} \leq k_ix(t) \leq u_{0(i)}; \\ -u_{0(i)}, & \text{if } k_ix(t) < u_{0(i)} \end{cases}$$

where, k_i is the ith row of the K gain matrix and the saturation is assumed to be symmetrical. The time-delay in (4.2), is time-varying and satisfies following conditions

$$0 \leq d(t) \leq d_u \tag{4.7}$$

$$0 \leq \dot{d}(t) \leq \mu \tag{4.8}$$

where, d_u is the maximum allowable delay upper bound and μ is the delay-derivative upper bound.

Definition 2 ([19]) Let the solution of the system defined in (4.2) with the initial condition $x_0 = \phi \in \mathcal{C}([-d_u, 0], \mathcal{R}^n)$ be $x(t, \phi)$, then the domain of attraction (DOA) of the origin of the system (4.2) is defined as

$$\Im = \left\{\phi \in \mathcal{C}[-d_u, 0] : \lim_{t\to\infty} x(t, \phi) = 0\right\} \tag{4.9}$$

where, $\mathcal{C}[-d_u, 0]$ denotes the space of continuously differentiable vector function ϕ over $[-d_u, 0]$. In sequel, the estimate of the domain of attraction (DOA) is defined in [15] as,

$$\mathcal{X}_\delta = \left\{\phi \in \mathcal{C}[-d_u, 0] : \max_{[-d_u,0]} \|\phi\|_2 \leq \delta_1, \max_{[-d_u,0]} \|\dot{\phi}\|_2 \leq \delta_2\right\} \tag{4.10}$$

Definition 3 Local stabilization means that, under the influence of designed state feedback gain, the closed-loop state trajectory of system (4.2) initiated in $\mathcal{X}_\delta$ will converge to origin without leaving at any time the convex set $\mathcal{E}(P, 1) \subset \mathcal{S}$, where $\mathcal{S}$ is a set of all the values of $x(t)$ defined within the sector of the nonlinearity.

The problem here is to design a state-feedback controller for the system (4.2) such that the closed loop system is asymptotically stable satisfying (4.7) and (4.8) for maximum delay upper bound and simultaneously obtain the estimate of DOA.

It is observed from all recent literatures [14, 25–29] that the delay-dependent stability analysis of linear TDS provides the sufficient conditions while replacing the $\dot{x}(t)$ term (arising out of the LK functional derivative) by current as well as delayed state vector that yields less conservative delay upper bound results, but it must be mentioned here that direct extension of these conditions to solve state feedback stabilization problem leads to nonlinear matrix inequality (NLMI) condition [30, 31].

To alleviate this difficulty, in [19], the term associated with $\dot{x}(t)$ has been retained and a suitable transformation is proposed that helped to solve stabilization condition in LMI framework while dealing with the actuator saturation problem via polytopic approach.

A new and improved delay-dependent stabilization condition for the system in (4.2) is presented adopting the sector nonlinearities approximation of the saturation function and subsequently the stabilization condition by adopting polytopic approach has also been considered for the comparative assessment of both the approaches.

4.4 Main Result on Stabilization of TDS with Actuator Saturation

In this section stabilization condition of a nominal time-delay system subjected to actuator saturation using two different approximations of saturation function mentioned above are presented in the form of following theorems.

4.4.1 Stabilization Using Sector Nonlinearities Approximation

In this section, an improved stabilization method is proposed by introducing two scalar variables which are associated with a free weighting matrix, also $\dot{x}(t)$ term is retained in the derivative of the LK functional and this, in turn allows us to obtain LMI condition thereby avoiding any transformation of matrix variables unlike in [19]. The proposed method is simpler and computationally efficient as the introduced scalar variables are associated with a common free matrix instead of using two different free matrices in the formulation of stabilization problem. Here, the actuator saturation term in (4.2) using dead-zone nonlinearity function following [18] is written as,

$$Sat(Kx(t)) = Kx(t) - \Psi(Kx(t)) \tag{4.11}$$

where, $\Psi(Kx(t))$ is deadzone nonlinearity function. So, in view of (4.11) the closed loop system (4.2) can be rewritten as

$$\dot{x}(t) = A_c x(t) + A_d x(t - d(t)) - B\psi(Kx(t)) \tag{4.12}$$

where, $A_c = (A + BK)$. For handling the dead-zone nonlinearity $\Psi(Kx(t))$, we consider an auxiliary feedback matrix $G \in \mathcal{R}^{m\times n}$ satisfying the condition $|(k_i - g_i)x| \le u_{0(i)}, i = 1, \dots m$ and hence define a set such that

$$\mathbf{S} = \{x \in \mathcal{R}^n : |(k_i - g_i)x| \le u_{0(i)}, i = 1, \dots m\} \tag{4.13}$$

We choose, the set S to be a compact convex set of the form of an ellipsoid,

$$\mathcal{E}(P, 1) = \{x : x^T Px \le 1\} \tag{4.14}$$

satisfying the condition that, $\mathcal{E}(P, 1) \subset \mathbf{S}$.

Theorem 4.1 ([32]) *Given scalars α and β, the time-delay saturating actuator system (4.12) is asymptotically stable for allowable delay upper bound d_u via memory less state feedback controller $K = YZ^{-T}$, if there exist matrices $\bar{P} = \bar{P}^T > 0$, $\bar{Q}_j = \bar{Q}_j^{\,T} > 0$, $\bar{R}_1 = \bar{R}_1^{\,T} > 0$, any free matrices Y, W, Z, $\bar{M}_j$, $\bar{N}_j$, $j = 1, 2$ and diagonal matrix $L = L^T > 0$ of appropriate dimensions such that following LMIs hold:*

$$\begin{bmatrix} \Sigma & \varphi_1 \\ \star & -\bar{R}_1 \end{bmatrix} < 0 \tag{4.15}$$

$$\begin{bmatrix} \Sigma & \varphi_2 \\ \star & -\bar{R}_1 \end{bmatrix} < 0 \tag{4.16}$$

$$\begin{bmatrix} \bar{P} & (y_i - w_i)^T \\ \star & u_i^2 \end{bmatrix} \ge 0, i = 1, 2, \dots m \tag{4.17}$$

where, y_i and w_i are the ith row of Y and W matrices respectively, and .

$$\varphi_1 = \begin{bmatrix} \bar{M}_1^{\,T} \; \bar{N}_1^{\,T} \; 0 \; 0 \; 0 \end{bmatrix}^T$$

$$\varphi_2 = \begin{bmatrix} 0 \; \bar{M}_2^{\,T} \; \bar{N}_2^{\,T} \; 0 \; 0 \end{bmatrix}^T$$

$$\Sigma = \begin{bmatrix} \Sigma_{11} & \Sigma_{12} & 0 & \Sigma_{14} & \Sigma_{15} \\ \star & \Sigma_{22} & \Sigma_{23} & \Sigma_{24} & \Sigma_{25} \\ 0 & \star & \Sigma_{33} & 0 & 0 \\ \star & \star & 0 & \Sigma_{44} & \Sigma_{45} \\ \star & \star & 0 & \star & \Sigma_{55} \end{bmatrix}$$

with, $\Sigma_{11}=AZ^T + ZA^T + BY + Y^TB^T + \bar{Q}_1 + \bar{Q}_2 + d_u^{-1}(\bar{M}_1 + \bar{M}_1^T)$
$\Sigma_{12}=A_dZ^T + \beta ZA^T + \beta Y^TB^T + d_u^{-1}(-\bar{M}_1 + \bar{N}_1^T)$
$\Sigma_{14}=-Z^T + \alpha ZA^T + \alpha Y^TB^T + \bar{P}$, $\Sigma_{15}=-BL + W^T$
$\Sigma_{22}=-(1-\mu)\bar{Q}_2 + \beta(A_dZ^T + ZA_d^T) + d_u^{-1}(\bar{M}_2 + \bar{M}_2^T - \bar{N}_1 - \bar{N}_1^T)$
$\Sigma_{23}=d_u^{-1}(-\bar{M}_2 + \bar{N}_2^T)$, $\Sigma_{24}=-\beta Z^T + \alpha ZA_d^T$, $\Sigma_{25}=-\beta BL$
$\Sigma_{33}=-\bar{Q}_1 + d_u^{-1}(-\bar{N}_2 - \bar{N}_2^T)$, $\Sigma_{44}=\bar{R}_1 - \alpha(Z + Z^T)$, $\Sigma_{45}=-\alpha BL$
$\Sigma_{55}=-2L.$

The corresponding estimate of domain of attraction $\mathcal{X}_\delta \leq 1$ is given as,

$$\begin{aligned}\mathcal{X}_\delta = \delta_1^2 \{\lambda_{max}(Z^{-1}\bar{P}Z^{-T}) + d_u\lambda_{max}(Z^{-1}\bar{Q}_1Z^{-T}) \\ d_u\lambda_{max}(Z^{-1}\bar{Q}_2Z^{-T})\} \\ +\delta_2^2\{\tfrac{1}{2}d_u\lambda_{max}(Z^{-1}\bar{R}_1Z^{-T})\}\end{aligned} \tag{4.18}$$

where, $\delta_1 = \max_{[-d_u,0]} \|\phi\|_2$, *and* $\delta_2 = \max_{[-d_u,0]} \|\dot{\phi}\|_2$

Proof The detailed proof of this theorem can be found as Theorem 3 in [32]. However we give a brief highlight of the proof here for easy understanding of the readers. Consider a LK functional for the system (4.12) as,

$$\begin{aligned}V(x_t) = x^T(t)Px(t) + \int_{t-d_u}^{t} x^T(\theta)Q_1x(\theta)d\theta \\ +d_u^{-1}\int_{t-d_u}^{t}\int_{\theta}^{t} \dot{x}^T(\phi)R_1\dot{x}(\phi)d\phi d\theta \\ +\int_{t-d(t)}^{t} x^T(\theta)Q_2x(\theta)d\theta\end{aligned} \tag{4.19}$$

The time derivative of (4.19) is obtained as,

$$\begin{aligned}\dot{V}(x_t) \leq 2x^T(t)P\dot{x}(t) + x^T(t)Q_1x(t) + x^T(t)Q_2x(t) \\ -(1-\mu)x^T(t-d(t))Q_2x(t-d(t)) \\ -x^T(t-d_u)Q_1x(t-d_u) + \dot{x}^T(t)R_1\dot{x}(t) \\ -d_u^{-1}\int_{t-d_u}^{t} \dot{x}^T(\theta)R_1\dot{x}(\theta)d\theta\end{aligned} \tag{4.20}$$

Using the integral inequality described in Lemma 3 of [32] one can approximate the integral terms in 4.20 to write $\dot{V}$ as,

$$\dot{V}(x_t) \leq \xi^T(t) \left\{ \begin{bmatrix} \theta_{11} & \theta_{12} & 0 & \theta_{14} & \theta_{15} \\ \star & \theta_{22} & \theta_{23} & \theta_{24} & \theta_{25} \\ 0 & \star & \theta_{33} & 0 & 0 \\ \star & \star & 0 & \theta_{44} & \theta_{45} \\ \star & \star & 0 & \star & 0 \end{bmatrix} + (1-\rho) \begin{bmatrix} 0 \\ M_2 \\ N_2 \\ 0 \\ 0 \end{bmatrix} R_1^{-1} \right.$$

$$\left. \times \begin{bmatrix} 0 \\ M_2 \\ N_2 \\ 0 \\ 0 \end{bmatrix}^T + \rho \begin{bmatrix} M_1 \\ N_1 \\ 0 \\ 0 \\ 0 \end{bmatrix} R_1^{-1} \begin{bmatrix} M_1 \\ N_1 \\ 0 \\ 0 \\ 0 \end{bmatrix}^T \right\} \xi(t)$$

where, $\theta_{11}{=}SA_c + A_c^T S^T + Q_1 + Q_2 + M_1 + M_1^T$; $\theta_{12}{=}SA_d + \beta A_c^T S^T + d_u^{-1}(-M_1 + N_1^T)$

$\theta_{14}{=}-S + \alpha A_c^T S^T + P$; $\theta_{15}{=}-SB$

$\theta_{22}{=}\beta(SA_d + A_d^T S^T) - (1-\mu)Q_2 + d_u^{-1}(-N_1 - N_1^T + M_2 + M_2^T)$

$\theta_{23}{=}d_u^{-1}(-M_2 + N_2^T)$; $\theta_{24}{=}\alpha A_d^T S^T - \beta S$; $\theta_{25}{=}-\beta SB$

$\theta_{33}{=}-Q_1 + d_u^{-1}(-N_2 + N_2^T)$; $\theta_{44}{=}R_1 - \alpha(S + S^T)$; $\theta_{45}{=}-\alpha SB$

If $x(t) \in \mathbf{S}$ then from Lemma 4.1, one can write

$$\dot{V}(x_t) \leq \dot{V}(x_t) - 2\psi^T(Kx(t))D[\psi(Kx(t)) - Gx(t)] \tag{4.21}$$

Thus in view of (4.21) one can write,

$$\begin{aligned} \dot{V}(x_t) \quad &\leq \xi^T(t) \left\{ \Theta + \rho\vartheta_1 R_1^{-1} \vartheta_1^T \right. \\ &\left. + (1-\rho)\vartheta_2 R_1^{-1} \vartheta_2^T \right\} \xi(t) \end{aligned} \tag{4.22}$$

From (4.22) one can conclude that the asymptotic stability of the system in (4.12) is guaranteed if following condition is satisfied

$$\{\Theta + \rho\vartheta_1 R_1^{-1} \vartheta_1^T + (1-\rho)\vartheta_2 R_1^{-1} \vartheta_2^T\} < 0 \tag{4.23}$$

This can be rewritten as

$$\rho(\Theta + \vartheta_1 R_1^{-1} \vartheta_1^T) + (1-\rho)(\Theta + \vartheta_2 R_1^{-1} \vartheta_2^T) < 0 \tag{4.24}$$

From (4.24) one can write,

$$\Theta + \vartheta_1 R_1^{-1} \vartheta_1^T < 0 \tag{4.25}$$
$$\Theta + \vartheta_2 R_1^{-1} \vartheta_2^T < 0 \tag{4.26}$$

Using Schur-complement (4.25)and (4.26) can be rewritten as

$$\begin{bmatrix} \Theta & \vartheta_1 \\ \star & -R_1 \end{bmatrix} < 0 \tag{4.27}$$
$$\begin{bmatrix} \Theta & \vartheta_2 \\ \star & -R_1 \end{bmatrix} < 0 \tag{4.28}$$

where, $\vartheta_1 = \begin{bmatrix} M_1^T \ N_1^T \ 0 \ 0 \ 0 \end{bmatrix}^T$ and $\vartheta_2 = \begin{bmatrix} 0 \ M_2^T \ N_2^T \ 0 \ 0 \end{bmatrix}^T$ are defined in Theorem 4.2.

Now, pre- and post multiplying (4.27) and (4.28) by $diag\{S^{-1}, S^{-1}, S^{-1}, S^{-1}, D^{-1}, S^{-1}\}$ and $diag\{S^{-T}, S^{-T}, S^{-T}, S^{-T}, D^{-1}, S^{-T}\}$ respectively and adopting following changes in variables

$$S^{-1} = Z;\ S^{-T} = Z^T;\ D^{-1} = L, KZ^T = Y;\ ZK^T = Y^T;\ GZ^T = W, ZG^T = W^T;$$
$$S^{-1}R_1S^{-T} = \bar{R}_1;\ S^{-1}PS^{-T} = \bar{P};\ S^{-1}Q_1S^{-T} = \bar{Q}_1;\ S^{-1}Q_2S^{-T} = \bar{Q}_2$$

may obtain the following LMIs

$$\begin{bmatrix} \Sigma_{11} & \Sigma_{12} & 0 & \Sigma_{14} & \Sigma_{15} & \bar{M}_1 \\ \star & \Sigma_{22} & \Sigma_{23} & \Sigma_{24} & \Sigma_{25} & \bar{N}_1 \\ 0 & \star & \Sigma_{33} & 0 & 0 & 0 \\ \star & \star & 0 & \Sigma_{44} & \Sigma_{45} & 0 \\ \star & \star & 0 & \star & \Sigma_{55} & 0 \\ \star & \star & 0 & 0 & 0 & -\bar{R}_1 \end{bmatrix} < 0 \tag{4.29}$$

$$\begin{bmatrix} \Sigma_{11} & \Sigma_{12} & 0 & \Sigma_{14} & \Sigma_{15} & 0 \\ \star & \Sigma_{22} & \Sigma_{23} & \Sigma_{24} & \Sigma_{25} & \bar{M}_2 \\ 0 & \star & \Sigma_{33} & 0 & 0 & \bar{N}_2 \\ \star & \star & 0 & \Sigma_{44} & \Sigma_{45} & 0 \\ \star & \star & 0 & \star & \Sigma_{55} & 0 \\ \star & \star & 0 & 0 & 0 & -\bar{R}_1 \end{bmatrix} < 0 \tag{4.30}$$

The elements $\Sigma_{i,j}$ $(i = 1, \ldots, 5\ and\ j = 1, 2, \ldots, 5)$ in (4.29) and (4.30) are defined in the Theorem 4.1.

To ensure that, $\dot{V}(x_t) < 0$, one must satisfy (4.29) and (4.30) for any $x(t) \in \mathcal{E}(P, 1)$, satisfying the condition $\mathcal{E}(P, 1) \subset \mathcal{S}$. The ellipsoid set along with the condition that guarantees $|(k_i - g_i)x| \leq u_{0(i)}, i = 1, \ldots m, \forall x(t) \in \mathcal{E}(P, 1)$ is necessary.

Following facts leads to formulate the equivalent LMI constraints as,

$$\mathcal{E}(P,1) = \{x \in \mathcal{R}^n | x^T P x\} \tag{4.31}$$

$$S = \{x \in \mathcal{R}^n || (k_i - g_i)x| \leq u_{O(i)}\} \tag{4.32}$$

and,

$$\mathcal{E}(P,1) \subset S \tag{4.33}$$

Using (4.31) one can write,

$$u_{O(i)}(1 + x^T P x) \leq 2u_{O(i)} \tag{4.34}$$

Using (4.32) one can write,

$$2|(k_i - g_i)x| \leq 2u_{O(i)} \tag{4.35}$$

Now, any point on the hyperplane $(k_i - g_i)x(t) = \pm u_{O(i)}$ makes the expression $x^T(t)Px(t) \geq 1$, which implies the condition (4.33), thus in view of this fact one can write,

$$u_{O(i)}(1 + x^T(t)Px(t)) \geq 2|(k_i - g_i)x(t)| \tag{4.36}$$

Thus in view of inequalities expressed in (4.34)–(4.36) the inequality holds,

$$2u_{O(i)} \geq u_{O(i)}(1 + x^T(t)Px(t)) \geq 2|(k_i - g_i)x(t)| \tag{4.37}$$

The latter inequality in (4.37) can be rewritten as,

$$u_{O(i)}(1 + x^T(t)Px(t)) - 2|(k_i - g_i)x(t)| \geq 0 \tag{4.38}$$

For any values of $x(t)$ taken from $\mathcal{R}^n$ space, one can rewrite (4.38) as,

$$\begin{bmatrix} 1 \pm x^T(t) \end{bmatrix} \begin{bmatrix} u_{O(i)} & (k_i - g_i) \\ \star & u_{O(i)}P \end{bmatrix} \begin{bmatrix} 1 \\ \pm x(t) \end{bmatrix} \geq 0 \tag{4.39}$$

The satisfaction of (4.39) gives the LMI condition,

$$\begin{bmatrix} u_{O(i)} & (k_i - g_i) \\ \star & u_{O(i)}P \end{bmatrix} \geq 0 \tag{4.40}$$

The LMI condition in (4.40) can be equivalently written using Schur-complement as,

$$\begin{bmatrix} P & (k_i - g_i)^T \\ \star & u_{0(i)}^2 \end{bmatrix} \geq 0 \tag{4.41}$$

Now, pre- and post multiplying (4.41) with $diag\{S^{-1}, I_n\}$ and $diag\{S^{-T}, I_n\}$ respectively, one can obtain

$$\begin{bmatrix} \bar{P} & (y_i - w_i)^T \\ \star & u_{0(i)}^2 \end{bmatrix} \geq 0 \tag{4.42}$$

The LMIs (4.29), (4.30) and (4.42) ensures that $\dot{V}(x_t) < -\eta \parallel x_t \parallel^2$ for a sufficiently small $\eta > 0$, thus $V(x_t) < V(x_{t_0})$ which consequently implies
$x^T(t)Px(t) \leq V(x_t) \leq V(x_{t_0})$

$$\begin{aligned} x^T(t)Px(t) \leq & \max_{\theta \in [-d_u, 0]} \|\phi(\theta)\|_2^2 \{\lambda_{max}(Z^{-1}\bar{P}Z^{-T}) \\ & + d_u\lambda_{max}(Z^{-1}\bar{Q}_1Z^{-T}) + d_u\lambda_{max}(Z^{-1}\bar{Q}_2Z^{-T})\} \\ & + \max_{\theta \in [-d_u, 0]} \|\dot{\phi}(\theta)\|_2^2 \left\{\frac{d_u^2}{2} d_u^{-1}\lambda_{max}(Z^{-1}\bar{R}_1Z^{-T})\right\} \end{aligned} \tag{4.43}$$

$$\begin{aligned} x^T(t)Px(t) \leq & \delta_1^2 \{\lambda_{max}(Z^{-1}\bar{P}Z^{-T}) \\ & + d_u\lambda_{max}(Z^{-1}\bar{Q}_1Z^{-T}) + d_u\lambda_{max}(Z^{-1}\bar{Q}_2Z^{-T})\} \\ & + \delta_2^2 \left\{\frac{d_u}{2}\lambda_{max}(Z^{-1}\bar{R}_1Z^{-T})\right\} \leq 1 \end{aligned} \tag{4.44}$$

The set $\mathcal{X}_\delta$ is a set of initial conditions and is defined in (4.10). If the condition $\mathcal{X}_\delta \leq 1$ belongs to $\mathcal{E}(P, 1)$, this implies that all the trajectories of $x(t)$ starting from $\mathcal{X}_\delta \leq 1$ remains within $x^T(t)Px(t) \leq 1$, and thereby constraint on the control $|(k_i - h_i)x(t)| \leq u_{0(i)}, i = 1, 2, \ldots m$ is also satisfied due to (4.42). This completes the proof. ■

4.4.2 Optimization Algorithm

To obtain the maximum estimate of the domain of attraction (DOA) for a particular delay value an LMI based optimization routine is proposed following [19, 23]. The tuning (or design) parameters α and β are determined in an adhoc fashion iteratively or one can use "fminsearch" algorithm available in MATLAB optimization toolbox. One can select $\delta_1 = \delta_2 = \delta$ in (4.44) [15, 19].

$$\text{Minimize } \kappa$$
$$\text{subject to (4.15)–(4.17)}, \begin{bmatrix} e_1 I_n & I_n \\ \star & Z \end{bmatrix} \geq 0$$
$$e_2 I_n - \bar{P} \geq 0 \; e_3 I_n - \bar{Q}_1 \geq 0; \; e_4 I_n - \bar{Q}_2 \geq 0; \; e_5 I_n - \bar{R}_1 \geq 0$$

where, $\kappa = \sigma * e_1 + e_2 + d_u e_3 + d_u e_4 + \frac{d_u}{2} e_5$

In minimization problem $e_i, i = 1, 2, \ldots 5$ and σ are the weights introduced for optimization.

The maximum domain of attraction is obtained by

$$\delta_{max} = \frac{1}{\sqrt{\varrho}}$$

where,

$$\varrho = \lambda_{max}(Z^{-1}\bar{P}Z^{-T}) + d_u \lambda_{max}(Z^{-1}\bar{Q}_1 Z^{-T})$$
$$+ d_u \lambda_{max}(Z^{-1}\bar{Q}_2 Z^{-T}) + \frac{d_u}{2}\lambda_{max}(Z^{-1}\bar{R}_1 Z^{-T})$$

Numerical Example 4.1 ([15]) Consider a linear state-delayed system with actuator saturation with the following constant matrices,

$$A = \begin{bmatrix} 0.5 & -1 \\ 0.5 & -0.5 \end{bmatrix}; \; A_d = \begin{bmatrix} 0.6 & 0.4 \\ 0 & -0.5 \end{bmatrix}; \; B = \begin{bmatrix} 1 \\ 1 \end{bmatrix}; \; \textit{and } u_{0(i)} = 5$$

We solve this problem for constant delay case ($\mu = 0$), for a considered delay bound (d_u) an appropriate choice of weight σ is introduced into the minimization problem. The scalar variables α and β are tuned such that less conservative delay upper bound is obtained compared to then the existing results and simultaneously domain of attraction is also estimated for a given delay. The comparative studies are made in Table 4.1. Subsequently, we have solved this problem for different delay derivatives ($\mu = 0.3$ and $\mu = 1.5$), the results are presented in Tables 4.2 and 4.3 respectively. It is observed that, as the delay derivative increases the maximum delay upper bound decreases, which is an expected result. It may be noted that, the matrix $(A + A_d)$ is unstable (for this numerical example), thus violating the necessary condition of delay-dependent stability analysis. Stabilization of TDS in presence of actuator saturation (4.2) using the proposed technique has been illustrated with this example.

Table 4.1 Computation result of Example 4.1 for $\mu = 0$

Methods	d_u	(α, β)	δ	K
Theorem 4.1	0.35	(0.7, −0.4)	2.7972	[−1.5743, 0.4505]
[15]	0.35	'(α, β)' not used as parameters	2.852	–
[14]	0.35	'(α, β)' not used as parameters	0.968	–
Theorem 4.1	1	(0.91, −0.59)	1.6141	[−2.8851, 0.6638]
[15]	1	'(α, β)' not used as parameters	1.7442	–
[14]	1	'(α, β)' not used as parameters	Infeasible	–
Theorem 4.1	1.854	(1.71, −0.74)	0.6626	[−1.7008, 0.2776]
[15]	1.854	'(α, β)' not used as parameters	0.091	[−25.8809, −4.9315]
[14]	1.854	'(α, β)' not used as parameters	Infeasible	–
Theorem 4.1	2.5	(2.34, −0.88)	0.2295	[−3.6360, −0.9380]
[15]	2.5	'(α, β)' not used as parameters	Infeasible	–
[14]	2.5	'(α, β)' not used as parameters	Infeasible	–
Theorem 4.1	3.310	(9, −0.998)	0.0024	[−605.9023, −401.8906]
[15]	3.310	'(α, β)' not used as parameters	Infeasible	–
[14]	3.310	'(α, β)' not used as parameters	Infeasible	–

'–' means result is not available in the reference

Table 4.2 Computation result of Example 4.1 for $\mu = 0.3$

d_u	(α, β)	δ	K
3.170	(13.9, −0.83)	0.0105	[−247.2298, −159.5089]

Table 4.3 Computation result of Example 4.1 for $\mu = 1.5$

d_u	(α, β)	δ	K
2.3721	(15.6, −0.38)	0.0437	[−39.7363, −23.1264]

Numerical Example 4.2 ([18]) Consider system (4.2) with following constant matrices,

$$A = \begin{bmatrix} 1 & 1.5 \\ 0.3 & -2 \end{bmatrix}; \quad A_d = \begin{bmatrix} 0 & -1 \\ 0 & 0 \end{bmatrix}; \quad B = \begin{bmatrix} 10 \\ 1 \end{bmatrix}; \quad and\ u_{0(i)} = 15$$

With the proper choice of weight (σ) on the optimization problem, we obtain the maximum estimate of DOA for a given delay upper bound ($d_u = 1$) and delay-derivative $\mu = 0$. The comparative results of DOA and controller gains are shown in Table 4.4. Similarly one may note, that the matrix $(A + A_d)$ is unstable.

Table 4.4 Computation result of Example 4.2 for $d_u = 1$ for $\mu = 0$

Methods	α	β	δ	K
Theorem 4.1	0.45	−0.021	106.2856	[−0.6646, −0.0239]
[18]	*	*	83.55	[−0.1950, 0.0649]
[15]	*	*	79.43	[−7.913, 0.7323]
[13]	*	*	58.395	–

'*' indicated parameters not used and '–' means result is not available in the reference

4.4.3 Simulation Results

The results obtained by solving LMIs in Theorem 4.1 have been used to simulate the system considered in Numerical Example 4.1 for $\mu = 0$, $d_u = 1.854$ s (corresponding to radius of DOA $\delta = 0.6626$). Figure 4.7 shows that the estimated domain of attraction is inside the ellipsoid set $\mathcal{E}(P, 1)$, Fig. 4.8 (enlarged part of estimated DOA region in Fig. 4.7) depicts that the state trajectories starting from the periphery of the estimated DOA finally converge to origin (inside the DOA), finally Figs. 4.9 and 4.10 show the response of state trajectories with an initial condition of $x(0) = [-0.57, 0.3327]^T$. The control input signal applied to the system is shown in Fig. 4.11 is well within the saturation limits.

Remark 4.2 A new delay-dependent LMI based stability criteria has been derived for local (regional) stabilization of linear time-delay systems with actuator saturation. The saturation nonlinearity is treated using dead-zone nonlinearity [18] which in turn, reduces the number of solvable LMIs compared to polytopic differential inclusion representation of the saturation nonlinearity [4].

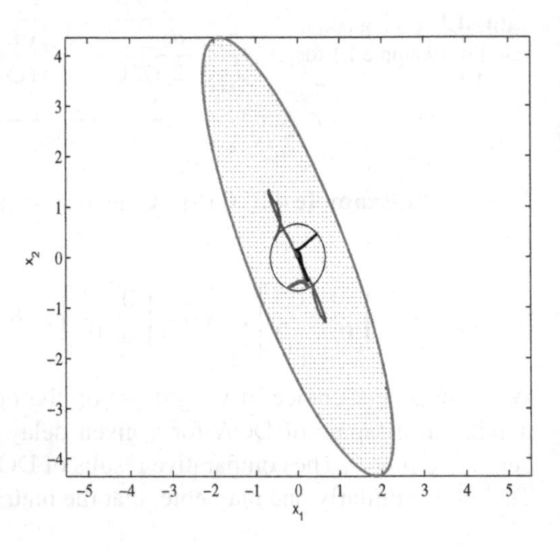

Fig. 4.7 Estimated DOA inside the ellipsoid

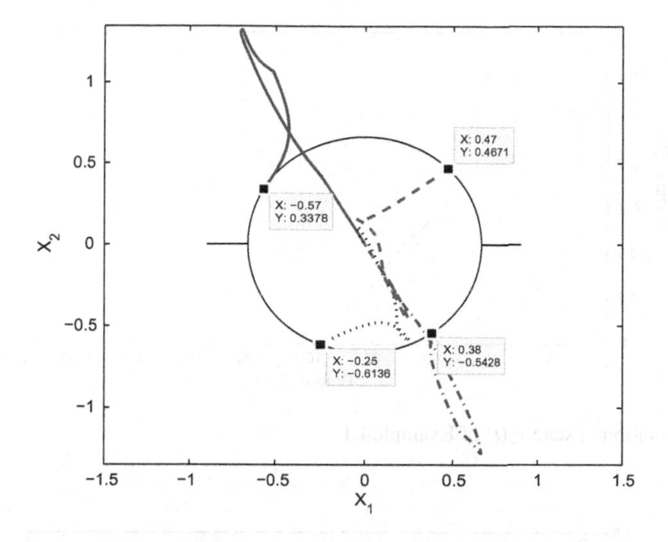

Fig. 4.8 Convergence of state trajectories inside the DOA

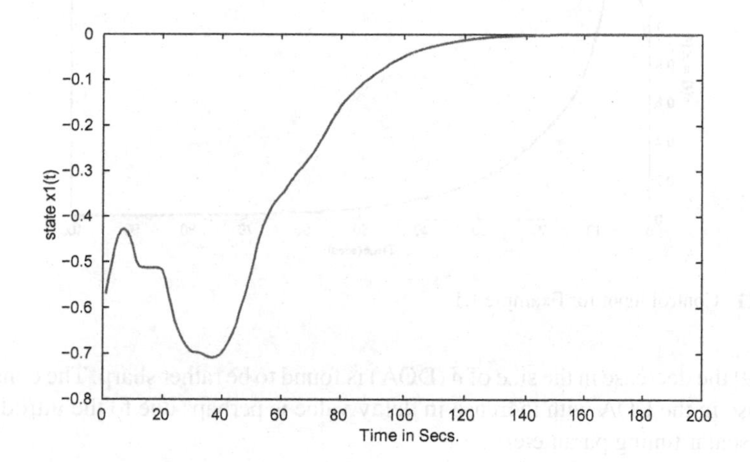

Fig. 4.9 Solution of state $x_1(t)$ of Example 4.1

Remark 4.3 The proposed method is new in following respect (i) it uses new bounding technique to approximate the integral inequality arising out of the LK functional derivative, (ii) it introduces two scalar variables α and β that are treated as design parameters and allows to obtain improved stabilization region compared to existing methods. It has also been observed that the merits of delay upper bound and the estimate of the domain of attraction are inversely related. It is to be mentioned here that, the decrease in the estimate of domain of attraction with increase in delay bound is quite uniform in case of proposed method, whereas in case of other existing results

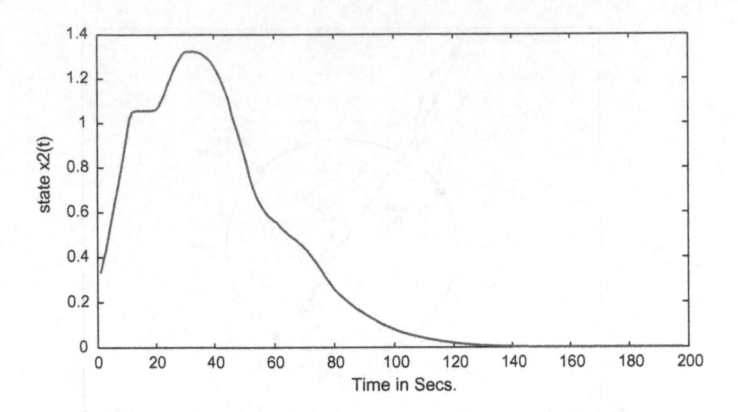

Fig. 4.10 Solution of state $x_2(t)$ of Example 4.1

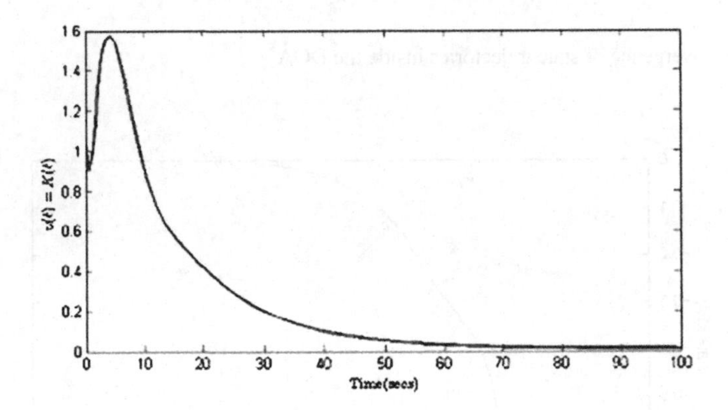

Fig. 4.11 Control input for Example 4.1

[15, 19] the decrease in the size of δ (DOA) is found to be rather sharp. The consistent decrease in the DOA with increase in delay value is perhaps due to the introduction of the scalar tuning parameters.

Remark 4.4 The choice of the scalar variables (α and β) to obtain the feasibility of the LMI conditions and subsequently maximizing the domain of attraction is done in an adhoc manner. One of the suggested tuning rule can be the implementation of iterative "*fminsearch*" algorithm to satisfy the obtained LMI conditions.

4.4.4 Stabilization Using Polytopic Approximation

In addition to the Lemma 1.1, following lemmas will be useful in solving the stabilization problems.

Lemma 4.2 ([15]) *Assume an auxiliary matrix feedback $H \in \mathcal{R}^{m\times n}$ such that $|h_i x| \leq u_{0(i)}$ $\forall x \in \mathbf{S}$, where $\mathbf{S}$ is any convex compact set defined as $\mathbf{S} = \{x \in \Re^n : |h_i x| \leq u_{0(i)}\}$, $i = 1, 2, \ldots, m$. Now for any $x(t) \in \mathbf{S}$ the system (4.2) can be represented as,*

$$\dot{x}(t) = \sum_{j=1}^{2^m} \lambda_j(t) A_j x(t) + A_d x(t - d(t)) \tag{4.45}$$

with,

$$\sum_{j=1}^{2^m} \lambda_j(t) = 1, \, and \, \lambda_j(t) \geq 0 \tag{4.46}$$

Let us assume that, $\bar{A} = \sum_{j=1}^{2^m} \lambda_j(t) A_j$, thus one can write (4.45) as,

$$\dot{x}(t) = \bar{A}x(t) + A_d x(t - d(t)) \tag{4.47}$$

Note that, the matrix A_j are the vertices of a convex polyhedron of matrices and for $x(t) \in \mathcal{S}$ using Lemma 1.1 one can get,

$$A_j = (A + B(D_j K + D_j^- H)) \in co\{A_1, \ldots, A_{2^m}\} \tag{4.48}$$

A modified delay-dependent stabilization of the system (4.47) is derived by following the procedures (i) congruence transformation for constructing LK functional (ii) estimate of domain of attraction and (ii) optimization algorithm as given in [19]. But unlike the introduction of $W \in \mathcal{R}^{m\times n}$ matrix in [19], we adopt the Lemma 4.2 of [15] to place the saturation function in the convex hull of group of linear feedbacks K and H. The purpose of this derivation is to show that a similar LMI structure can be obtained as in Corollary 1 of [19] for ascertaining the stabilization of (4.47). The modified delay-dependent stabilization criteria is presented below in the form of lemma.

Lemma 4.3 ([33]) *The time-delay saturating actuator system (4.47) is asymptotically stable via memory less state feedback controller $K = FX^{-1}$ for an allowable delay, if there exist symmetric matrices $X > 0$, $\bar{Q}_1 > 0$, $\bar{Q}_2 > 0$, $\bar{P} > 0$ and any free matrices $\bar{N}_k$, $k = 1, 2, 3$, F and G, of appropriate dimensions such that following LMIs hold,*

$$\bar{\Theta}_j = \begin{bmatrix} \bar{\theta}_{11} & \bar{\theta}_{12} & \bar{\theta}_{13} & \bar{N}_1 \\ \star & \bar{\theta}_{22} & \bar{\theta}_{23} & \bar{N}_2 \\ \star & \star & \bar{\theta}_{33} & \bar{N}_3 \\ \star & \star & \star & \bar{\theta}_{44} \end{bmatrix} < 0, j = 1, 2, \ldots 2^m \tag{4.49}$$

$$\begin{bmatrix} u_{0(i)} & g_i \\ \star & u_{0(i)}\bar{P} \end{bmatrix} \geq 0, i = 1, \ldots, m \tag{4.50}$$

where, g_i and denote the ith row of matrix G and,

$$\bar{\theta}_{11}{=}AX + XA^T + B(D_jK + D_j^- H)X + X(D_jK + D_j^- H)^T B^T + \bar{Q}_1 + \bar{N}_1 + \bar{N}_1^T$$
$$\bar{\theta}_{12}{=}A_dX - \bar{N}_1 + \bar{N}_2^T;\ \bar{\theta}_{13}{=}{-X} + XA^T + \bar{P} + X(D_jK + D_j^- H)^T B^T$$
$$\bar{\theta}_{22}{=}{-(1-\mu)}\bar{Q}_1 - \bar{N}_2 - \bar{N}_2^T;\ \bar{\theta}_{23}{=}XA_d^T$$
$$\bar{\theta}_{33}{=}d_u\bar{Q}_2 - 2X;\ \bar{\theta}_{44}{=}{-d_u^{-1}}\bar{Q}_2$$

then the closed-loop system (4.47) is asymptotically stable and the estimate of the domain of attraction is given by $\mathcal{X}_\delta \leq 1$, where,

$$\mathcal{X}_\delta = \delta_1^2 \left\{\lambda_{max}(X^{-1}\bar{P}X^{-1}) + d_u\lambda_{max}(X^{-1}\bar{Q}_1X^{-1})\right\} + \delta_2^2 \left\{\frac{d_u^2}{2}\lambda_{max}(X^{-1}\bar{Q}_2X^{-1})\right\} \tag{4.51}$$

Proof The detailed proof can be found in [33]. Here we present the brief idea (outline of the proof). Consider LK functional for the system in (4.2) or equivalently for (4.47) as,

$$V(t) = x^T Px(t) + \int_{t-d(t)}^{t} x^T(\alpha)Q_1x(\alpha)d\alpha + \int_{-d_u}^{0}\int_{t+\alpha}^{t} \dot{x}^T(\alpha)Q_2\dot{x}(\alpha)d\alpha ds \tag{4.52}$$

The matrix variables in (4.52) is defined as in [19],

$$P = X^{-1}\bar{P}X^{-1},\ \ Q_1 = X^{-1}\bar{Q}_1X^{-1},\ \ Q_2 = X^{-1}\bar{Q}_2X^{-1}$$

Now, finding the time-derivative of (4.52), one can get,

$$\dot{V}(t) \leq \dot{x}^T Px(t) + x^T P\dot{x}(t) + x^T Q_1x(t) + d_u\dot{x}^T(t)Q_2\dot{x}(t) - (1-\mu)x^T(t-d(t))Q_1x(t-d(t)) - \int_{t-d_u}^{t} \dot{x}^T(s)Q_2\dot{x}(s)ds \tag{4.53}$$

Defining an augmented state vector $\eta(t) = [x^T(t), x^T(t-d(t)), \dot{x}^T(t)]^T$ one can write the integral term in (4.53) along with the introduction of free matrices N_k, $(k = 1, 2, 3)$ and using integral inequality (one can refer Lemma 2 of [31]) as,

$$-\int_{t-d_u}^{t} \dot{x}^T(s)Q_2\dot{x}(s)ds \le \eta^T(t)\left\{\begin{bmatrix} N_1+N_1^T & -N_1+N_2^T & N_3^T \\ \star & -N_2-N_2^T & -N_3^T \\ \star & \star & 0 \end{bmatrix} + \begin{bmatrix} N_1 \\ N_2 \\ N_3 \end{bmatrix} d_u Q_2^{-1}\left[N_1^T\ N_2^T\ N_3^T\right]\right\}\eta(t) \quad (4.54)$$

As $\dot{x}(t)$ is introduced in the augmented state vector so to introduce system matrices in the LMI condition introducing following equality,

$$2\left[x^T(t)T+\dot{x}^T(t)T\right]\times\left[-\dot{x}(t)+\bar{A}_jx(t)+A_dx(t-d(t))\right]=0 \quad (4.55)$$

where matrix $T=T^T>0$. Carrying out algebraic manipulations, substitutions and finally using Schur-complement one may obtain,

$$\dot{V}(t)\le\sum_{j=1}^{2^m}\lambda_j(t)\{\eta^T(t)\Theta_j\eta(t)\} \quad (4.56)$$

where,

$$\Theta_j=\begin{bmatrix} \theta_{11} & \theta_{12} & \theta_{13} & N_1 \\ \star & \theta_{22} & \theta_{23} & N_2 \\ \star & \star & \theta_{33} & N_3 \\ \star & \star & \star & -d_u^{-1}Q_2 \end{bmatrix} \quad (4.57)$$

where, $\theta_{11}=TA+A^TT+\sum_{j=1}^{2^m}\lambda_j(t)\{TB(D_jK+D_j^-H)+(D_jK+D_j^-H)^TB^TT\}+Q_1+N_1+N_1^T$, $\theta_{12}=TA_d-N-1+N_2^T$, $\theta_{13}=-T+A^TT+\sum_{j=1}^{2^m}\lambda_j(t)\{(D_jK+D_j^-H)^TB^TT\}+P+N_3^T$
$\theta_{22}=-(1-\mu)Q_1-N_2-N_2^T$, $\theta_{23}=A_d^TT-N_3^T$, $\theta_{33}=-2T+d_uQ_2$.

Set, $X=T^{-1}$ and performing pre- and post-multiplication of matrix Θ_j by $diag\{X,X,X,X\}$, one can obtain the following,

$$\dot{V}(t)\le\sum_{j=1}^{2^m}\lambda_j(t)\left\{\eta^T(t)\bar{\Theta}_j\eta(t)\right\} \quad (4.58)$$

where,

$$\bar{\Theta}_j=\begin{bmatrix} \bar{\theta}_{11} & \bar{\theta}_{12} & \bar{\theta}_{13} & \bar{N}_1 \\ \star & \bar{\theta}_{22} & \bar{\theta}_{23} & \bar{N}_2 \\ \star & \star & \theta_{33} & \bar{N}_3 \\ \star & \star & \star & -d_u^{-1}\bar{Q}_2 \end{bmatrix} \quad (4.59)$$

the change in variables after the pre- and post-multiplications are, $XN_kX = \bar{N}_k$, $(k = 1, 2, 3)$, $KX = F$ and $HX = G$. The elements of the matrix $\bar{\Theta}_j$ are already defined in (4.49) (see Lemma 4.3).

Asymptotic stability of the system (4.47) implies that, $\dot{V}(t) < 0$ in (4.58), for which $\bar{\Theta}_j < 0$ for any $x(t) \in \mathcal{S}$, where $\mathcal{S}$ denotes any compact convex set. Here we consider this set to be ellipsoid ($\mathcal{E}(P, 1)$) such that, $\mathcal{E}(P, 1) \subset \mathcal{S}$ and thus guaranteeing the condition $|h_i x(t)| \leq u_{0(i)}, \forall x(t) \in \mathcal{E}(P, 1)$. The set $\mathcal{E}(P, 1)$ is defined as, $\mathcal{E}(P, 1) \equiv \{x(t) : x^T(t)Px(t) \leq 1\}$. This condition results from the fact that, when $x(t) \in \mathcal{E}(P, 1)$ then following inequalities hold as discussed in Theorem 4.1 and [15],

$$2u_{0(i)} \geq u_{0(i)}(1 + x^T(t)Px(t)) \geq 2|h_i x(t)|, i = 1, 2, \ldots m$$

which implies that, $|h_i x(t)| \leq u_{0(i)}$, this is interpreted in terms of LMI as,

$$\begin{bmatrix} u_{0(i)} & h_i \\ \star & u_{0(i)}P \end{bmatrix} \geq 0 \tag{4.60}$$

Pre- and post-multiplying (4.60) with $diag\{I, X\}$ one can get,

$$\begin{bmatrix} u_{0(i)} & g_i \\ \star & u_{0(i)}\bar{P} \end{bmatrix} \geq 0 \tag{4.61}$$

where, g_i is the ith row of the G matrix and $X = T^{-1}$. As $\dot{V}(x_t) < 0$, if the matrix $\bar{\Theta}_j < 0$ in (4.58), which implies that, $V(x_t) < V(x_{t_0})$ and in this view one can infer that,

$$x^T(t)Px(t) \leq V(x_t) \leq V(x_{t_0})$$

$$\begin{aligned} x^T(t)Px(t) \leq & \max_{\theta \in [-d_u, 0]} \|\phi(\theta)\|_2^2 \{\lambda_{max}(X^{-1}\bar{P}X^{-1}) + d_u\lambda_{max}(X^{-1}\bar{Q}_1X^{-1})\} \\ & + \max_{\theta \in [-d_u, 0]} \|\dot{\phi}(\theta)\|_2^2 \left\{\frac{d_u^2}{2}\lambda_{max}(X^{-1}\bar{Q}_2X^{-1})\right\} \end{aligned}$$

$$\begin{aligned} x^T(t)Px(t) \leq & \delta_1^2 \{\lambda_{max}(X^{-1}\bar{P}X^{-1}) + d_u\lambda_{max}(X^{-1}\bar{Q}_1X^{-1})\} \\ & + \delta_2^2 \left\{\frac{d_u^2}{2}\lambda_{max}(X^{-1}\bar{Q}_2X^{-1})\right\} \leq 1 \end{aligned} \tag{4.62}$$

If $\dot{V}(x_t) < 0$ then the inequality (4.62) gives the estimate of the domain of attraction and guarantees that for all the initial functions $\phi \in \mathcal{X}_\delta$, the trajectories of $x(t)$

remain within $\mathcal{E}(P, 1)$, which in turn implies that the stabilization of polytopic time-delay system (4.47) is equivalent to the time-delay systems with actuator saturation in (4.2). ■

Optimization algorithm:
Here, the domain of attraction for a particular delay value is maximized following the optimization algorithm of [13, 19], for which one selects $\delta_1 = \delta_2 = \delta$ in (4.62) or (4.65).

$$\text{Min. } \kappa$$

$$\text{s.t. (4.49), (4.50) and } \begin{bmatrix} w_1 I & I \\ \star & X \end{bmatrix} \geq 0,$$

$$w_2 I - \bar{P} \geq 0, \quad w_3 I - \bar{Q}_1 \geq 0, \quad w_4 I - \bar{Q}_2 \geq 0,$$

where, $\kappa = \epsilon \times w_1 + w_2 + d_u w_3 + \frac{d_u^2}{2} w_4$.

The variables $w_i, i = 1, \ldots 4$ are introduced for multi-objective optimization problem and ϵ is the weight in the optimization routine for obtaining the maximized estimate of DOA. The maximized DOA can be obtained by $\delta_{max} = \frac{1}{\sqrt{\Lambda}}$, where, $\Lambda = \lambda_{max}(X^{-1}\bar{P}X^{-1}) + d_u\lambda_{max}(X^{-1}\bar{Q}_1X^{-1}) + \frac{d_u^2}{2}\lambda_{max}(X^{-1}\bar{Q}_2X^{-1})$ with $X > 0, \bar{P} > 0, \bar{Q}_1 > 0$ and $\bar{Q}_2 > 0$.

Remark 4.5 The LMIs in (4.49) and (4.50) are same as in Corollary 1 of [19] for $\mu = 0$ as we adopt the same congruence transformation. Next, considering the matrix $T = 0$ and expressing $\dot{x}(t)$ in terms of the current as well as delayed states, a new stabilization condition using polytopic representation for actuator saturation of time-delay system (4.47) with the same choice of LK functional considered in Lemma 4.3 is presented below in the form of theorem.

Theorem 4.2 ([33]) *Consider the time-delay systems with actuator saturation (4.47), for any delay $d(t)$ satisfying the conditions (4.7) and (4.8), if there exist symmetric matrices $Y > 0, \bar{Q}_1 > 0, \bar{Q}_2 > 0$ and any free matrices $\bar{N}_k, (k = 1, 2)$, F, G and scalar $\epsilon > 0$, such that the following LMIs hold:*

$$\Pi_j = \begin{bmatrix} \pi_{11} & \pi_{12} & d_u\bar{N}_1 & \pi_{14} \\ \star & \pi_{22} & d_u\bar{N}_2 & d_u Y A_d^T \\ \star & \star & -d_u\alpha_1^{-1}Y & 0 \\ \star & \star & 0 & -d_u\alpha_1 Y \end{bmatrix} < 0 \tag{4.63}$$

where, $\pi_{11} = YA^T + AY + \bar{Q}_1 + \bar{N}_1 + \bar{N}_1^T + Y(D_jK + D_j^- H)^T B^T + B(D_jK + D_j^- H)Y$,

$$\pi_{12} = A_d Y - \bar{N}_1 + \bar{N}_2^T, \quad \pi_{14} = d_u Y A^T + d_u Y(D_jK + D_j^- H)^T B^T$$

$$\pi_{22} = -(1 - \mu)\bar{Q}_1 - \bar{N}_2 - \bar{N}_2^T$$

$$\begin{bmatrix} u_{0(i)} & g_i \\ \star & u_{0(i)}Y \end{bmatrix} \geq 0 \tag{4.64}$$

then the closed-loop system (4.47) is asymptotically stable and the estimate of the domain of attraction is given by $\mathcal{X}_\delta \leq 1$, where,

$$\begin{aligned} \mathcal{X}_\delta = &\ \delta_1^2\{\lambda_{max}(Y^{-1}) + d_u\lambda_{max}(Y^{-1}\bar{Q}_1Y^{-1})\} \\ &+\delta_2^2\left\{\frac{d_u^2}{2}\lambda_{max}(Y^{-1})\right\} \end{aligned} \tag{4.65}$$

Further the feedback gain is given by $K = FY^{-1}$.

Proof The detailed proof can be found in [33]. It is proved with same LK functional as in Lemma 4.3 with $T = 0$. □

Optimization algorithm:
Here, the domain of attraction for a particular delay value is maximized following the optimization algorithm of [13, 19], for which one selects $\delta_1 = \delta_2 = \delta$ in (4.62) or (4.65).

$$\begin{aligned} &\text{Min. } \kappa \\ &\text{s.t. } (4.63), (4.64), \begin{bmatrix} w_1 I & I \\ \star & Y \end{bmatrix} \geq 0, \\ &w_2 I - \bar{Q}_1 \geq 0, \begin{bmatrix} w_3 I & I \\ \star & \alpha_1 Y \end{bmatrix} \geq 0 \end{aligned}$$

where, $\kappa = \epsilon \times w_1 + d_u w_2 + \frac{d_u^2}{2} w_3$.

Remark 4.6 The scalar variable ϵ in the objective function is treated as a tuning parameter whose appropriate selection leads to improved delay upper bound value with corresponding larger estimate of DOA.

Remark 4.7 The result proposed in Lemma 4.3 is examined with a Numerical Example 4.1 and it is observed that, maximum delay bound $d_u = 1.3608$ and δ_{max} =0.0091 (with $\epsilon = 10^3$). It must be mentioned here that, the LMI conditions obtained in Lemma 4.3 is exactly same as in Corollary 1 of [19], but it is claimed in [19] that the maximum delay bound $d_u = 2.248$ and δ_{max} =0.3272 for same value of ϵ using the same optimization algorithm.

Furthermore, Example 4.1 is considered to simulate the TDS with actuator saturation with the same values of $d_u = 2.248$, gain matrix $K = [-2.82, 0.21]$, $u_{0(i)} = \pm 5$ and the initial condition of $x(0) = [-0.27, 0.16]^T$ (which lies within the DOA $\delta_{max} = 0.3272$) as in [19]. It is found that, (i) the system response is unbounded (as shown in Figs. 4.12 and 4.13) and (ii) the phase plane plot is not enclosed inside the estimated DOA (as shown in the Fig. 4.14).

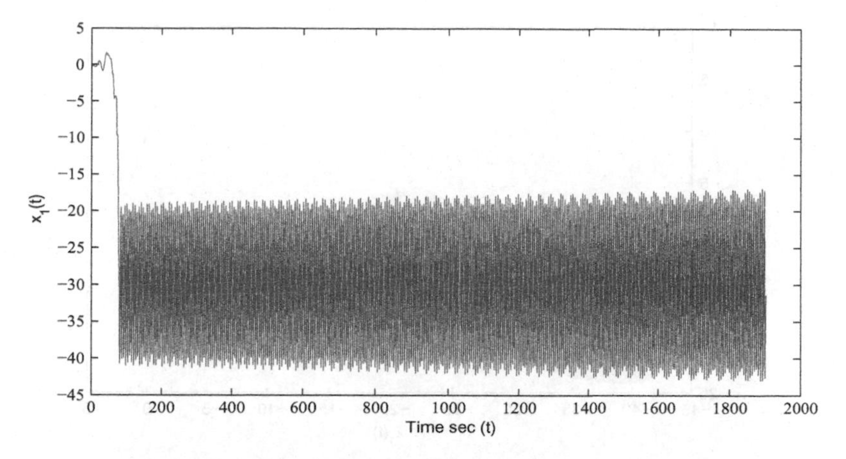

Fig. 4.12 Solution of state $x_1(t)$ of Example 4.1 by Corollary 1 of [19]

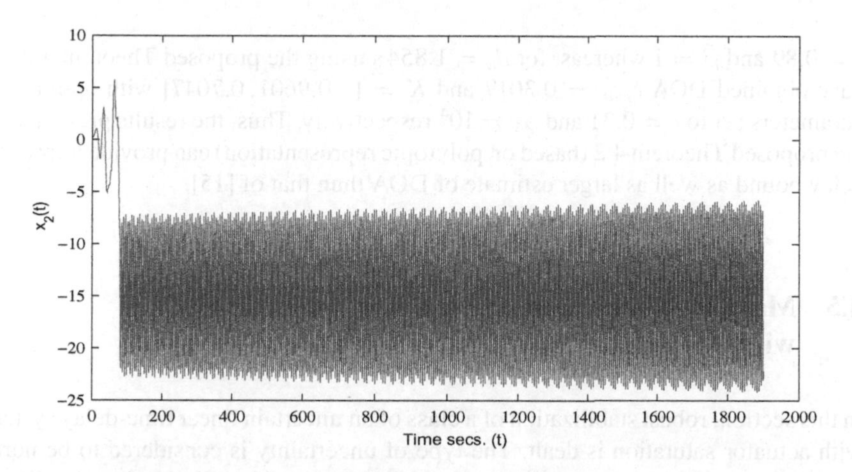

Fig. 4.13 Solution of state $x_2(t)$ of Example 4.1 by Corollary 1 of [19]

In view of the above, results of [19] appears to be erroneous and thus it has not been included for comparison in the preceding sections.

Remark 4.8 Results of Theorem 4.2 is demonstrated through the Numerical Example 4.1. The solution of LMIs (4.63) and (4.64) provide the maximum delay bound $d_u = 1.9555$, DOA $\delta_{max} = 0.0157$ and $K = [-0.9404, 0.7065]$ (with $\alpha_1 = 0.33$, and $\epsilon = 10^3$) for $\mu = 0$. As the stabilization condition obtained in Theorem 4.2 is NLMI, so one can transform it to an LMI condition using an assumption $X = \alpha_1 Y$, but with this transformation the stabilization result is expected to be more conservative.

In [15] the same Numerical Example (Example 4.1) for $\mu = 0$ is considered to obtain the upper delay bound as $d_u = 1.854$, DOA $\delta_{max} = 0.091$ and stabilizing gain is $K = -[25.8809, 4.9315]$ by adopting some optimizing parameter values

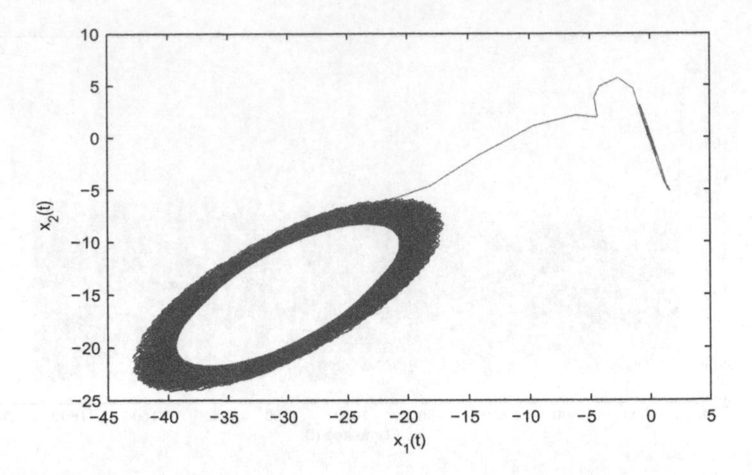

Fig. 4.14 Phase plane trajectory obtained for Example 4.1 by Corollary 1 of [19]

$\epsilon = 0.89$ and $\beta = 1$ whereas, for $d_u = 1.854$ s using the proposed Theorem 4.2 we have obtained DOA $\delta_{max} = 0.3017$ and $K = [-0.9601, 0.7047]$ with optimizing parameters set to $\epsilon = 0.31$ and $\alpha_1 = 10^4$ respectively. Thus, the results presented in the proposed Theorem 4.2 (based on polytopic representation) can provide improved delay bound as well as larger estimate of DOA than that of [15].

4.5 Main Result on Robust Stabilization of TDS with Actuator Saturation

In this section, robust stabilization of a class of an uncertain linear time-delay system with actuator saturation is dealt. The type of uncertainty is considered to be norm bounded type. The robust stabilization condition derived is a direct extension of the stabilization condition presented in Sect. 4.4.1. Before presenting the theorem on robust stabilization, we introduce the system description below.

4.5.1 Uncertain TDS with Actuator Saturation

Consider the following uncertain time-delay systems with actuator saturation,

$$\dot{x}(t) = A(t)x(t) + A_d(t)x(t - d(t)) + BSat(u(t)) \tag{4.66}$$

$$x(t) = \phi(t), \forall t \in [-d_u, 0] \tag{4.67}$$

The uncertain matrices $A(t)$ and $A_d(t)$ are defined below,

$$A(t) = A + \Delta A(t), \quad A_d(t) = A_d + \Delta A_d(t) \tag{4.68}$$

The uncertainties are assumed only in system matrices. A and A_d are constant known matrices and $\Delta A(t)$ and $\Delta A_d(t)$ are time-varying uncertain matrices of appropriate dimensions with the following uncertainty structure defined below:

$$\Delta A(t) = D_a F_a(t) E_a, \quad \Delta A_d(t) = D_d F_d(t) E_d \tag{4.69}$$

In (4.69), the $F_a(t)$ and $F_d(t)$ are unknown time-varying real matrices with Lebesgue measure elements satisfying the conditions,

$$F_a^T F_a(t) \leq I, \quad F_d^T(t) F_d(t) \leq I \tag{4.70}$$

where, D_a, D_d, E_a and E_d are known real constant matrices, $d(t)$ in (4.66) is the time-varying delay which satisfies the condition,

$$0 \leq d(t) \leq d_u, \quad 0 \leq \dot{d}(t) \leq \mu \tag{4.71}$$

where, $d_u > 0$ and $\mu < 1$.

Using the state feedback control law $u(t) = Kx(t)$ in (4.66), the closed-loop system under dead-zone nonlinearities representation becomes,

$$\dot{x}(t) = A_c(t)x(t) + A_d(t)x(t - d(t)) - B\psi(Kx(t)) \tag{4.72}$$

where, $A_c(t) = A(t) + BK$.

4.5.2 Robust Stabilization Using Sector Nonlinearities Approach

The control input saturation function (or called as actuator saturation) in (4.66) is already described in the Sect. 4.4.1.

In this section, we derive a delay-dependent robust stabilization condition to find feedback gains that can stabilize the uncertain time-delay systems with actuator saturation described in (4.72) with a view to achieve the maximum delay bound and simultaneously estimate the domain of attraction.

Theorem 4.3 *Given scalars α and β, the time-delay saturating actuator system (4.66) is asymptotically stable for allowable delay upper bound d_u via memory less state feedback controller $K = YZ^{-T}$, if there exist matrices $\bar{P} = \bar{P}^T > 0, \bar{Q}_j = \bar{Q}_j^T > 0, (j = 1, 2)$ $\bar{R}_1 = \bar{R}_1^T > 0$, any free matrices $Y, W, Z, \bar{M}_j, \bar{N}_j, j = 1, 2$ and diagonal matrix $L = L^T > 0$ of appropriate dimensions such that following LMIs hold*

$$\begin{bmatrix} \bar{\Pi} & \phi_1 \\ \star & -\bar{R}_1 \end{bmatrix} < 0 \tag{4.73}$$

$$\begin{bmatrix} \bar{\Pi} & \phi_2 \\ \star & -\bar{R}_1 \end{bmatrix} < 0 \tag{4.74}$$

$$\begin{bmatrix} \bar{P} & (y_i - w_i)^T \\ \star & u_{0(i)}^2 \end{bmatrix} \geq 0,\ i = 1, 2, \ldots m \tag{4.75}$$

y_i and w_i are the ith row of Y and W matrix respectively and the state feedback gain is given by $K = YZ^{-T}$. where,

$$\phi_1 = \begin{bmatrix} \bar{M_1}^T & \bar{N_1}^T & 0\,0\,0\,0\,0 \end{bmatrix}^T$$

$$\phi_2 = \begin{bmatrix} 0 & \bar{M_2}^T & \bar{N_2}^T & 0\,0\,0\,0 \end{bmatrix}^T$$

$$\bar{\Pi} = \begin{bmatrix} \Pi & E_1^T & E_2^T \\ \star & -\epsilon_1 I & 0 \\ \star & 0 & -\epsilon_2 I \end{bmatrix}$$

where,

$$E_1 = \begin{bmatrix} E_a Z^T & 0\,0\,0\,0 \end{bmatrix}$$
$$E_2 = \begin{bmatrix} E_d Z^T & 0\,0\,0\,0 \end{bmatrix}$$

and,

$$\Pi = \begin{bmatrix} \Pi_{11} & \Pi_{12} & 0 & \Pi_{14} & \Pi_{15} \\ \star & \Pi_{22} & \Pi_{23} & \Pi_{24} & \Pi_{25} \\ 0 & \star & \Pi_{33} & 0 & 0 \\ \star & \star & 0 & \Pi_{44} & \Pi_{45} \\ \star & \star & 0 & \star & \Pi_{55} \end{bmatrix}$$

where, $\Pi_{11} = AZ^T + ZA^T + BY + Y^T B^T + \bar{Q}_1 + \bar{Q}_2 + d_u^{-1}(\bar{M}_1 + \bar{M}_1^T) + \epsilon_1 D_a D_a^T + \epsilon_2 D_d D_d^T$,

$\Pi_{12} = A_d Z^T + \beta Z A^T + \beta Y^T B^T + d_u^{-1}(-\bar{M}_1 + \bar{N}_1^T) + \beta(\epsilon_1 D_a D_a^T + \epsilon_2 D_d D_d^T)$,

$\Pi_{14} = -Z^T + \alpha Z A^T + \alpha Y^T B^T + \bar{P} + \alpha(\epsilon_1 D_a D_a^T + \epsilon_2 D_d D_d^T)$,

$\Pi_{15} = -BL + W^T$,

$\Pi_{22} = -(1-\mu)\bar{Q}_2 + \beta(A_d Z^T + Z A_d^T) + d_u^{-1}(\bar{M}_2 + \bar{M}_2^T - \bar{N}_1 - \bar{N}_1^T) + \beta^2(\epsilon_1 D_a D_a^T + \epsilon_2 D_d D_d^T)$,

$$\Pi_{23}=d_u^{-1}(-\bar{M}_2+\bar{N}_2^T),\ \Pi_{24}=-\beta Z^T+\alpha ZA_d^T+\alpha\beta(\epsilon_1 D_a D_a^T+\epsilon_2 D_d D_d^T),$$
$$\Pi_{25}=-\beta BL$$
$$\Pi_{33}=-\bar{Q}_1+d_u^{-1}(-\bar{N}_2-\bar{N}_2^T),\ \Pi_{44}=\bar{R}_1-\alpha(Z+Z^T)+\alpha^2(\epsilon_1 D_a D_a^T$$
$$+\epsilon_2 D_d D_d^T),$$
$$\Pi_{45}=-\alpha BL,\ \Pi_{55}=-2L$$

the corresponding estimate of the domain of attraction is given by $\mathcal{X}_\delta \leq 1$, where $\mathcal{X}_\delta$ is as represented in (4.18).

Proof The stabilization conditions obtained in (4.15)–(4.17) (refer Theorem 4.1) in an LMI framework can directly be extended to solve the robust stabilization problem. In (4.15) and (4.16) the matrix Σ contains the nominal matrices A and A_d, which is now replaced by uncertain matrices $A(t)$ and $A_d(t)$ respectively. One can rewrite matrix Σ in (4.15) and (4.16) as,

$$\tilde{\Sigma}=\begin{bmatrix}\tilde{\Sigma}_{11} & \tilde{\Sigma}_{12} & 0 & \tilde{\Sigma}_{14} & \tilde{\Sigma}_{15}\\ \star & \tilde{\Sigma}_{22} & \tilde{\Sigma}_{23} & \tilde{\Sigma}_{24} & \tilde{\Sigma}_{25}\\ 0 & \star & \tilde{\Sigma}_{33} & 0 & 0\\ \star & \star & 0 & \tilde{\Sigma}_{44} & \tilde{\Sigma}_{45}\\ \star & \star & 0 & \star & \tilde{\Sigma}_{55}\end{bmatrix}$$

where, $\tilde{\Sigma}_{11}=A(t)Z^T+ZA^T(t)+BY+Y^TB^T+\bar{Q}_1+\bar{Q}_2+d_u^{-1}(\bar{M}_1+\bar{M}_1^T)$
$\tilde{\Sigma}_{12}=A_d(t)Z^T+\beta ZA^T(t)+\beta Y^TB^T+d_u^{-1}(-\bar{M}_1+\bar{N}_1^T)$
$\tilde{\Sigma}_{14}=-Z^T+\alpha ZA^T(t)+\alpha Y^TB^T+\bar{P}$, $\tilde{\Sigma}_{15}=-BL+W^T$
$\tilde{\Sigma}_{22}=-(1-\mu)\bar{Q}_2+\beta(A_d(t)Z^T+ZA_d^T(t))+d_u^{-1}(\bar{M}_2+\bar{M}_2^T-\bar{N}_1-\bar{N}_1^T)$
$\tilde{\Sigma}_{23}=d_u^{-1}(-\bar{M}_2+\bar{N}_2^T)$, $\tilde{\Sigma}_{24}=-\beta Z^T+\alpha ZA_d^T(t)$, $\tilde{\Sigma}_{25}=-\beta BL$
$\tilde{\Sigma}_{33}=-\bar{Q}_1+d_u^{-1}(-\bar{N}_2-\bar{N}_2^T)$, $\tilde{\Sigma}_{44}=\bar{R}_1-\alpha(Z+Z^T)$, $\tilde{\Sigma}_{45}=-\alpha BL$
$\tilde{\Sigma}_{55}=-2L$

Now, substituting $A(t)$ and $A_d(t)$ matrices from (4.68) and (4.69) into the matrix $\tilde{\Sigma}$ and subsequently $\tilde{\Sigma}$ is decomposed into nominal and uncertain matrices as indicated below,

$$\tilde{\Sigma}=\tilde{\Sigma}_{nom}+\tilde{\Sigma}_{unc} \tag{4.76}$$

where,

$$\tilde{\Sigma}_{nom}=\begin{bmatrix}\tilde{\Sigma}_{11}|_{\Delta=0} & \tilde{\Sigma}_{12}|_{\Delta=0} & 0 & \tilde{\Sigma}_{14}|_{\Delta=0} & \tilde{\Sigma}_{15}\\ \star & \tilde{\Sigma}_{22}|_{\Delta=0} & \tilde{\Sigma}_{23} & \tilde{\Sigma}_{24}|_{\Delta=0} & \tilde{\Sigma}_{25}\\ 0 & \star & \tilde{\Sigma}_{33} & 0 & 0\\ \star & \star & 0 & \tilde{\Sigma}_{44} & \tilde{\Sigma}_{45}\\ \star & \star & 0 & \star & \tilde{\Sigma}_{55}\end{bmatrix} \tag{4.77}$$

and

$$\tilde{\Sigma}_{unc} = \begin{bmatrix} D_aF_aE_aZ^T + ZE_a^TF_a^TD_a^T & D_dF_dE_dZ^T + \beta ZE_a^TF_a^TD_a^T & 0 & \alpha ZE_a^TF_a^TD_a^T & 0 \\ \star & \beta(D_dF_dE_dZ^T + ZE_d^TF_d^TD_d^T) & 0 & \alpha ZE_d^TF_d^TD_d^T & 0 \\ 0 & 0 & 0 & 0 & 0 \\ \star & \star & 0 & 0 & 0 \\ 0 & 0 & 0 & 0 & 0 \end{bmatrix}, \tag{4.78}$$

The uncertain matrix in (4.78) can be expressed in a compact form as,

$$\tilde{\Sigma}_{unc} = D_1F_a(t)E_1 + D_2F_d(t)E_2 + E_1^TF_a^T(t)D_1^T + E_2^TF_d^T(t)D_2^T \tag{4.79}$$

where, $D_1 = [D_a^T, \beta D_a^T, 0, \alpha D_a^T, 0]^T$, $D_2 = [D_d^T, \beta D_d^T, 0, \alpha D_d^T, 0]^T$, $E_1 = [E_aZ^T, 0, 0, 0, 0]^T$ and $E_2 = [E_dZ^T, 0, 0, 0, 0]^T$. the uncertain time-varying matrices in (4.79) is eliminated using Lemma 2.6,[2] thus one can write,

$$\tilde{\Sigma}_{unc} \leq \epsilon_1 D_1D_1^T + \epsilon_2 D_2D_2^T + \epsilon_1^{-1}E_1^TE_1 + \epsilon_2^{-1}E_2^TE_2 \tag{4.80}$$

In view of (4.80), one can express (4.76) as matrix inequality form,

$$\tilde{\Sigma} \leq \Pi + E_1^T\epsilon_1^{-1}E_1 + E_2^T\epsilon_2^{-1}E_2 \tag{4.81}$$

where,

$$\begin{bmatrix} \Pi_{11} & \Pi_{12} & 0 & \Pi_{14} & \Pi_{15} \\ \star & \Pi_{22} & \Pi_{23} & \Pi_{24} & \Pi_{25} \\ 0 & \star & \Pi_{33} & 0 & 0 \\ \star & \star & 0 & \Pi_{44} & \Pi_{45} \\ \star & \star & 0 & \star & \Pi_{55} \end{bmatrix} \tag{4.82}$$

and the elements of matrix Π are given below,

where, $\Pi_{11}{=}AZ^T + ZA^T + BY + Y^TB^T + \bar{Q}_1 + \bar{Q}_2 + d_u^{-1}(\bar{M}_1 + \bar{M}_1^T)$
$+\epsilon_1 D_aD_a^T + \epsilon_2 D_dD_d^T$,
$\Pi_{12}{=}A_dZ^T + \beta ZA^T + \beta Y^TB^T + d_u^{-1}(-\bar{M}_1 + \bar{N}_1^T)$
$+\beta(\epsilon_1 D_aD_a^T + \epsilon_2 D_dD_d^T)$,
$\Pi_{14}{=}-Z^T + \alpha ZA^T + \alpha Y^TB^T + \bar{P} + \alpha(\epsilon_1 D_aD_a^T + \epsilon_2 D_dD_d^T)$,
$\Pi_{15}{=}-BL + W^T$,
$\Pi_{22}{=}-(1-\mu)\bar{Q}_2 + \beta(A_dZ^T + ZA_d^T) + d_u^{-1}(\bar{M}_2 + \bar{M}_2^T - \bar{N}_1 - \bar{N}_1^T)$
$+\beta^2(\epsilon_1 D_aD_a^T + \epsilon_2 D_dD_d^T)$,
$\Pi_{23}{=}d_u^{-1}(-\bar{M}_2 + \bar{N}_2^T)$, $\Pi_{24}{=}-\beta Z^T + \alpha ZA_d^T + \alpha\beta(\epsilon_1 D_aD_a^T + \epsilon_2 D_dD_d^T)$,
$\Pi_{25}{=}-\beta BL$

[2] [29, 34], Let A, D, E and F be real matrices of appropriate dimensions with $\| F \| \leq 1$, then we have for any scalar $\epsilon > 0$, $DFE + E^TF^TD^T \leq \frac{1}{\epsilon}DD^T + \epsilon E^TE$.

$$\Pi_{33}{=}-\bar{Q}_1 + d_u^{-1}(-\bar{N}_2 - \bar{N}_2^T),\ \Pi_{44}{=}\bar{R}_1 - \alpha(Z+Z^T) + \alpha^2(\epsilon_1 D_a D_a^T + \epsilon_2 D_d D_d^T),$$
$$\Pi_{45}{=}-\alpha BL,\ \ \Pi_{55}{=}-2L$$

Now, using Schur-complement one can rewrite, (4.81) as,

$$\bar{\Pi} = \begin{bmatrix} \Pi & E_1^T & E_2^T \\ \star & -\epsilon_1 I & 0 \\ \star & 0 & -\epsilon_2 I \end{bmatrix} \tag{4.83}$$

So, in view of (4.83), one can obtain the LMI conditions (4.73) and (4.74) following the stabilization conditions presented in (4.15) and (4.16). This completes the proof. ■

Remark 4.9 If $D_a = D_d = D$ and $F_a(t) = F_d(t) = F(t)$, such that $F^T(t)F(t) \leq I$, then the matrix $\bar{\Pi}$, and in turn Π in (4.73) and (4.74) in Theorem 4.3 gets modified and the robust stabilization condition is presented in the form of following corollary,

Corollary 4.1 *Given scalars α and β, the time-delay saturating actuator system (4.66) is asymptotically stable for allowable delay upper bound d_u via memory less state feedback controller $K = YZ^{-T}$, if there exist matrices $\bar{P} = \bar{P}^T > 0, \bar{Q}_j = \bar{Q}_j^T > 0, (j = 1, 2)$ $\bar{R}_1 = \bar{R}_1^T > 0$, any free matrices $Y, W, Z, M_j, N_j, j = 1, 2$ and diagonal matrix $L = L^T > 0$ of appropriate dimensions such that following LMIs hold,*

$$\begin{bmatrix} \tilde{\Pi} & \phi_1 \\ \star & -\bar{R}_1 \end{bmatrix} < 0 \tag{4.84}$$

$$\begin{bmatrix} \tilde{\Pi} & \phi_2 \\ \star & -\bar{R}_1 \end{bmatrix} < 0 \tag{4.85}$$

$$\begin{bmatrix} \bar{P} & (y_i - w_i)^T \\ \star & u_{0(i)}^2 \end{bmatrix} \geq 0, i = 1, 2, \ldots m \tag{4.86}$$

$$\tilde{\Pi} = \begin{bmatrix} \hat{\Pi} & E^T \\ \star & -\epsilon I \end{bmatrix} \tag{4.87}$$

where,

$$E = \begin{bmatrix} E_a Z^T & E_d Z^T & 0 & 0 & 0 \end{bmatrix} \tag{4.88}$$
$$D = \begin{bmatrix} D^T & \beta D^T & 0 & \alpha D^T & 0 \end{bmatrix}^T \tag{4.89}$$

$$\hat{\Pi} = \begin{bmatrix} \hat{\Pi}_{11} & \hat{\Pi}_{12} & 0 & \hat{\Pi}_{14} & \hat{\Pi}_{15} \\ \star & \hat{\Pi}_{22} & \hat{\Pi}_{23} & \hat{\Pi}_{24} & \hat{\Pi}_{25} \\ 0 & \star & \hat{\Pi}_{33} & 0 & 0 \\ \star & \star & 0 & \hat{\Pi}_{44} & \hat{\Pi}_{45} \\ \star & \star & 0 & \star & \hat{\Pi}_{55} \end{bmatrix} \tag{4.90}$$

where, $\hat{\Pi}_{11}=AZ^T+ZA^T+BY+Y^TB^T+\bar{Q}_1+\bar{Q}_2+d_u^{-1}(\bar{M}_1+\bar{M}_1^T)$
$+\epsilon DD^T$,
$\hat{\Pi}_{12}=A_dZ^T+\beta ZA^T+\beta Y^TB^T+d_u^{-1}(-\bar{M}_1+\bar{N}_1^T)$
$+\beta(\epsilon DD^T)$,
$\hat{\Pi}_{14}=-Z^T+\alpha ZA^T+\alpha Y^TB^T+\bar{P}+\alpha(\epsilon DD^T)$,
$\Pi_{15}=-BL+W^T$,
$\hat{\Pi}_{22}=-(1-\mu)\bar{Q}_2+\beta(A_dZ^T+ZA_d^T)+d_u^{-1}(\bar{M}_2+\bar{M}_2^T-\bar{N}_1-\bar{N}_1^T)$
$+\beta^2(\epsilon DD^T)$,
$\hat{\Pi}_{23}=d_u^{-1}(-\bar{M}_2+\bar{N}_2^T)$, $\hat{\Pi}_{24}=-\beta Z^T+\alpha ZA_d^T+\alpha\beta(\epsilon DD^T)$,
$\hat{\Pi}_{25}=-\beta BL$
$\hat{\Pi}_{33}=-\bar{Q}_1+d_u^{-1}(-\bar{N}_2-\bar{N}_2^T)$, $\hat{\Pi}_{44}=\bar{R}_1-\alpha(Z+Z^T)+\alpha^2(\epsilon DD^T)$,
$\hat{\Pi}_{45}=-\alpha BL$, $\hat{\Pi}_{55}=-2L$

Numerical Example 4.3 Consider system (4.66) with the following nominal matrices as considered in Example 4.1,

$$A=\begin{bmatrix} 0.5 & -1 \\ 0.5 & -0.5 \end{bmatrix}; \; A_d=\begin{bmatrix} 0.6 & 0.4 \\ 0 & -0.5 \end{bmatrix}; \; B=\begin{bmatrix} 1 \\ 1 \end{bmatrix};$$

. The parametric uncertainty is of norm bounded type (4.69) with the following matrices,

$$\Delta A(t) = D_aF_a(t)E_a$$
$$\Delta A_d(t) = D_dF_d(t)E_a$$

with

$$D_a = D_d = D = \begin{bmatrix} 0.2 & 0 \\ 0 & 0.2 \end{bmatrix}; \; E_a = E_d = E = \begin{bmatrix} 1 & 0 \\ 0 & 1 \end{bmatrix}; \; \textit{and } u_{0(i)} = 5, (i = 1)$$

In this problem, we want to find the maximum delay upper bound for $\mu = 0$ and estimate the corresponding domain of attraction. The value of the weight σ associated with optimization algorithm ϵ is chosen appropriately.

Table 4.5 Computation result of Example 4.3 for $\mu = 0$

Methods	α	β	d_u	δ	K
Corollary 4.1	0.64	−0.002	0.1	0.9376	[−15.4196, 6.8430]
	0.69	−0.003	0.2	0.2232	[−27.3300, 12.4036]
	0.711	−0.003	0.2322	0.0091	[−473.9933, 221.3437]

Remark 4.10 The numerical Example 4.3 is now solved by reducing the amount of parametric uncertainties with $D = \begin{bmatrix} 0.1 & 0 \\ 0 & 0.1 \end{bmatrix}$ and the corresponding results obtained are presented in Table 4.6 for $\mu = 0$ with appropriate value of σ (weight on optimizing algorithm).

The simulation results are presented in the Figs. 4.15, 4.16, 4.17 and 4.18 for these values: $D = \begin{bmatrix} 0.1 & 0 \\ 0 & 0.1 \end{bmatrix}$, $E = \begin{bmatrix} 1 & 0 \\ 0 & 1 \end{bmatrix}$ and $F(t)^T F(t) \leq I$. Considering the maximum limiting value of the uncertain matrix $F(t) = \begin{bmatrix} 1 & 0 \\ 0 & 1 \end{bmatrix}$, makes $A(t) = A + DE$ and $A_d(t) = A_d + DE$. The delay value is taken as $d_u = 0.8$ s, input

Table 4.6 Computation result of Example 4.3 for $\mu = 0$

Methods	α	β	d_u	δ	K
Corollary 4.1	0.9	−0.004	0.1	2.3222	[−6.3454, 1.8892]
	2.5	−0.01	0.8	0.5535	[−9.4117, 0.4357]
	2.5	−0.01	0.874	0.0045	1e+3*[−1.0533, −0.0431]

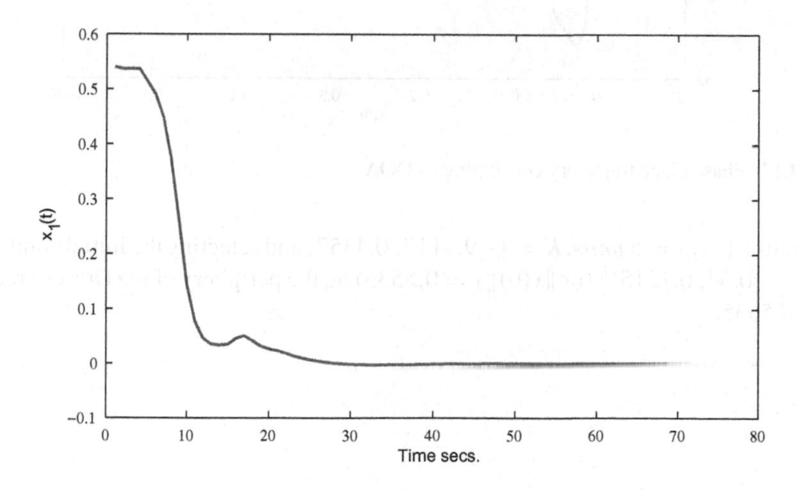

Fig. 4.15 Solution of state $x_1(t)$ of Example 4.3

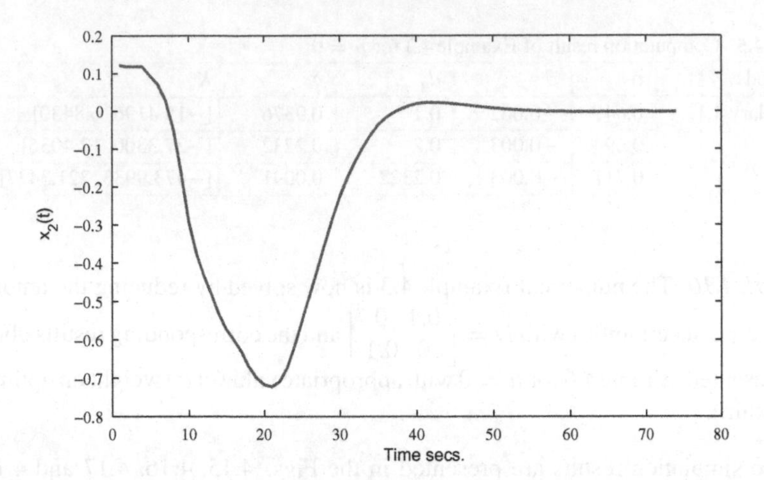

Fig. 4.16 Solution of state $x_2(t)$ of Example 4.3

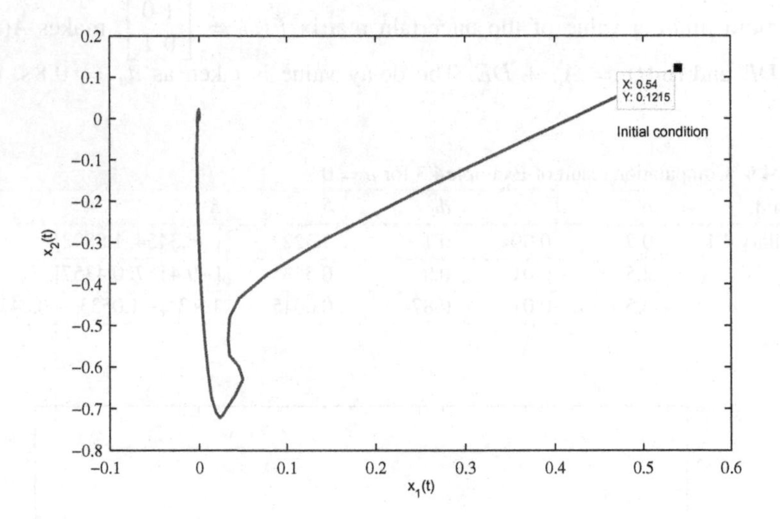

Fig. 4.17 Phase plane trajectory converging to DOA

saturation $|u_{0(i)}| = 5\, units, K = [-9.4117, 0.4357]$ and selecting the initial condition $x(0) = [0.54, 0.1215]^T$ (or $\|x(0)\|_2 = 0.5535$) on the periphery of the DOA of radius $\delta = 0.5535$.

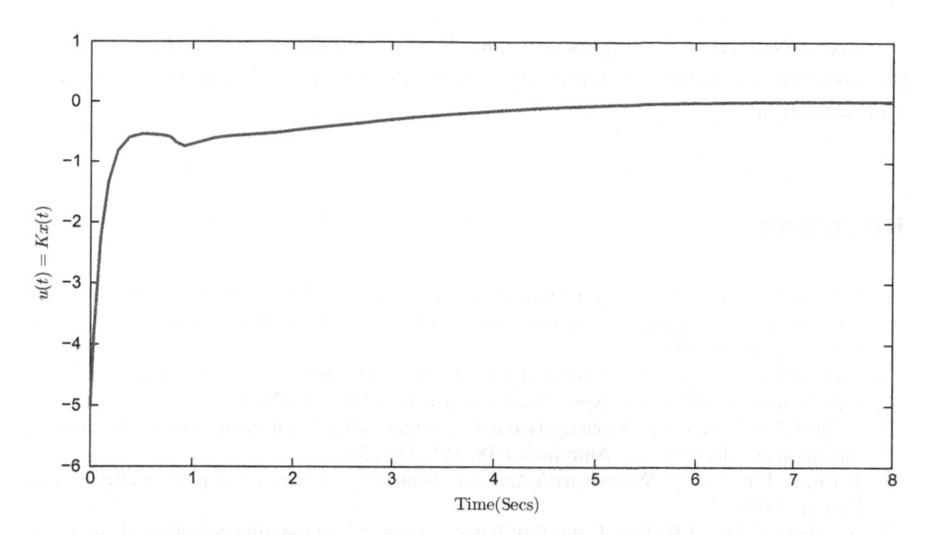

Fig. 4.18 Control input of Example 4.3

4.6 Conclusion

In this chapter, local stabilization for a class of linear time-varying delay system in presence of saturating inputs has been addressed. Improved delay-dependent stabilizing conditions have been derived using LK functional approach in an LMI framework for two different representations of the actuator saturation function, namely (i) sector nonlinearities approach and (ii) polytopic approach. The obtained stabilization conditions are solved to find the controller gains that asymptotically stabilizes the given unstable open-loop system for a certain delay upper bound and also the corresponding estimates of the domain of attraction (DOA) are obtained. The result of the work of [19] are investigated and it appears to be erroneous, which has been highlighted in the Sect. 4.4.4. It is worth mentioning at this stage that, the use of the scalar tuning parameters α and β introduced in the formulation of stabilization condition helps to maintain a trade-off among the delay upper bound (d_u), estimate of DOA (δ_{max}) and the statc feedback controller gains (K). The values of 'd_u', 'δ_{max}' and 'K' obtained using the two different proposed methods are compared with the existing results, the result of the proposed methods are presented in Tables 4.1 and 4.4.

Next, the improved delay-dependent stabilizing condition obtained with sector nonlinearities approximation is extended to solve the robust stabilization problem for norm bounded parametric uncertain time-delay system with actuator saturation. The results of the proposed robust stabilization algorithm are presented in Tables 4.5 and 4.6 for two different bounds of uncertainties. By reducing the bound of uncertainty, one may expect more delay bound and larger estimate of DOA.

Several Numerical Examples are considered to validate the effectiveness of the proposed stabilizing/robust stabilizing control algorithms for linear TDS with actuator saturation.

References

1. C. Briat, O. Sename, J.F. Lafay, Parameter-dependent sate feedback control of LPV time-delay systems with time varying delay using projection approach, in *17th IFAC World Congress 2008* (2008), pp. 4946–4951
2. S. Oucheriah, Synthesis of controllers for time-delay systems subject to actuator saturation and disturbances. ASME J. Dyn. Syst. Meas. Control **125**, 244–249 (2003)
3. T. Hu, Z. Lin, B.M. Chen, An analysis and design method for linear systems subject to actuator saturation and disturbances. Automatica **38**, 351–359 (2002)
4. T. Hu, Z. Lin, *Control Systems with Actuator Saturation: Analysis and Design* (Birkhauser, Boston, 2001)
5. A. Saberi, Z. Lin, A.R. Teel, Control of linear systems with saturating actuators. IEEE Trans. Autom. Control **41**, 368–378 (1996)
6. J.M.G.D. Silva, A. Fischman, S. Tarbouriech, J.M. Dion, L. Dugard, Synthesis of state feedback control for linear systems subject to control saturation by an LMI approach, in *Proceedings of ROCOND, Hungary* (1997) pp. 219–234
7. Y.Y. Cao, K. Fang, Parameter-dependent Lyapunov function appraoch to satbility analysis and design of polytopic system with input saturation. Asian J. Control **9**, 1–10 (2007)
8. V.F. Montagner, R.C.L.F. Oliveria, P.L.D. Peres, S.Tarbouriech, I.Queinnec, in Gain scheduled controller for LPV systems with saturating actuators: LMI based approach, in *2007 Proceedings of the ACC, New York City* (2007), pp. 6067–6072
9. B.S. Chen, S.S. Wang, H.C. Lu, Stabilization of time-delay systems containing saturating actuators. Int. J. Control **47**, 867–881 (1988)
10. J. Chou, I. Hong, B. Chen, Dynamical feedback compensators for uncertain time-delay systems containing saturating actuators. Int. J. Control **49**, 961–968 (1989)
11. E. Tissir, A. Hmamed, Further results on stabilization of time-delay systems containing saturating actuators. Int. J. Syst. Sci. **23**, 615–622 (1992)
12. S. Oucheriah, Global stabilization of a class of linear continous time-delay systems with saturating controls. IEEE Trans. Circ. Syst. I Fundam. Theory Appl. **43**, 1012–1015 (1996)
13. S. Tarbouriech, J.M.G.D. Silva, Synthesis of controllers for continuous time-delay system with saturating control via LMIs. IEEE Trans. Autom. Control **45**, 105–111 (2000)
14. Y.Y. Cao, Z. Lin, T. Hu, Stability analysis of linear time-delay systems subjected to input saturation. IEEE Trans. Circ. Syst. I Fundam. Theory Appl. **49**, 233–240 (2002)
15. E. Fridman, A. Pila, U. Shaked, Regional stabilization and H_∞ control of time-delay systems with saturating actuators. Int. J. Robust Nonlinear Control **13**, 885–907 (2003)
16. T.J. Sun, P.L. Liu, J.T. Tsay, Stabilization of delay-dependence for saturating actuator systems, in *30th Proceedings of IEEE CDC, Brighton* (1991) pp. 2891–2892
17. H.Y. Su, F. Liu, J. Chu, Robust stabilization of uncertain time-delay systems containing saturating actuators. IEE Proc. Control Theory Appl. **148**, 323–328 (2001)
18. J.M.G.D. Silva, A. Seuret, E. Fridman, J.P. Richard, Stabilization of neutral systems with saturating control inputs. Int. J. Syst. Sci. **42**(7), 1093–1103 (2011)
19. L. Zhang, E.-K. Boukas, A. Haider, Delay-range-dependent control synthesis for time-delay systems with actuator saturation. Automatica **44**, 2691–2695 (2008)
20. P.-L. Liu, Stabilization criteria for neutral time-delay systems with saturating actuators. J. Franklin Inst. **347**, 1577–1588 (2010)
21. S.I. Niclescu, J.M. Dion, L. Dugard, Robust stabilization of uncertain time-delay systems containing saturating actuators. IEEE Trans. Autom. Control **41**, 742–747 (1996)

22. S.I. Niclescu, J.M. Dion, L. Dugard, Delay-dependent stabilization for linear time-delay uncertain systems with saturating actuators. Int. J. Gen. Syst. **40**(3), 301–312 (2011)
23. S. Tarbouriech, J.M.G. da Silva, G. Garcia, Delay-dependent anti-windup strategy for linear system with saturating inputs and delayed outputs. Int. J. Robust Nonlinear Control **14**, 665–682 (2004)
24. Y. Zhao, Y. Ou, L. Zhang, H. Gao, H_∞ control of uncertain seat suspension systems subject to input delay and actuator saturation, in *IEEE 48th Conference on Decision and Control*, vol. 14 (2009), pp. 5164–5169
25. Y. He, Q.G. Wang, C. Lin, M. Wu, Delay-range-dependent stability for systems with time-varying delay. Automatica **43**, 371–376 (2007)
26. Y. He, Q.-G. Wang, L. Xie, C. Lin, Further improvements of free weighting matrices technique for systems with time-varying delay. IEEE Trans. Autom. Control **52**, 293–299 (2007)
27. Y. He, Q.-G. Wang, L. Xie, C. Lin, New delay-dependent stability criteria for systems with interval delay. Automatica **45**, 744–749 (2009)
28. H. Shao, Improved delay-dependent stability criteria for systems with a delay varying in range. Automatica **44**, 3215–3218 (2008)
29. M. Wu, Y. He, J.H. She, G.P. Liu, Delay-dependent criteria for robust stability of time-varying delay systems. Automatica **40**, 1435–1439 (2004)
30. M.N.A. Parlakci, Improved robust stability criteria and design of robust stabilizing controller for uncertain linear time-delay system. Int. J. Robust Nonlinear Control **16**, 599–636 (2006)
31. R. Dey, S. Ghosh, G. Ray, A. Rakshit, State feedback stabilization of uncertain linear time-delay systems: a nonlinear matrix inequality appraoch. Numer. Linear Algebra Appl. **18**(3), 351–361 (2011)
32. R. Dey, S. Ghosh, G. Ray, A. Rakshit, Improved delay-dependent stabilization of time-delay systems with actuator saturation. Int. J. Robust. Nonlinear Control **24**, 902–917 (2012)
33. R. Dey, A. Rakshit, G. Ray, S. Ghosh, New delay-dependent stabilization result for linear time-delay system with actuator saturation, in *IEEE conference EPSICON* (2012), pp. 1–6
34. X. Li, C.E. de Souza, Criteria for robust stability and stabilization of uncertain linear systems with state delays. Automatica **33**, 1657–1662 (1997)

Chapter 5
Fuzzy Time-Delay System

5.1 Introduction

Almost all the physical systems and processes in the real world are nonlinear [1] in nature, so over the years a considerable attention has been paid to the investigation of the dynamic behaviour of nonlinear systems. Development of nonlinear control technique has serious demerits of difficult analytical framework, computational issues and specific design for specific nonlinearity. Thus in recent times fuzzy logic control of nonlinear system [2], particularly Takagi-Sugeno (T-S) fuzzy model [3] based control has attracted attention among the researchers. It has been proved that Takagi-Sugeno (T-S) fuzzy control technique can approximate the complex nonlinear systems. T-S fuzzy modelling technique can provide a suitable representation of nonlinear systems in terms of fuzzy sets and fuzzy reasoning applied to a set of linear sub-models such that it takes advantage of the advances in linear system theories. Hence the resulting stability criteria can be recast in linear matrix inequality (LMI) framework [4] which can be solved efficiently by using existing software [5].

Time delay occurs in many industrial systems such as mechanical transmissions, fluid transmissions, metallurgical processes, and networked control systems [6]. Further, the presence of time-delay in a system model induces the infinite roots in its characteristic equation which imposes thc exact analysis that is computationally difficult. Also, time-delay is one of the key sources to instability and performance deterioration of the physical systems. Time delay is usually categorized as time constant and time-varying. Thus in recent years it is gradually accepted as an important issue among the researchers and several methods [7, 8], (references are therein) have been reported to analyze the time-delay systems via Lyapunov second method.

In the recent years, the topic of stability analysis and stabilization of time-varying delay systems in T-S fuzzy modeling framework has become an interesting research field [9]. Stabilization of fuzzy time delay systems is studied via parallel distributed compensation (PDC) technique, which employs multiple linear controller corresponding to the locally linear sub-models with fuzzy rules [10, 11]. Several LMI based less conservative stability and stabilization condition for T-S fuzzy time delay

R. Dey et al., *Stability and Stabilization of Linear and Fuzzy Time-Delay Systems*, Intelligent Systems Reference Library 141, https://doi.org/10.1007/978-3-319-70149-3_5

system have been proposed in [9, 12–15] by constructing an appropriate Lyapunov-Krasovskii (L-K) functional and using a tighter bounding integral inequality for approximating the integral term arising out of the derivative of L-K functional. Robust stability and stabilization condition for delayed T-S fuzzy system has been studied in [16] by using the Newton-Leibnitz formula together with free-weighting matrix technique. Utilizing Jensens inequality [6] approach, a delay dependent robust stability and stabilization criteria was proposed in [13, 17] where as in [9, 18, 19] affine form of Jensens inequality has been used to obtain the improved stability condition for fuzzy time delay systems. A new stability and stabilization condition was established in [20] by using Wirtinger inequality [21]. It is pertinent to mention here that the improvement due to the earlier methods is still limited and further obtain the less conservative stability results is an important and challenging issue.

In this chapter, a new and improved delay-dependent stability and stabilization condition for T-S fuzzy time delay system is presented. The stability and stabilization condition is derived in this chapter by choosing an appropriate augmented L-K functional and utilizing Wirtinger inequality combined with reciprocal convex lemma. State feedback controllers are designed via the PDC scheme for stabilization and free weighting matrices are taken to convert bilinear matrix inequality (BMI) problem into LMI one. Finally, three numerical examples are given to demonstrate the effectiveness and advantages of the proposed method.

5.2 Problem Formulation and Preliminaries

Consider a nonlinear time-varying delay system as,

$$\dot{x}(t) = f\,(t, x(t), x(t-\tau(t)), u(t))\,, \qquad t \geq 0, \tag{5.1}$$

$$x(t) = \phi(t), \qquad -\tau_0 \leq t \leq 0$$

where, $x(t) \in \mathbb{R}^n$ is the state vector of the system, $u(t) \in \mathbb{R}^m$ is the control input, $'f'$ is a smooth continuous non-linear function, $x(t) = \phi(t)$ is the vector valued initial condition on $[-\tau_0, 0]$ and $\tau(t)$ is the time-varying differentiable function.

Nonlinear system given in (5.1) can be represented by the T-S fuzzy model and it is expressed in terms of fuzzy IF-THEN rules in the following form:
Rule i: IF $z_1(t)$ is M_{i1} and … and $z_p(t)$is M_{ip} THEN

$$\dot{x}(t) = A_i x(t) + A_{\tau_i} x(t-\tau(t)) + B_i u(t), \quad t \geq 0, \tag{5.2}$$

$$x(t) = \phi(t), \quad -\tau_0 \leq t \leq 0,$$

where, $z_1(t), z_2(t), \ldots, z_p(t)$ are the premise variables, M_{ij} are the fuzzy membership functions with $i = 1, 2, 3, \ldots, r$, $j = 1, 2, 3, \ldots, p$, the scalars r and p indicates the number of fuzzy IF-THEN rules and number of premise variables respectively.

A_i, A_{τ_i} are system matrices with appropriate dimensions. The time-varying delay differential function $\tau(t)$ is considered to satisfy the following conditions:

$$0 \le \tau(t) \le \tau_0, \tag{5.3}$$

and

$$\dot{\tau}(t) \le \mu < 1, \tag{5.4}$$

where, τ_0 and μ are positive constants representing the upper delay bound $\tau(t)$ and $\dot{\tau}(t)$, respectively. If $z_j(t) = z_j^0$ given, where z_j^0 are singletons, then for each ith fuzzy rule, the truth values of $\dot{x}(t)$ is calculated as,

$$h_i(z(t)) = \left(M_{i1}(z_1(t)) \wedge \ldots \wedge M_{ip}(z_p(t))\right), i = 1, 2, \ldots, r, \tag{5.5}$$

where,
$z(t) = \left[z_1(t)\ z_2(t) \ldots z_p(t)\right]^T$, $M_{i1}(z_1(t)), \ldots, M_{ip}(z_p(t))$ is the grade of the membership of $z_1(t), \ldots, z_p(t)$ in M_{ij} and $\wedge$ denote the 'min' operation.

By the centroid method for defuzzification, product inference and singleton fuzzifier, the final output of the fuzzy system (5.2) is calculated as,

$$\begin{aligned}\dot{x}(t) &= \frac{\sum_{i=1}^{r} h_i(z(t)) \left\{A_i x(t) + A_{\tau_i} x(t - \tau(t)) + B_i u(t)\right\}}{\sum_{i=1}^{r} h_i(z(t))}, \\ &= \sum_{i=1}^{r} w_i(z(t)) \left\{A_i x(t) + A_{\tau_i} x(t - \tau(t)) + B_i u(t)\right\}, \end{aligned} \tag{5.6}$$

where, $w_i(z(t)) = \dfrac{h_i(z(t))}{\sum_{i=1}^{r} h_i(z(t))}$, $\forall\, t$ and $i = 1, 2, \ldots, r$, is called the fuzzy weighting function. By definition, fuzzy weighting function $w_i(z(t)) \ge 0$ and $\sum_{i=1}^{r} w_i(z(t)) = 1$.

In this chapter, a TS fuzzy-model-based controller will be designed via parallel distributed compensation (PDC) to stabilize the system (5.6). The PDC control structure utilizes a nonlinear state feedback controller which is associated with the structure of T?S fuzzy model. The gains of the controller can be obtained by using a linear matrix inequality (LMI) formulation [1].

Consider that the following controller rules are given:
Rule j: IF $z_1(t)$ is M_{j1} and … and $z_p(t)$is M_{jp} THEN

$$u(t) = K_j x(t), \quad t \ge 0, \; j = 1, 2, \ldots, r \tag{5.7}$$

The defuzzified output of the controller rule (5.7) is obtained as,

$$u(t) = \sum_{j=1}^{r} w_j(z(t)) K_j x(t) \tag{5.8}$$

where, K_j $(j = 1, 2, \ldots, r)$ are controller gains to be determined.

Substituting (5.8) into (5.6), one can obtain the following closed-loop fuzzy system as below:

$$\dot{x}(t) = \sum_{i=1}^{r}\sum_{j=1}^{r} w_i(z(t))w_j(z(t)) \left\{ \left(A_i + B_i K_j\right) x(t) + A_{\tau_i} x(t - \tau(t)) \right\}, \quad t \geq 0$$

$$= \overline{A}_j^i x(t) + \overline{A}_\tau^i x(t - \tau(t)), \quad t \geq 0 \tag{5.9}$$

$$x(t) = \phi(t), \qquad -\tau_0 \leq t \leq 0,$$

where, $\overline{A}_j^i = \sum_{i=1}^{r}\sum_{j=1}^{r} w_i(z(t))w_j z(t) \left(A_i + B_i K_j\right)$ and $\overline{A}_\tau^i = \sum_{i=1}^{r} \sum_{j=1}^{r} w_i(z(t))w_j z(t) A_{\tau_i}$. Furthermore, w_i is used to denote the fuzzy weighting function instead of $w_i(z(t))$ for notational simplicity.

Before proceeding next, we recall two integral inequality lemma and some existing stability and stabilization condition for T-S fuzzy time delay systems.

Lemma 5.1 [22], (Reciprocal convex lemma) *For given positive integers m, n, a scalar α in the interval* $(0, 1)$, *a given $n \times n$ matrix $R > 0$, two matrices W_1 and W_2 in $\mathbb{R}^{n\times m}$. Define for all vector ς in $\mathbb{R}^m$, the function $\Theta(\alpha, R)$ is given by:*

$$\Theta(\alpha, R) = \frac{1}{\alpha}\varsigma^T W_1^T R W_1 \varsigma + \frac{1}{1-\alpha}\varsigma^T W_2^T R W_2 \varsigma, \tag{5.10}$$

Then, if there exists a matrix X in $\mathbb{R}^{n\times n}$ such that $\begin{bmatrix} R & X \\ * & R \end{bmatrix} \geq 0$, *then the following inequality holds,*

$$\min_{\alpha\in(0,1)} \Theta(\alpha, R) \geq \begin{bmatrix} W_1\varsigma \\ W_2\varsigma \end{bmatrix}^T \begin{bmatrix} R & X \\ * & R \end{bmatrix} \begin{bmatrix} W_1\varsigma \\ W_2\varsigma \end{bmatrix}. \tag{5.11}$$

Lemma 5.2 [21], (Wirtinger inequality lemma) *For a given matrix $R > 0$, the following inequality holds for all continuously differentiable functions x from $[a, b] \rightarrow \mathbb{R}^n$:*

$$\int_a^b \dot{x}^T(s)R\dot{x}(s)ds \geqslant \frac{1}{b-a}\varsigma_1^T(t)R\varsigma_1(t) + \frac{3}{b-a}\varsigma_2^T(t)R\varsigma_2(t) \tag{5.12}$$

where, $\varsigma_1(t) = [x(b) - x(a)]$ and $\varsigma_2(t) = \left[x(b) + x(a) - \frac{2}{b-a}\int_a^b x(s)ds\right]$.

Lemma 5.3 [9] *Given a fuzzy time-delay system (5.9) satisfying the conditions (5.3)–(5.4) is asymptotically stable, if there exist real symmetric positive-definite matrices $P,\ Q_k,\ R_m > 0,\ k = 1, 2, 3, 4,\ m = 1, 2$ and any free matrices $\Phi_{li},\ l = 1, 2,\ i = 1, 2, \ldots, r$ of appropriate dimensions with scalars $d_1,\ d_2$ and $\overline{d} = d_2 - d_1$*

representing delay lower bound, delay upper bound and delay range respectively, such that the following LMIs are satisfied,

$$\Theta_i + \Phi_{li} R_2^{-1} \Phi_{li}^T, \quad l = 1, 2, \quad i = 1, 2, \ldots, r \tag{5.13}$$

where,

$$\Theta_i = \begin{bmatrix} \Theta_{11i} & R_1 & \Theta_{13i} & 0 \\ * & \Theta_{22i} & \Theta_{23i} & 0 \\ * & * & \Theta_{33i} & \Theta_{34i} \\ * & * & * & \Theta_{44i} \end{bmatrix}$$

$\Theta_{11i} = PA_i + A_i^T P + \sum_{k=1}^{3} Q_3 - R_1 + A_i^T(d_1^2 R_1 + R_2)A_i$, $\Theta_{13i} = PA_{di} + A_i^T(d_1^2 R_1 + R_2)A_{di}$,
$\Theta_{22i} = Q_4 - Q_1 - R_1 + \overline{d}^{-1}(T_{1i} + T_{1i}^T)$, $\Theta_{23i} = -\overline{d}^{-1}(-T_{1i} + T_{2i}^T)$
$\Theta_{33i} = -(1-\mu)(Q_3 + Q_4) + \overline{d}^{-1}(N_{1i} + N_{1i}^T - T_{2i} + T_{2i}^T) + A_{di}^T(d_1^2 R_1 + R_2)A_{di}$
$\Theta_{34i} = -\overline{d}^{-1}(-N_{1i} + N_{1i}^T)$, $\Theta_{44i} = -Q_2 - \overline{d}^{-1}(N_{2i} + N_{2i}^T)$,
$\Phi_{1i} = [0 \ T_{1i}^T \ T_{2i}^T \ 0]^T$, $\Phi_{2i} = [0 \ 0 \ N_{1i}^T \ N_{2i}^T]^T$.

Lemma 5.4 [29] *Consider the fuzzy system (5.9). For given scalars τ_0 and μ, if for some scalars $\lambda_1 \neq 0$, λ_2 and $\lambda > 0$, there exist matrices $\mathcal{P} = \mathcal{P}^T > 0$, $\mathcal{Q}_i = \mathcal{Q}_i^T \geq 0$, $\mathcal{R} = \mathcal{R}^T > 0$, G_i, χ_{ki}, $i \in \mathbb{S}$, $k = 1, 2, 3$ satisfying the following LMIs:*

$$\Phi_{\vartheta\varpi ii} < 0, \quad \vartheta, \varpi, i \in \mathbb{S} \tag{5.14}$$

$$\frac{1}{r-1}\Phi_{\vartheta\varpi ii} + \frac{1}{2}(\Phi_{\vartheta\varpi ij} + \Phi_{\vartheta\varpi ji}) < 0, \quad \vartheta, \varpi, i, j \in \mathbb{S} \tag{5.15}$$

where,

$$\Phi_{\vartheta\varpi ij} \triangleq \begin{bmatrix} \mathcal{Q}_i + \widetilde{\Upsilon}_{11,ij} & \widetilde{\Upsilon}_{12,ij} & \lambda_1\lambda_3\mathcal{P} + \widetilde{\Upsilon}_{13,ij} & \tau_0\lambda_1^{-1}\lambda_3\chi_{1i} \\ * & -(1-\mu)\mathcal{Q}_\vartheta + \widetilde{\Upsilon}_{22,i} & \widetilde{\Upsilon}_{23,i} & \tau_0\lambda_1^{-1}\lambda_3\chi_{2i} \\ * & * & \tau_0\mathcal{R}_i - 2\lambda_3\mathcal{P} & \tau_0\chi_{3i} \\ * & * & * & -\tau_0\mathcal{R}_\varpi \end{bmatrix},$$

$\widetilde{\Upsilon}_{11,ij} = \chi_{1i} + \chi_{1i}^T + \lambda_1 A_i\mathcal{P} + \lambda_1 B_i G_j + \lambda_1\mathcal{P}A_i^T + \lambda_1 G_j^T B_i^T$,
$\widetilde{\Upsilon}_{12,ij} = -\chi_{1i} + \chi_{2i}^T + \lambda_1\lambda_2\mathcal{P}A_i^T + \lambda_1\lambda_2 G_j^T B_i^T + \lambda_1 A_{di}\mathcal{P}$,
$\widetilde{\Upsilon}_{13,ij} = \lambda_1\lambda_3^{-1}\chi_{3i}^T + \lambda_1\mathcal{P}A_i^T + \lambda_1 G_j^T B_i^T - \lambda_3\mathcal{P}$, $\widetilde{\Upsilon}_{22,i} = -\chi_{2i} - \chi_{2i}^T + \lambda_1\lambda_2 A_{di}\mathcal{P} + \lambda_1\lambda_2\mathcal{P}A_{di}^T$
$\widetilde{\Upsilon}_{23,i} = -\lambda_1\lambda_3^{-1}\chi_{3i}^T + \lambda_1\mathcal{P}A_{di}^T - \lambda_2\lambda_3\mathcal{P}$.

then there exists a fuzzy controller of the form (5.8) such that the closed-loop fuzzy system (5.9) is asymptotically stable in the large. Furthermore, the state-feedback gain matrices are given by $K_j = G_j\mathcal{P}^{-1}$, $j \in \mathbb{S}$.

5.3 Stability Analysis

In this section, a new and improved delay-dependent stability condition for T-S fuzzy time delay system in (5.9) is presented here.

Theorem 5.1 *For given scalars $\tau_0 > 0$, μ and the state feedback gain matrices K_j, $j = 1, 2, \ldots, r$, the closed-loop T-S fuzzy system (5.9) subject to the conditions (5.3) and (5.4) is asymptotically stable, if there exist symmetric positive definite matrices $P \in \mathbb{R}^{3n\times 3n}$, $Q_l \in \mathbb{R}^{n\times n}$, $l = 1, 2$, $R \in \mathbb{R}^{n\times n}$ and any matrices $X \in \mathbb{R}^{2n\times 2n}$, M_k, $k = 1, 2$ with appropriate dimension such that the following LMIs are holds:*

$$\Psi = \begin{bmatrix} \widetilde{R} & X \\ * & \widetilde{R} \end{bmatrix} > 0 \tag{5.16}$$

$$\Omega_{ii}(0, 1) - \Gamma^T \Psi \Gamma < 0, \quad i = 1, 2, \ldots, r \tag{5.17}$$

$$\Omega_{ii}(1, 0) - \Gamma^T \Psi \Gamma < 0, \quad i = 1, 2, \ldots, r \tag{5.18}$$

and

$$\Omega_{ij}(0, 1) + \Omega_{ji}(\alpha, (1-\alpha)) - 2\Gamma^T \Psi \Gamma, \quad 1 \le i < j \le r \tag{5.19}$$

$$\Omega_{ij}(1, 0) + \Omega_{ji}(\alpha, (1-\alpha)) - 2\Gamma^T \Psi \Gamma, \quad 1 \le i < j \le r \tag{5.20}$$

where,

$$\begin{aligned}\Omega_{ij}(0, 1) = sym\left\{\Pi_{11,1}^T P \Pi_{22}\right\} + \sum_{l=1}^{2} e_1^T Q_l e_1 - e_3^T Q_1 e_3 - (1-\mu) e_2^T Q_2 e_2 + \tau_0^2 e_6^T R e_6 \\ + 2\left\{e_1^T M_1^T \left(\overline{A}_j^i e_1 + \overline{A}_\tau^i e_2 - e_6\right) + e_6^T M_2^T \left(\overline{A}_j^i e_1 + \overline{A}_\tau^i e_2 - e_6\right)\right\},\end{aligned}$$

$$\begin{aligned}\Omega_{ij}(1, 0) = sym\left\{\Pi_{11,2}^T P \Pi_{22}\right\} + \sum_{l=1}^{2} e_1^T Q_l e_1 - e_3^T Q_1 e_3 - (1-\mu) e_2^T Q_2 e_2 + \tau_0^2 e_6^T R e_6 \\ + 2\left\{e_1^T M_1^T \left(\overline{A}_j^i e_1 + \overline{A}_\tau^i e_2 - e_6\right) + e_6^T M_2^T \left(\overline{A}_j^i e_1 + \overline{A}_\tau^i e_2 - e_6\right)\right\},\end{aligned}$$

$\Pi_{11,1} = \left[e_1^T \ \ 0 \ \ \tau_0 e_5^T\right]$, $\quad \Pi_{11,2} = \left[e_1^T \ \ \tau_0 e_4^T \ \ 0\right]$, $\quad \Pi_{22} = \left[e_6^T \ \ e_1^T - (1-\mu)\right.$ $\left. e_2^T \ \ (1-\mu) e_2^T - e_3^T\right]^T$, $\widetilde{R} = diag\{R, 3R\}$
and $e_s = \left[e_{n\times(s-1)n} \ \ I_{n\times n} \ \ 0_{n\times(6-s)n}\right]^T$, $\quad s = 1, 2, \ldots, 6$ *denote the block entry matrices.*

Proof We define a Lyapunov-Krasovskii functional candidate as,

$$V(x_t) = V_1(x_t) + V_2(x_t) + V_3(x_t) \tag{5.21}$$

where,

$$V_1(x_t) = \varpi^T(t) P \varpi(t),$$

$$V_2(x_t) = \int_{t-\tau_0}^{t} x^T(s) Q_1 x(s) ds + \int_{t-\tau(t)}^{t} x^T(s) Q_2 x(s) ds,$$

$$V_3(x_t) = \tau_0 \int_{-\tau_0}^{0} \int_{t+\lambda}^{t} \dot{x}^T(s) R \dot{x}(s) ds d\lambda$$

with $\varpi(t) = \left[x^T(t) \;\; \int_{t-\tau(t)}^{t} x^T(s) ds \;\; \int_{t-\tau_0}^{t-\tau(t)} x^T(s) ds \right]^T$ and $P = \begin{bmatrix} P_{11} & P_{12} & P_{13} \\ * & P_{22} & P_{23} \\ * & * & P_{33} \end{bmatrix}$.

Taking the time derivative of (5.21) along with the conditions (5.3) and (5.4), one can obtain

$$\dot{V}(x_t) = \dot{V}_1(x_t) + \dot{V}_2(x_t) + \dot{V}_3(x_t) \tag{5.22}$$

where,

$$\dot{V}_1(x_t) = 2\varpi^T(t) P \dot{\varpi}(t) = 2 \begin{bmatrix} x(t) \\ \int_{t-\tau(t)}^{t} x(s) ds \\ \int_{t-\tau_0}^{t-\tau(t)} x(s) ds \end{bmatrix}^T \begin{bmatrix} P_{11} & P_{12} & P_{13} \\ * & P_{22} & P_{23} \\ * & * & P_{33} \end{bmatrix} \begin{bmatrix} \dot{x}(t) \\ x(t) - (1-\mu) x(t-\tau(t)) \\ (1-\mu) x(t-\tau(t)) - x(t-\tau_0) \end{bmatrix} \tag{5.23}$$

Here, we introduce an augmented state vector $\xi(t)$ and the block entry matrices e_s with,

$$\xi(t) = \left[x^T(t) \; x^T(t-\tau(t)) \; x^T(t-\tau_0) \; \chi_4^T(t) \; \chi_5^T(t) \; \dot{x}^T(t) \right]^T,$$

and

$$e_s = \left[e_{n\times(s-1)n} \;\; I_{n\times n} \;\; 0_{n\times(6-s)n} \right]^T, \quad s = 1, 2, \ldots, 6$$

where,

$$\chi_4(t) = \begin{cases} \frac{1}{\tau(t)} \int_{t-\tau(t)}^{t} x(s) ds; & if \;\; 0 < \tau(t), \\ x(t); & if \;\; \tau(t) = 0, \end{cases} \tag{5.24}$$

$$\chi_5(t) = \begin{cases} \frac{1}{\tau_0 - \tau(t)} \int_{t-\tau_0}^{t-\tau(t)} x(s) ds; & if \;\; \tau(t) < \tau_0, \\ x(t-\tau_0); & if \;\; \tau_0 = \tau(t), \end{cases} \tag{5.25}$$

Now, applying (5.4) in (5.23) and with the help of $\xi(t)$ and e_s, (5.23) can be rewritten in the following form as,

$$\dot{V}_1(x_t) \leq \xi^T(t) \left[sym \left\{ \Pi_{11}^T P \Pi_{22} \right\} \right] \xi(t) \tag{5.26}$$

where, $\Pi_{11} = \left[e_1^T \ \alpha e_4^T \ (1-\alpha)e_5^T\right]$, $\Pi_{22} = \left[e_6^T \ e_1^T - (1-\mu)e_2^T \ (1-\mu) e_2^T - e_3^T\right]^T$, $\alpha = \frac{\tau(t)}{\tau_0}$ and $\alpha \in [0 \ 1]$.
Now, we shall continue the differentiating of $V(x_t)$, $\dot{V}_2(x_t)$ and $\dot{V}_3(x_t)$ can be obtained as,

$$\dot{V}_2(x_t) = \sum_{l=1}^{2} x^T(t)Q_l x(t) - x^T(t-\tau_0)Q_1 x(t-\tau_0) - (1-\dot{\tau}(t))x^T(t-\tau(t))Q_2 x(t-\tau(t))$$
$$\leq \xi^T(t)\left\{\sum_{l=1}^{2} e_1^T Q_l e_1 - e_3^T Q_1 e_3 - (1-\mu)e_2^T Q_2 e_2\right\}\xi(t) \tag{5.27}$$

$$\dot{V}_3(x_t) = \tau_0^2 \dot{x}^T(t)R\dot{x}(t) - \tau_0 \int_{t-\tau_0}^{t} \dot{x}^T(s)R\dot{x}(s)ds$$
$$= \tau_0^2 \xi^T(t)e_6^T R e_6 \xi(t) - \tau_0 \int_{t-\tau_0}^{t} \dot{x}^T(s)R\dot{x}(s)ds \tag{5.28}$$

Next, consider the integral term in the right hand side (RHS) of (5.28) which contain uncertain limit of integration. To formulate the reciprocal convex function, first we splitted this integral term into two parts and then carrying out appropriate algebraic manipulation for the chosen L-K functional as described below,

$$-\tau_0 \int_{t-\tau_0}^{t} \dot{x}^T(s)R\dot{x}(s)ds = \frac{\tau_0}{\tau(t)}\left(-\tau(t)\int_{t-\tau(t)}^{t} \dot{x}^T(s)R\dot{x}(s)ds\right)$$
$$+ \frac{\tau_0}{\tau_0 - \tau(t)}\left(-(\tau_0 - \tau(t))\int_{t-\tau_0}^{t-\tau(t)} \dot{x}^T(s)R\dot{x}(s)ds\right)$$
$$= \frac{1}{\alpha}\left(-\tau(t)\int_{t-\tau(t)}^{t} \dot{x}^T(s)R\dot{x}(s)ds\right) + \frac{1}{1-\alpha}\left(-(\tau_0 - \tau(t))\int_{t-\tau_0}^{t-\tau(t)} \dot{x}^T(s)R\dot{x}(s)ds\right) \tag{5.29}$$

To approximate the integral terms in the RHS of (5.29) Lemma 1 is applied, that yields,

$$-\tau_0 \int_{t-\tau_0}^{t} \dot{x}^T(s)R\dot{x}(s)ds \leq -\frac{1}{\alpha}\xi^T(t)\begin{bmatrix}\zeta_1\\ \zeta_2\end{bmatrix}^T \begin{bmatrix}R & 0\\ * & 3R\end{bmatrix}\begin{bmatrix}\zeta_1\\ \zeta_2\end{bmatrix}\xi(t)$$
$$-\frac{1}{1-\alpha}\xi^T(t)\begin{bmatrix}\zeta_3\\ \zeta_4\end{bmatrix}^T \begin{bmatrix}R & 0\\ * & 3R\end{bmatrix}\begin{bmatrix}\zeta_3\\ \zeta_4\end{bmatrix}\xi(t) \tag{5.30}$$

where, $\zeta_1 = [e_1 - e_2]$, $\zeta_2 = [e_1 + e_2 - 2e_4]$, $\zeta_3 = [e_2 - e_3]$, $\zeta_4 = [e_2 + e_3 - 2e_5]$.
Now, to formulate the function defined in (5.10), one can write RHS of (5.30) as,

$$-\tau_0 \int_{t-\tau_0}^{t} \dot{x}^T(s)R\dot{x}(s)ds \leq -\frac{1}{\alpha}\xi^T(t)\Omega_1^T \widetilde{R}\Omega_1 - \frac{1}{1-\alpha}\xi^T(t)\Omega_2^T \widetilde{R}\Omega_2 \tag{5.31}$$

where, $\Omega_1 = \left[\zeta_1^T \ \zeta_2^T\right]^T$, $\Omega_2 = \left[\zeta_3^T \ \zeta_4^T\right]^T$ and $\widetilde{R}$ is defined after (5.19).
Further, using Lemma 1 in the RHS of (5.31), if there exist any matrix $X \in \mathbb{R}^{2n\times 2n}$, one can obtain that,

$$-\tau_0 \int_{t-\tau_0}^{t} \dot{x}^T(s)R\dot{x}(s)ds \leq -\xi^T(t)\begin{bmatrix}\Omega_1\\ \Omega_2\end{bmatrix}^T \begin{bmatrix}\widetilde{R} & X\\ * & \widetilde{R}\end{bmatrix}\begin{bmatrix}\Omega_1\\ \Omega_2\end{bmatrix}\xi(t)$$
$$\leq -\xi^T(t)\Gamma^T\Psi\Gamma\xi(t) \tag{5.32}$$

where, $\Gamma = \left[\Omega_1^T \ \Omega_2^T\right]^T$ and Ψ is defined in (5.16).
Substituting (5.3) in the RHS of (5.28) yields,

$$\dot{V}_3(x_t) \leq \tau_0^2\xi^T(t)e_6^T Re_6\xi(t) - \xi^T(t)\Gamma^T\Psi\Gamma\xi(t) \tag{5.33}$$

Moreover, introducing some free matrices M_k, $k = 1, 2$, the quadratic form of the system dynamics can be written as,

$$2\left[x^T(t)M_1^T + \dot{x}^T(t)M_2^T\right]\left[\overline{A}_j^i x(t) + \overline{A}_\tau^i x(t-\tau(t)) - \dot{x}(t)\right] = 0$$

$$\Rightarrow 2\xi^T(t)\left[e_1^T M_1^T\left(\overline{A}_j^i e_1 + \overline{A}_\tau^i e_2 - e_6\right) + e_6^T M_2^T\left(\overline{A}_j^i e_1 + \overline{A}_\tau^i e_2 - e_6\right)\right]\xi(t) = 0 \tag{5.34}$$

Finally, considering (5.26), (5.27), (5.33) and (5.34) together, we have

$$\begin{aligned}
\dot{V}(x_t) \leq \sum_{i=1}^{r}\sum_{j=1}^{r} w_i w_j \xi^T(t)\Big[& sym\left\{\Pi_{11}^T P\Pi_{22}\right\} + \sum_{l=1}^{2} e_1^T Q_l e_1 - e_3^T Q_1 e_3 \\
& - (1-\mu)e_2^T Q_2 e_2 + \tau_0^2 e_6^T Re_6 - \Gamma^T\Psi\Gamma \\
& + 2\Big\{e_1^T M_1^T\left(\overline{A}_j^i e_1 + \overline{A}_\tau^i e_2 - e_6\right) \\
& + e_6^T M_2^T\left(\overline{A}_j^i e_1 + \overline{A}_\tau^i e_2 - e_6\right)\Big\}\Big]\xi(t) \\
= \sum_{i=1}^{r}\sum_{j=1}^{r} & w_i w_j \xi^T(t)\left\{\Omega_{ij}(\alpha,(1-\alpha)) - \Gamma^T\Psi\Gamma\right\}\xi(t) \\
\leq \sum_{i=1}^{r} & w_i^2 \xi^T(t)\left[\Omega_{ii}(\alpha,(1-\alpha)) - \Gamma^T\Psi\Gamma\right]\xi(t) \\
+ \sum_{i=1}^{r}\sum_{i<j}^{r} & w_i w_j \xi^T(t)\left[\Omega_{ij}(\alpha,(1-\alpha)) + \Omega_{ji}(\alpha,(1-\alpha)) - 2\Gamma^T\Psi\Gamma\right]\xi(t)
\end{aligned} \tag{5.35}$$

where,

$$\Omega_{ij}(\alpha,(1-\alpha)) = sym\left\{\Pi_{11}^T P \Pi_{22}\right\} + \sum_{l=1}^{2} e_1^T Q_l e_1 - e_3^T Q_1 e_3 - (1-\mu) e_2^T Q_2 e_2 + \tau_0^2 e_6^T R e_6$$
$$+ 2\left\{e_1^T M_1^T\left(\overline{A}_j^i e_1 + \overline{A}_\tau^i e_2 - e_6\right) + e_6^T M_2^T\left(\overline{A}_j^i e_1 + \overline{A}_\tau^i e_2 - e_6\right)\right\}.$$

Since, $\Omega_{ij}(\alpha,(1-\alpha))$ is depend on the convex parameter $\alpha \in [0\ 1]$. Therefore, the stability requirement is that the inequality

$$\dot{V}(x_t) \leq \sum_{i=1}^{r} w_i^2 \xi^T(t)\left[\Omega_{ii}(\alpha,(1-\alpha)) - \Gamma^T \Psi \Gamma\right]\xi(t)$$
$$+ \sum_{i=1}^{r}\sum_{i<j}^{r} w_i w_j \xi^T(t)\left[\Omega_{ij}(\alpha,(1-\alpha)) + \Omega_{ji}(\alpha,(1-\alpha)) - 2\Gamma^T \Psi \Gamma\right]\xi(t) < 0 \tag{5.36}$$

should be satisfied once for $\alpha = 0$ and again for $\alpha = 1$. That is,

$$\Omega_{ii}(0,1) - \Gamma^T \Psi \Gamma < 0, \quad for\ i = 1, 2, \ldots, r \tag{5.37}$$

$$\Omega_{ii}(1,0) - \Gamma^T \Psi \Gamma < 0, \quad for\ i = 1, 2, \ldots, r \tag{5.38}$$

and

$$\Omega_{ij}(0,1) + \Omega_{ji}(0,1) - 2\Gamma^T \Psi \Gamma, \quad for\ 1 \leq i < j \leq r \tag{5.39}$$

$$\Omega_{ij}(1,0) + \Omega_{ji}(1,0) - 2\Gamma^T \Psi \Gamma, \quad for\ 1 \leq i < j \leq r \tag{5.40}$$

where, $\Omega_{ij}(0,1)$ and $\Omega_{ij}(1,0)$ are defined after (5.19). Hence, $\dot{V}(x_t) < 0$ means that $\dot{V}(x_t) < -\epsilon\|x_t\|^2$, for sufficiently small $\epsilon > 0$, and which ensures the asymptotic stability of system (5.9) as per Lyapunov-Krasovskii Theorem. This completes the proof of the Theorem 5.1.

5.4 State Feedback Stabilization

In this section, our objective is to design the state feedback controller that guaranteeing the asymptotic stability of the system (5.9).

Theorem 5.2 *For given scalars $\tau_0 > 0$, ρ and μ, the T-S fuzzy time delay system (5.9) satisfying (5.3) and (5.4), is asymptotically stable with feedback gains $K_j = N_j Y^{-T}$, $(j = 1, 2, \ldots, r)$, if there exist symmetric positive matrices $\widehat{P} \in \mathbb{R}^{3n\times 3n}$, $\widehat{Q}_l \in \mathbb{R}^{n\times n}$, $l = 1, 2$, $\widehat{R} \in \mathbb{R}^{n\times n}$ and any matrices $\widehat{X} \in \mathbb{R}^{2n\times 2n}$, N_j, $(j = 1, 2, \ldots, r)$ and Y with appropriate dimension such that the following LMIs are holds:*

$$\widehat{\Psi} = \begin{bmatrix} \overline{R} & \widehat{X} \\ * & \overline{R} \end{bmatrix} > 0 \tag{5.41}$$

$$\widehat{\Omega}_{ii}(0,1) - \widehat{\Gamma}^T \widehat{\Psi} \widehat{\Gamma} < 0, \quad i = 1, 2, \ldots, r \tag{5.42}$$

$$\widehat{\Omega}_{ii}(1,0) - \widehat{\Gamma}^T \widehat{\Psi} \widehat{\Gamma} < 0, \quad i = 1, 2, \ldots, r \tag{5.43}$$

and

$$\widehat{\Omega}_{ij}(0,1) + \widehat{\Omega}_{ji}(0,1) - 2\widehat{\Gamma}^T \widehat{\Psi} \widehat{\Gamma}, \quad \le i < j \le r \tag{5.44}$$

$$\widehat{\Omega}_{ij}(1,0) + \widehat{\Omega}_{ji}(1,0) - 2\widehat{\Gamma}^T \widehat{\Psi} \widehat{\Gamma}, \quad \le i < j \le r \tag{5.45}$$

where,

$$\widehat{\Omega}_{ij}(0,1) = sym\left\{\widehat{\Pi}_{11,1}^T \widehat{P} \widehat{\Pi}_{22}\right\} + \sum_{l=1}^{2} e_1^T \widehat{Q}_l e_1 - e_3^T \widehat{Q}_1 e_3 - (1-\mu) e_2^T \widehat{Q}_2 e_2 + \tau_0^2 e_6^T \widehat{R} e_6$$
$$+ 2\left\{e_1^T M_1^T \left(\overline{A}_j^i e_1 + \overline{A}_\tau^i e_2 - e_6\right) + e_6^T M_2^T \left(\overline{A}_j^i e_1 + \overline{A}_\tau^i e_2 - e_6\right)\right\},$$

$$\widehat{\Omega}_{ij}(1,0) = sym\left\{\widehat{\Pi}_{11,2}^T \widehat{P} \widehat{\Pi}_{22}\right\} + \sum_{l=1}^{2} e_1^T \widehat{Q}_l e_1 - e_3^T \widehat{Q}_1 e_3 - (1-\mu) e_2^T \widehat{Q}_2 e_2 + \tau_0^2 e_6^T \widehat{R} e_6$$
$$+ 2\left\{e_1^T M_1^T \left(\overline{A}_j^i e_1 + \overline{A}_\tau^i e_2 - e_6\right) + e_6^T M_2^T \left(\overline{A}_j^i e_1 + \overline{A}_\tau^i e_2 - e_6\right)\right\},$$

$\widehat{\Pi}_{11,1} = \left[e_1^T \ 0 \ \tau_0 e_5^T\right]$, $\widehat{\Pi}_{11,2} = \left[e_1^T \ \tau_0 e^T \ 0\right]$, $\widehat{\Pi}_{22} = \left[e_6^T \ e_1^T - (1-\mu)e_2^T \right.$ $\left.(1-\mu)e_2^T - e_3^T\right]^T$, $\overline{R} = diag\left\{\widehat{R}, \ 3\widehat{R}\right\}$.

Proof The proof is based on the conditions of Theorem 5.1. Note that the block (6, 6) in (5.17) is $\left\{\tau_0^2 e_6^T R e_6 - (M_2 + M_2^T)\right\}$. Since, $R > 0$ and $\tau_0 \neq 0$, $-(M_2 + M_2^T)$ must be negative definite, which implies that M_2 is non-singular. Choosing $M_2^{-1} = Y$ and $M_1 = \rho M_2$, where ρ is a scalar parameter. Next, pre and post multiply both sides of (5.16) with $diag\{Y, Y, Y, Y\}$ and its transpose, yields (5.41) with changes of variables $\widehat{R} = YRY^T$ and $\widehat{X} = YXY^T$. Again, pre and post multiply of (5.17) and (5.19) with $diag\{Y, Y, Y, Y, Y, Y\}$ and its transpose, respectively. Defining the variables $\widehat{P}_{11} = YP_{11}Y^T$, $\widehat{P}_{12} = YP_{12}Y^T$, $\widehat{P}_{13} = YP_{13}Y^T$, $\widehat{P}_{22} = YP_{22}Y^T$, $\widehat{P}_{23} = YP_{23}Y^T$, $\widehat{P}_{33} = YP_{33}Y^T$, $\widehat{Q}_1 = YQ_1Y^T$, $\widehat{Q}_2 = YQ_2Y^T$ and $N_j = K_jY^T$, one can obtain the LMIs in (5.42) and (5.44) respectively. This complete the proof of Theorem 5.2.

5.5 Numerical Examples

To justify the efficiency of the proposed stability and stabilization condition let us demonstrate the following list of examples.

Example 1 Consider the following nonlinear system in [23] with time varying delay:

$$\begin{cases} \dot{x}_1(t) = 0.5\left(1-\sin^2(\theta(t))\right)x_2(t) - x_1(t-\tau(t)) - \left(1+\sin^2(\theta(t))\right)x_1(t) \\ \dot{x}_2(t) = \mathrm{sgn}\left(\mid \theta(t)\mid -\frac{\pi}{2}\right)\left(0.9\cos^2(\theta(t))-1\right)x_1(t-\tau(t)) - x_2(t-\tau(t)) - \left(0.9+0.1\cos^2(\theta(t))\right)x_2(t) \end{cases} \tag{5.46}$$

The above system (5.46) can be rewritten in the following state-space form as,

$$\begin{aligned} \dot{x}(t) &= \begin{bmatrix} -1-\sin^2(\theta(t)) & 0.5-0.5\sin^2(\theta(t)) \\ 0 & -0.9-0.1\cos^2(\theta(t)) \end{bmatrix} x(t) \\ &+ \begin{bmatrix} -1 & 0 \\ \mathrm{sgn}\left(\mid\theta(t)\mid-\frac{\pi}{2}\right)\left(0.9\cos^2(\theta(t))-1\right) & -1 \end{bmatrix} x(t-\tau(t)), \end{aligned} \tag{5.47}$$

where, $x(t) = [x_1^T(t) \;\; x_2^T(t)]^T$, $x(t-\tau(t)) = [x_1^T(t-\tau(t)) \;\; x_2^T(t-\tau(t))]^T$ and the nonlinear terms in (5.47) are $\sin^2(\theta(t))$ and $\cos^2(\theta(t))$ and assume that $-\frac{\pi}{2} \leq \theta(t) \leq \frac{\pi}{2}$.

Generally, two approaches namely (i) identification using input-output data and (ii) derivation from given nonlinear system equation are there to approximate the nonlinear dynamical systems by T-S fuzzy modelling [3]. The first approach mainly consists of two parts: structure identification and parameter identification and this approach is suitable for plants that are unable or difficult to be represented by analytical methods. In the second approach, nonlinear dynamic models for physical systems can be obtained by the Lagrange methods and the Newton-Euler method and this approach uses the idea of sector nonlinearity, local approximation or a combination of them.

To approximate the nonlinear system described in (5.47) by T-S fuzzy modelling, we take the local approximation method because it uses the less number of fuzzy rules for fuzzy models compared as to the sector nonlinearity method.

Therefore, the nonlinear system (5.47) is modelled by the following fuzzy rules as,

Rule 1: **IF** $\theta(t)$ is about $\pm(\dfrac{\pi}{2})$ **THEN** $\dot{x}(t) = A_1 x(t) + A_{\tau_1} x(t-\tau(t))$
Rule 2: **IF** $\theta(t)$ is about '0' **THEN** $\dot{x}(t) = A_2 x(t) + A_{\tau_2} x(t-\tau(t))$
where the subsystem matrices are calculated as,

$$A_1 = \begin{bmatrix} -2 & 0 \\ 0 & -0.9 \end{bmatrix},\; A_2 = \begin{bmatrix} -1 & 0.5 \\ 0 & -1 \end{bmatrix},\; A_{\tau_1} = \begin{bmatrix} -1 & 0 \\ -1 & -1 \end{bmatrix},\; A_{\tau_2} = \begin{bmatrix} -1 & 0 \\ 0.1 & -1 \end{bmatrix},$$

and the membership functions are defined as,

Table 5.1 Maximum time-delay τ_0 for Example 1 with $\mu = 0$

Method	[24]	[25]	[18]	[26]	**Theorem** 5.1
Maximum τ_0	1.5974	1.5974	1.6341	2.0290	**2.5503**

$$M_1(\theta(t)) = \frac{1}{1+e^{-2\theta(t)}}, \quad M_2(\theta(t)) = 1 - M_1(\theta(t))$$

Since, M_1 and M_2 are two fuzzy sets, so according to fuzzy mathematics,

$$M_1(\theta(t)) + M_2(\theta(t)) = 1$$

Now, the truth values of $\dot{x}(t)$ is calculated by using (5.5) as,

$$h_1(\theta(t)) = M_1(\theta(t)) \wedge M_2(\theta(t)), \quad h_2(\theta(t)) = M_1(\theta(t)) \wedge M_2(\theta(t))$$

By using (5.9) with $u(t) = 0$, the final output is obtained as,

$$\dot{x}(t) = \sum_{i=1}^{2} w_i(\theta(t)) \left\{A_i x(t) + A_{\tau_i} x(t - \tau(t))\right\}, \quad t \geq 0 \tag{5.48}$$

where, $w_1(\theta(t)) = \frac{h_1(\theta(t))}{h_1(\theta(t))+h_2(\theta(t))}$ and $w_2(\theta(t)) = \frac{h_2(\theta(t))}{h_1(\theta(t))+h_2(\theta(t))}$ and A_1, A_2, A_{τ_1}, A_{τ_2} are given in above.

Using Theorem 5.1, the maximum delay bound τ_0 is calculated for $\mu = 0$ and the results are presented in Table 5.1. From Table 5.1, it is observed that Theorem 5.1 gives the larger time-delay τ_0 than some existing stability criterion.

Example 2 Consider a time-delayed T-S fuzzy system given in [9] as,
Rule 1: **IF** $x(t)$ is M_{11} **THEN** $\dot{x}(t) = A_1 x(t) + A_{\tau_1} x(t - \tau(t))$
Rule 2: **IF** $x(t)$ is M_{11} **THEN** $\dot{x}(t) = A_2 x(t) + A_{\tau_2} x(t - \tau(t))$

The membership function for rule 1 and 2 are,

$$M_1(x(t)) = \left\{1 - \frac{1}{1+\exp^{-5(x_1(t)-\frac{\pi}{6})}}\right\} \frac{1}{1+\exp^{-5(x_1(t)-\frac{\pi}{6})}}, \quad M_2 = 1 - M_1(x(t)).$$

where, $x(t) = [x_1(t) \; x_2(t)]^T$ and

$$A_1 = \begin{bmatrix} -2.0 & 0.0 \\ 0.0 & -0.9 \end{bmatrix}, A_2 = \begin{bmatrix} -1.5 & 1.0 \\ 0.0 & -0.75 \end{bmatrix}, A_{\tau_1} = \begin{bmatrix} -1.0 & 0.0 \\ -1.0 & -1.0 \end{bmatrix} A_{\tau_2} = \begin{bmatrix} -1.0 & 0.0 \\ 1.0 & -0.85 \end{bmatrix} \tag{5.49}$$

For Example 2, maximum delay bound τ_0 is calculated by using Theorem 5.1 with different values of μ. The obtained delay bound results are given in Table 5.2. It is

Table 5.2 Maximum time-delay τ_0 for Example 1 with different μ

Method	$\mu = 0.2$	$\mu = 0.4$	$\mu = 0.6$
[19]	0.9119	0.9793	1.0639
[27]	1.1410	1.1500	1.1720
[17]	1.1639	1.1734	1.1994
[28]	1.4618	1.4202	1.4005
[26]	1.7805	1.5339	1.4082
Theorem 5.1	**1.9607**	**1.7027**	**1.5612**

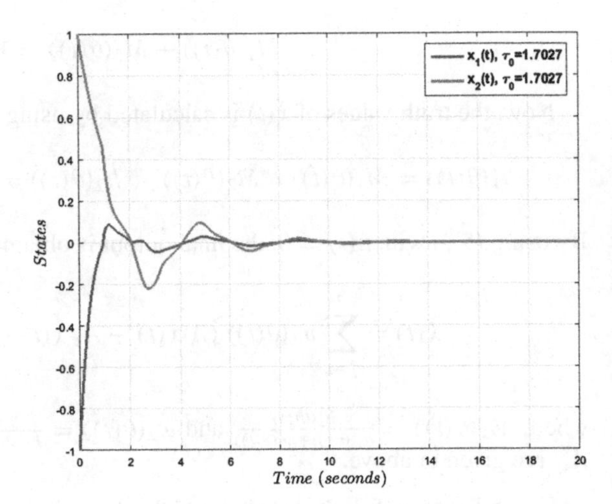

Fig. 5.1 State responses of the system given in Example 2 for $\tau(t) \in [0 \ \ 1.7027]$ and $\mu = 0.4$

observed from Table 5.2 that our proposed Theorem 5.1 gives the less conservative delay bound result over the some recent existing methods. Numerical simulation is carried out for the system given in Example 2 with the initial state $x(0) = [-1 \ \ 1]^T$ and $0 \leq \tau(t) \leq 1.7027$, $\mu = 0.4$. Also, it is considered that $\tau(t)$ is a slow time-varying sine signal following the expression $\tau(t) = 0.8513 \sin(\omega t) + 0.8513$, where $\omega = 3.14 \ red/sec$. The state responses of the T-S fuzzy system given in Example 2 are shown in Fig. 5.1 and time-varying delay $\tau(t)$ takes all the values within $0 \leq \tau(t) \leq 1.7027$ as depicted in Fig. 5.2.

Example 3 Consider the following T-S fuzzy model in [29] with

$$\dot{x}(t) = \sum_{i=1}^{r} w_i(z(t)) \left\{ A_i x(t) + A_{\tau_i} x(t - \tau(t)) + B_i u(t) \right\} \tag{5.50}$$

where, $x(t) = [x_1(t) \ \ x_2(t)]^T$ and

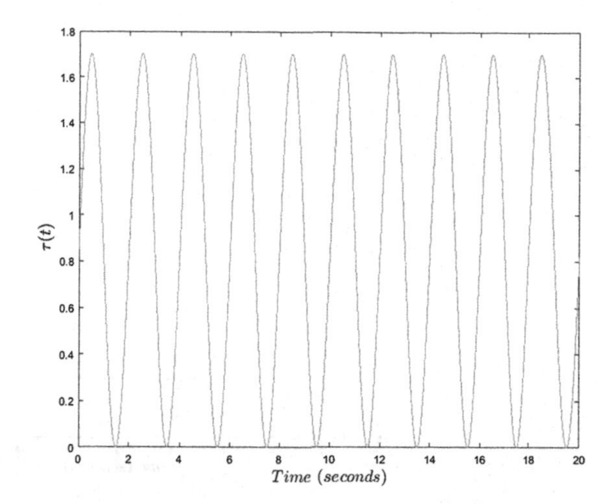

Fig. 5.2 State responses of the system given in Example 1 for $\tau(t) \in [1\ \ 1.9624]$ and $0 < \mu < 1$

Table 5.3 Comparison between various stabilization methods

Method	Corollary 2 of [29]	Theorem 5.1 of [29]	Theorem 5.1 of [12]	**Theorem** 5.2
Maximum allowable τ_0	0.2574	0.2664	0.4909	0.8152

$$A_1 = \begin{bmatrix} 0.0 & 0.6 \\ 0.0 & 1.0 \end{bmatrix},\ A_2 = \begin{bmatrix} 1.0 & 0.0 \\ 1.0 & 0.0 \end{bmatrix},\ A_{\tau 1} = \begin{bmatrix} 0.5 & 0.9 \\ 0.0 & 2.0 \end{bmatrix}\ A_{\tau 2} = \begin{bmatrix} 0.9 & 0.0 \\ 1.0 & 1.6 \end{bmatrix},\ B_1 = B_2 = \begin{bmatrix} 1 \\ 1 \end{bmatrix}. \tag{5.51}$$

The membership function are defined as,

$$M_1(x(t)) = \frac{1}{1 + \exp(-2x_1(t) + 0.5)},\ M_2 = 1 - M_1(x(t)).$$

For Example 3, delay upper bound τ_0 is calculated by using Theorem 5.2 with the choice of $\mu = 0$, and $\rho = 1.00$ and the results are given in Table 5.3. Further, Table 5.3 shows that obtained delay bound result is less conservative than the other existing approaches. Also, with the choice of $\mu = 0,\ \rho = 1.00$ and $\tau_0 = 0.8152$ following state feedback gain matrices are obtained by using Theorem 5.2 as,

$$K_1 = \begin{bmatrix} 16.0226 & -50.2955 \end{bmatrix},\ K_2 = \begin{bmatrix} 15.8894 & -53.8924 \end{bmatrix} \tag{5.52}$$

Figures 5.3 and 5.4 shows the state responses and the control signal for stabilize the system (5.50) via the fuzzy controller (5.8) with gain matrices (5.52) under the initial condition $x(0) = [2\ \ 0]^T$.

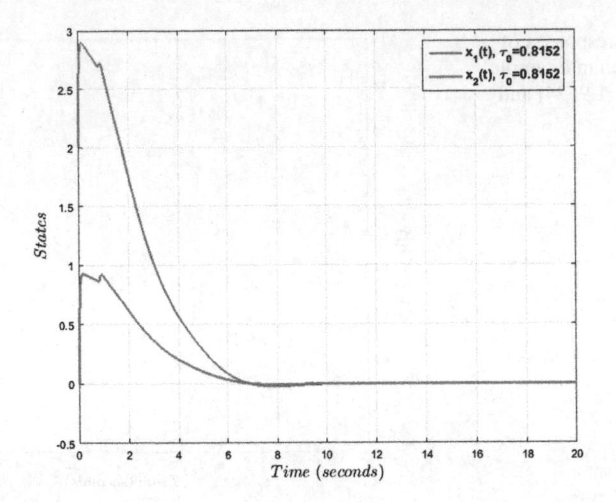

Fig. 5.3 State responses of the system given in Example 3

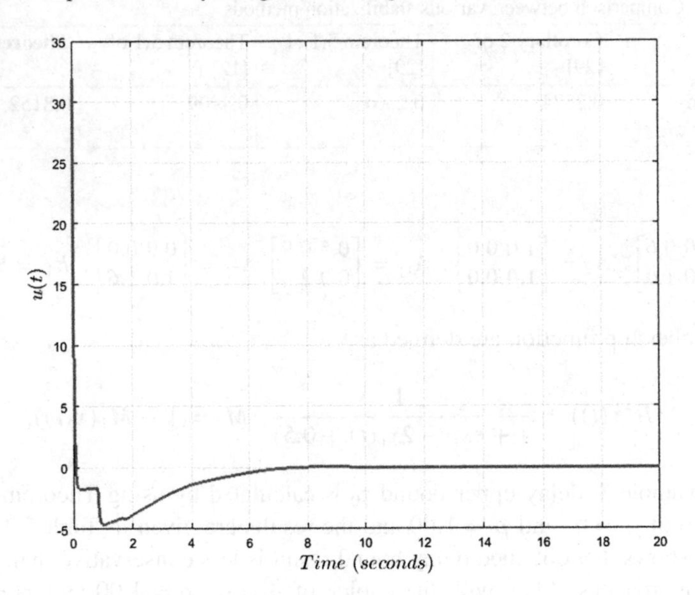

Fig. 5.4 Control signal for stabilization of Example 3

5.6 Conclusion

To obtain a less conservative delay-dependent stability and stabilization condition for T-S fuzzy time-delay system, Wirtinger inequality with reciprocal convex is used to estimate the quadratic integral integral term coming out from the derivative of L-K functional. Improvement in the delay bound results over some existing methods are illustrated by solving three different numerical examples.

References

1. H.K. Khalil, *Nonlinear systems* (Prentice-Hall, New Jersey, 1996), pp. 1–5
2. K. Tanaka, H.O. Wang, *Fuzzy Control Systems Design and Analysis: A Linear Matrix Inequality Approach* (Wiley, New York, 2004)
3. T. Takagi, M. Sugeno, Fuzzy identification of systems and its applications to modeling and control. IEEE Trans. Syst. Man Cybern. **1**, 116–132 (1985)
4. S. Boyd, L. El Ghaoui, E. Feron, V. Balakrishnan, *Linear Matrix Inequalities in System and Control Theory* (SIAM, 1994)
5. P. Gahinet, A. Nemirovski, A. Laub, M. Chilali, LMI Control Toolbox-for Use with Matlab, Natick, MA: The MATH Works (1995)
6. K. Gu, J. Chen, V.L. Kharitonov, *Stability of Time-Delay Systems* (Springer Science & Business Media, 2003)
7. R. Dey, S. Ghosh, G. Ray, A. Rakshit, State feedback stabilization of uncertain linear time-delay systems: a nonlinear matrix inequality approach. Numer. Linear Algebra Appl. **18**(3), 351–361 (2011)
8. R. Dey, S. Ghosh, G. Ray, A. Rakshit, Improved delay-dependent stabilization of time-delay systems with actuator saturation. Int. J. Robust Nonlinear Control **24**(5), 902–917 (2014)
9. R. Dey, V.E. Balas, T. Lin, G. Ray, Improved delay-range-dependent stability analysis of a T-S fuzzy system with time varying delay, in *2013 IEEE 14th International Symposium on Computational Intelligence and Informatics (CINTI)* (IEEE, 2013), pp. 173–178
10. C. Peng, D. Yue, T.-C. Yang, E.G. Tian, On delay-dependent approach for robust stability and stabilization of T-S fuzzy systems with constant delay and uncertainties. IEEE Trans. Fuzzy Syst. **17**(5), 1143–1156 (2009)
11. L. Wu, X. Su, P. Shi, *Fuzzy Control Systems with Time-Delay and Stochastic Perturbation* (Springer, Cham, 2015)
12. H. Gassara, A. El Hajjaji, M. Chaabane, Robust control of T-S fuzzy systems with time-varying delay using new approach. Int. J. Robust Nonlinear Control **20**(14), 1566–1578 (2010)
13. L. Li, X. Liu, New results on delay-dependent robust stability criteria of uncertain fuzzy systems with state and input delays. Inf. Sci. **179**(8), 1134–1148 (2009)
14. F. Liu, M. Wu, Y. He, R. Yokoyama, New delay-dependent stability criteria for T-S fuzzy systems with time-varying delay. Fuzzy Sets Syst. **161**(15), 2033–2042 (2010)
15. S. Mobayen, An lmi-based robust controller design using global nonlinear sliding surfaces and application to chaotic systems. Nonlinear Dyn. **79**(2), 1075–1084 (2015)
16. Y. Zhao, H. Gao, J. Lam, B. Du, Stability and stabilization of delayed T-S fuzzy systems: a delay partitioning approach. IEEE Trans. Fuzzy Syst. **17**(4), 750–762 (2009)
17. F.O. Souza, V.C. Campos, R.M. Palhares, On delay-dependent stability conditions for Takagi-Sugeno fuzzy systems. J. Franklin Inst. **351**(7), 3707–3718 (2014)
18. C. Peng, Y.-C. Tian, E. Tian, Improved delay-dependent robust stabilization conditions of uncertain T-S fuzzy systems with time-varying delay. Fuzzy Sets Syst. **159**(20), 2713–2729 (2008)

19. C. Peng, L.-Y. Wen, J.-Q. Yang, On delay-dependent robust stability criteria for uncertain T-S fuzzy systems with interval time-varying delay. Int. J. Fuzzy Syst. **13**(1) (2011)
20. Z. Zhang, C. Lin, B. Chen, New stability and stabilization conditions for T-S fuzzy systems with time delay. Fuzzy Sets Syst. **263**, 82–91 (2015)
21. A. Seuret, F. Gouaisbaut, Wirtinger-based integral inequality: application to time-delay systems. Automatica **49**(9), 2860–2866 (2013)
22. P. Park, J.W. Ko, C. Jeong, Reciprocally convex approach to stability of systems with time-varying delays. Automatica **47**(1), 235–238 (2011)
23. C. Peng, Q.-L. Han, Delay-range-dependent robust stabilization for uncertain t-s fuzzy control systems with interval time-varying delays. Inf. Sci. **181**(19), 4287–4299 (2011)
24. E. Tian, C. Peng, Delay-dependent stability analysis and synthesis of uncertain T-S fuzzy systems with time-varying delay. Fuzzy Sets Syst. **157**(4), 544–559 (2006)
25. C.-H. Lien, K.-W. Yu, W.-D. Chen, Z.-L. Wan, Y.-J. Chung, Stability criteria for uncertain Takagi-Sugeno fuzzy systems with interval time-varying delay. IET Control Theory Appl. **1**(3), 764–769 (2007)
26. Z. Lian, Y. He, C.-K. Zhang, M. Wu, Stability analysis for T-S fuzzy systems with time-varying delay via free-matrix-based integral inequality. Int. J. Control Autom. Syst. **14**(1), 21–28 (2016)
27. E. Tian, D. Yue, Y. Zhang, Delay-dependent robust H_∞ control for T-S fuzzy system with interval time-varying delay. Fuzzy Sets Syst. **160**(12), 1708–1719 (2009)
28. X. Xia, R. Li, J. An, On delay-fractional-dependent stability criteria for Takagi-Sugeno fuzzy systems with interval delay. Math. Probl. Eng. **2014** (2014)
29. H.-N. Wu, H.-X. Li, New approach to delay-dependent stability analysis and stabilization for continuous-time fuzzy systems with time-varying delay. IEEE Trans. Fuzzy Syst. **15**(3), 482–493 (2007)

Appendix

Here we present some representative MATLAB codes for solving the LMI conditions reported in this book using LMI toolbox of MATLAB. Two solvers of LMI toolbox have been presented here (i) 'feasp' and (ii) 'mincx'. The readers are advised to refer LMI toolbox user guide [1] of MATLAB in order to succinctly understand the usage of various commands used in the codes.

A1: MATLAB code to simulate uncertain linear time-delay system (Retarded system).

The MATLAB code presented here describes closed loop simulation of an uncertain linear time-delay system. The code is developed using standard RK4 routine. This code is developed for the system described in Example 7 [2]. This code can be modified to carry out open loop simulation of a nominal time-delay system. This is left to the readers as an exercise.

```
1  %%%%%%%%%%%%%%%%%%%%%%%%%%%%%%%%%%%%%%%%%%%%%%%%%%%%%%%%%%%%%%%%%%%%%%%%%%%%%%%%
2  %Closed loop simulation of uncertain time-delay system using the parameters
3  %given in Example 7 of paper "Improved robust stability criteria and design
4  %of robust stabilizing controller for uncertain linear time-delay systems"
5  %M.N. Alpaslan Paralkci,Int J of Robust and Nonlinear %Control,2006:16,599 -636.
6  %%%%%%%%%%%%%%%%%%%%%%%%%%%%%%%%%%%%%%%%%%%%%%%%%%%%%%%%%%%%%%%%%%%%%%%%%%%%%%%%
7  clc
8  clear all
9  A=[0 0;0 1];Ad=[-1 -1;0 -0.9];B=[0;1];D=[0.2 0;0 0.2];Ea=eye(2);Ed=eye(2);
10 %A=[0 0;0 1];Ad=[-2 -0.5;0 -1];B=[0;1];D=0.2*eye(2);Ea=eye(2);Ed=eye(2);%Ex2.1
11 h=input('input the value of(steps)"h"=');
12 tau=input('input the value of delay,=');
13 t0=input('input the value of intial time"t0"=');
14 tf=input('input the value of final time"tf"=');
15 m=round(tau/h);
16 t(1)=t0;
17 N=(tf-t0)/h;
18 x_int=[1,0];
19 x(1,:)=x_int(:)';
20 % K=[-4.0313 -10.1156];%d=2.0,iteration=50
21 % K=[-0.8035 -2.7533];%d=1.4,iteration=10
22 % K=[-2.5419 -6.4414];%d=1.8,iteration=20
23 % K=[-3.4953 -8.8552];%d=2,iteration=38
24 % K=[-0.6390 -2.3261];%d=1.3,iteration=07
```


R. Dey et al., *Stability and Stabilization of Linear and Fuzzy Time-Delay Systems*,
Intelligent Systems Reference Library 141, https://doi.org/10.1007/978-3-319-70149-3

```
25 % K=[-0.5638 -2.0496];%d=1.2,iteration=05
26 K=1e3*[-6.0105 -4.0486];%d=1.8,iteration=18
27 Ac=A+B*K;
28  for k=1:N
29      t(k+1)=t(1)+(k*h);
30      F=[cos(t(k)),0;0,sin(t(k))];
31      if (k-m)≤1
32          c_1=h*(x(k,:)*(Ac+D*F*Ea)'+x(1,:)*(Ad+D*F*Ed)');
33          c_1=c_1(:)';
34          c_2=h*((x(k,:)+(c_1)/2)*(Ac+D*F*Ea)'+(x(1,:)+(c_1)/2)*(Ad+D*F*Ed)');
35          c_2=c_2(:)';
36          c_3=h*((x(k,:)+(c_2)/2)*(Ac+D*F*Ea)'+(x(1,:)+(c_2)/2)*(Ad+D*F*Ed)');
37          c_3=c_3(:)';
38          c_4=h*((x(k,:)+(c_3))*(Ac+D*F*Ea)'+(x(1,:)+(c_3))*(Ad+D*F*+Ed)');
39          c_4=c_4(:)';
40          x(k+1,:)=x(k,:)+((c_1)+2*(c_2+c_3)+c_4)*(1/6);
41      elseif (k-m)>1
42          c_1=h*(x(k,:)*(Ac+D*F*Ea)'+x(k-m,:)*(Ad+D*F*Ed)');
43          c_1=c_1(:)';
44          c_2=h*((x(k,:)+(c_1)/2)*(Ac+D*F*Ea)'+(x(k-m,:)+(c_1)/2)*(Ad+D*F*Ed)');
45          c_2=c_2(:)';
46          c_3=h*((x(k,:)+(c_2)/2)*(Ac+D*F*Ea)'+(x(k-m,:)+(c_2)/2)*(Ad+D*F*Ed)');
47          c_3=c_3(:)';
48          c_4=h*((x(k,:)+(c_3))*(Ac+D*F*Ea)'+(x(k-m,:)+(c_3))*(Ad+D*F*Ed)');
49          c_4=c_4(:)';
50          x(k+1,:)=x(k,:)+(c_1+2*(c_2+c_3)+c_4)*(1/6);
51           u(k+1,:)=K*x(k+1,:)';
52      end
53  end
54 plot(t,x(:,1),'b',t,x(:,2),'r')
55 title(' x1(blue),x2(red)')
56 % plot(t,u)
57 % title('plot for x1')
```

A2: MATLAB code for Delay-dependent stabilization with actuator saturation—Theorem 4.1

```
1 clc
2 clear all
3 A=[0.5 -1;0.5 -0.5];Ad=[0.6 0.4;0 -0.5];B=[1;1];u=5;mu=0;
4 % A=[1 1.5;0.3 -2];Ad=[0 -1;0 0];B=[10;1];u=15;
5 %A=[-2,0;0,-0.9];Ad=[-1,0;-1,-1];B=[1;0]; u=1;
6 mu=0;
7 tau_bar=3;
8 % alpha=0.34;
9 beta=0.4;
10 sigma=1e+17; %weight on the optimization algorithm
11
12 %LMI variables
13 setlmis([]);
14   P=lmivar(1,[2,1]);
15   Q1=lmivar(1,[2,1]);
16   Q2=lmivar(1,[2,1]);
17   R1=lmivar(1,[2,1]);
```

```
18    Y=lmivar(2,[1,2]);
19    W=lmivar(2,[1,2]);%%%%%%%G in derivation
20    M1=lmivar(2,[2,2]);
21    N1=lmivar(2,[2,2]);
22    M2=lmivar(2,[2,2]);
23    N2=lmivar(2,[2,2]);
24    S=lmivar(2,[2,2]);%S_bar in the derivation
25    e1=lmivar(1,[1,0]);
26    e2=lmivar(1,[1,0]);
27    e3=lmivar(1,[1,0]);
28    e4=lmivar(1,[1,0]);
29    e5=lmivar(1,[1,0]);
30
31 % LMI index i=1,j=1
32 lmiterm([1 1 1 Q1],1,1);
33 lmiterm([1 1 1 Q2],1,1);
34 lmiterm([1 1 1 S],1,A','s');
35 lmiterm([1 1 1 M1],(1/tau_bar),1,'s');
36 lmiterm([1 1 1 Y],B,1,'s');
37
38 lmiterm([1 1 2 -S],Ad,1);
39 lmiterm([1 1 2 S],beta,A');
40 lmiterm([1 1 2 M1],-(1/tau_bar),1);
41 lmiterm([1 1 2 -N1],(1/tau_bar),1);
42 lmiterm([1 1 2 -Y],beta,B');
43
44 lmiterm([1 1 4 P],1,1);
45 lmiterm([1 1 4 -S],1,-1);
46 lmiterm([1 1 4 S],alpha,A');
47 lmiterm([1 1 4 -Y],alpha,B');
48
49 lmiterm([1 2 2 -S],beta*Ad,1,'s');
50 lmiterm([1 2 2 Q2],-(1-mu),1);
51 lmiterm([1 2 2 M2],(1/tau_bar),1,'s');
52 lmiterm([1 2 2 N1],-(1/tau_bar),1,'s');
53
54 lmiterm([1 2 3 M2],-(1/tau_bar),1);
55 lmiterm([1 2 3 -N2],(1/tau_bar),1);
56
57 lmiterm([1 2 4 -S],-beta,1);
58 lmiterm([1 2 4 S],alpha,Ad');
59
60 lmiterm([1 3 3 Q1],-1,1);
61 lmiterm([1 3 3 N2],-(1/tau_bar),1,'s');
62
63 lmiterm([1 4 4 R1],1,1);
64 lmiterm([1 4 4 S],-1,alpha,'s');
65
66 lmiterm([1 1 5 M1],1,1);
67 lmiterm([1 2 5 N1],1,1);
68 lmiterm([1 5 5 R1],-1,1);
69
70 % LMI index i=1(for convex combination),j=2(for j=1,2^m)
71 lmiterm([2 1 1 Q1],1,1);
72 lmiterm([2 1 1 Q2],1,1);
```

```
73  lmiterm([2 1 1 S],1,A','s');
74  lmiterm([2 1 1 M1],(1/tau_bar),1,'s');
75  lmiterm([2 1 1 W],B,1,'s');
76
77  lmiterm([2 1 2 -S],Ad,1);
78  lmiterm([2 1 2 S],beta,A');
79  lmiterm([2 1 2 M1],-(1/tau_bar),1);
80  lmiterm([2 1 2 -N1],(1/tau_bar),1);
81  lmiterm([2 1 2 -W],beta,B');
82
83  lmiterm([2 1 4 P],1,1);
84  lmiterm([2 1 4 -S],1,-1);
85  lmiterm([2 1 4 S],alpha,A');
86  lmiterm([2 1 4 -W],alpha,B');
87
88  lmiterm([2 2 2 -S],beta*Ad,1,'s');
89  lmiterm([2 2 2 Q2],-(1-mu),1);
90  lmiterm([2 2 2 M2],(1/tau_bar),1,'s');
91  lmiterm([2 2 2 N1],-(1/tau_bar),1,'s');
92
93  lmiterm([2 2 3 M2],-(1/tau_bar),1);
94  lmiterm([2 2 3 -N2],(1/tau_bar),1);
95
96  lmiterm([2 2 4 -S],-beta,1);
97  lmiterm([2 2 4 S],alpha,Ad');
98
99  lmiterm([2 3 3 Q1],-1,1);
100 lmiterm([2 3 3 N2],-(1/tau_bar),1,'s');
101
102 lmiterm([2 4 4 R1],1,1);
103 lmiterm([2 4 4 S],-1,alpha,'s');
104
105 lmiterm([2 1 5 M1],1,1);
106 lmiterm([2 2 5 N1],1,1);
107 lmiterm([2 5 5 R1],-1,1);
108
109 % LMI index i=2,j=1
110 lmiterm([3 1 1 Q1],1,1);
111 lmiterm([3 1 1 Q2],1,1);
112 lmiterm([3 1 1 S],1,A','s');
113 lmiterm([3 1 1 M1],(1/tau_bar),1,'s');
114 lmiterm([3 1 1 Y],B,1,'s');
115
116 lmiterm([3 1 2 -S],Ad,1);
117 lmiterm([3 1 2 S],beta,A');
118 lmiterm([3 1 2 M1],-(1/tau_bar),1);
119 lmiterm([3 1 2 -N1],(1/tau_bar),1);
120 lmiterm([3 1 2 -Y],beta,B');
121
122 lmiterm([3 1 4 P],1,1);
123 lmiterm([3 1 4 -S],1,-1);
124 lmiterm([3 1 4 S],alpha,A');
125 lmiterm([3 1 4 -Y],alpha,B');
126
127 lmiterm([3 2 2 -S],beta*Ad,1,'s');
```

```
128 lmiterm([3 2 2 Q2],-(1-mu),1);
129 lmiterm([3 2 2 M2],(1/tau_bar),1,'s');
130 lmiterm([3 2 2 N1],-(1/tau_bar),1,'s');
131
132 lmiterm([3 2 3 M2],-(1/tau_bar),1);
133 lmiterm([3 2 3 -N2],(1/tau_bar),1);
134
135 lmiterm([3 2 4 -S],-beta,1);
136 lmiterm([3 2 4 S],alpha,Ad');
137
138 lmiterm([3 3 3 Q1],-1,1);
139 lmiterm([3 3 3 N2],-(1/tau_bar),1,'s');
140
141 lmiterm([3 4 4 R1],1,1);
142 lmiterm([3 4 4 S],-1,alpha,'s');
143
144 lmiterm([3 2 5 M2],1,1);
145 lmiterm([3 3 5 N2],-1,1);
146 lmiterm([3 5 5 R1],-1,1);
147
148
149 % LMI index i=2(for convex combination),j=2(for j=1,2^m)
150 lmiterm([4 1 1 Q1],1,1);
151 lmiterm([4 1 1 Q2],1,1);
152 lmiterm([4 1 1 S],1,A','s');
153 lmiterm([4 1 1 M1],(1/tau_bar),1,'s');
154 lmiterm([4 1 1 W],B,1,'s');
155
156 lmiterm([4 1 2 -S],Ad,1);
157 lmiterm([4 1 2 S],beta,A');
158 lmiterm([4 1 2 M1],-(1/tau_bar),1);
159 lmiterm([4 1 2 -N1],(1/tau_bar),1);
160 lmiterm([4 1 2 -W],beta,B');
161
162 lmiterm([4 1 4 P],1,1);
163 lmiterm([4 1 4 -S],1,-1);
164 lmiterm([4 1 4 S],alpha,A');
165 lmiterm([4 1 4 -W],alpha,B');
166
167 lmiterm([4 2 2 -S],beta*Ad,1,'s');
168 lmiterm([4 2 2 Q2],-(1-mu),1);
169 lmiterm([4 2 2 M2],(1/tau_bar),1,'s');
170 lmiterm([4 2 2 N1],-(1/tau_bar),1,'s');
171
172 lmiterm([4 2 3 M2],-(1/tau_bar),1);
173 lmiterm([4 2 3 -N2],(1/tau_bar),1);
174
175 lmiterm([4 2 4 -S],-beta,1);
176 lmiterm([4 2 4 S],alpha,Ad');
177
178 lmiterm([4 3 3 Q1],-1,1);
179 lmiterm([4 3 3 N2],-(1/tau_bar),1,'s');
180
181 lmiterm([4 4 4 R1],1,1);
182 lmiterm([4 4 4 S],-1,alpha,'s');
```

```
183
184 lmiterm([4 2 5 M2],1,1);
185 lmiterm([4 3 5 N2],1,1);
186 lmiterm([4 5 5 R1],-1,1);
187
188 %Control Constraint LMI
189 lmiterm([-5 1 1 0],u);                          % LMI #3: 1
190 lmiterm([-5 1 2 W],1,1);
191 lmiterm([-5 2 2 P],u,1);                        % LMI #3: P
192
193 lmiterm([-6 1 1 e1],1,1);                       % LMI #4: w1
194 lmiterm([-6 1 2 0],1);                          % LMI #4: 1
195 lmiterm([-6 2 2 S],1,1);                        % LMI #4: X
196
197 lmiterm([-7 1 1 e2],1,1);                       % LMI #4: w1
198 lmiterm([-7 1 1 P],1,-1);                          % LMI #4: 1
199
200 lmiterm([-8 1 1 e3],1,1);                       % LMI #4: w1
201 lmiterm([-8 1 1 Q1],1,-1);                         % LMI #4: 1
202
203 lmiterm([-9 1 1 e4],1,1);                       % LMI #4: w1
204 lmiterm([-9 1 1 Q2],1,-1);                         % LMI #4: 1
205
206 lmiterm([-10 1 1 e5],1,1);                       % LMI #4: w1
207 lmiterm([-10 1 1 R1],1,-1);                         % LMI #4: 1
208
209 lmiterm([-11 1 1 Q1],1,1);                      % LMI #11: Q1
210
211 lmiterm([-12 1 1 Q2],1,1);                      % LMI #11: Q1
212
213 lmiterm([-13 1 1 P],1,1);                      % LMI #14: Z1
214
215 lmiterm([-14 1 1 R1],1,1);                      % LMI #14: Z1
216
217 lmi_actuator_sat3050=getlmis;
218 [tmin xfeas]=feasp(lmi_actuator_sat3050,[.01,100,1e7,1,1]);
219  if tmin<0
220 n=decnbr(lmi_actuator_sat3050);
221 c=zeros(n,1);
222 for j1=1:n
223 [e1j,e2j,e3j,e4j,e5j]=defcx(lmi_actuator_sat3050,j1,e1,e2,e3,e4,e5);
224 c(j1,:)=sigma*e1j+e2j+tau_bar*e3j+tau_bar*e4j+(tau_bar/2)*e5j;
225 end
226 [copt,xopt] = mincx(lmi_actuator_sat3050,c);
227 Sopt=dec2mat(lmi_actuator_sat3050,xopt,S);
228 Q1_opt=dec2mat(lmi_actuator_sat3050,xopt,Q1);
229 Q2_opt=dec2mat(lmi_actuator_sat3050,xopt,Q2);
230 Yopt=dec2mat(lmi_actuator_sat3050,xopt,Y);
231 P_opt=dec2mat(lmi_actuator_sat3050,xopt,P);
232 R1_opt=dec2mat(lmi_actuator_sat3050,xopt,R1);
233
234 %%%%%%Estimating the domain of attraction of origin%%%%%%%%%%%%
235
236 S_inv=inv(Sopt);
237 S_inv_trans=S_inv';
```

```
238
239 P=(S_inv*P_opt*S_inv_trans);
240 Eigen_P=eig(P);
241 Max_eigen_P=max(Eigen_P);
242
243 Q1=(S_inv*Q1_opt*S_inv_trans);
244 Eigen_Q1=eig(Q1);
245 Max_eigen_Q1=max(Eigen_Q1);
246
247 Q2=(S_inv*Q2_opt*S_inv_trans);
248 Eigen_Q2=eig(Q2);
249 Max_eigen_Q2=max(Eigen_Q2);
250
251 R1=(S_inv*R1_opt*S_inv_trans);
252 Eigen_R1=eig(R1);
253 Max_eigen_R1=max(Eigen_R1);
254
255 gamma=Max_eigen_P+tau_bar*Max_eigen_Q1+tau_bar*Max_eigen_Q2+(tau_bar/2)*Max_eigen_R1;
256 omega=sqrt(gamma);
257 Δ_max=(1/omega)
258 K=(Yopt*S_inv_trans)
```

A3: SIMULINK diagram for simulating the time-delay system with actuator saturation.

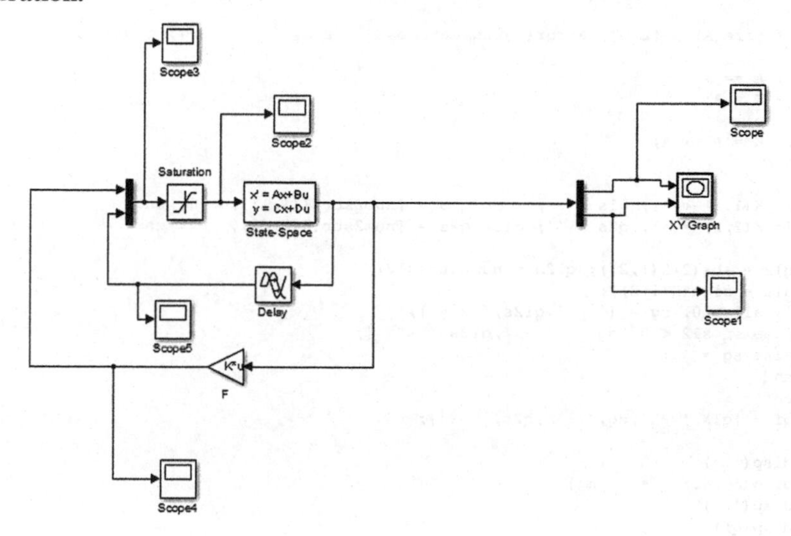

A4: MATLAB code to draw ellipse (for 2x2 system) convex compact set.

Here the input to the function is taken as symmetric positive definite matrix '*P*' solution after solving the appropriate LMI condition.

```
1  function [K,Q,D,W,l,qf] = ellipse1(A)
2  %
3  %  [K,Q,D,W,l,qf] = ellipse(A)
4  %
5  %   Draws the ellipse
6  %      x' * K * x = 1
7  %    corresponding to the positive definite symmetric 2 by 2 matrix
8  %    where K = A if A is symmetric; otherwise K = A' * A
9  %
10 %  K - symmetric positive definite matrix used
11 %  Q - orthogonal eigenvectors or principal axes of ellipse
12 %  D - diagonal matrix with eigenvalues, so K = Q * D * Q'
13 %  l = 1/sqrt(diag(D)) - lengths of semi-axes of ellipse
14 %  W = Q*D^(-1/2) - linear transformation mapping unit circle to ellipse
15 %  qf - string giving quadratic form x'*K*x
16 %
17 %  See also ELLIPSOID
18 %
19
20 if size(A) ≠ [2 2], error('argument must be 2 by 2 matrix'); end
21
22 if A == A'
23    K = A;
24    else
25    K = A' * A;
26 end
27
28 if K(1,1) == 1, q1s = ''; else q1s = [num2str(K(1,1)),' ']; end
29 if K(2,2) == 1, q2s = ''; else q2s = [num2str(K(2,2)),' ']; end
30
31 q12 = abs(2*K(1,2)); q12s = num2str(q12);
32 s12 = sign(K(1,2));
33 if s12 > 0, sg = [' + ',q12s,' x y'];
34 else if s12 < 0, sg = [' - ',q12s,' x y'];
35 else sg = '';
36 end
37 end
38 qf = [q1s,'x^2',sg,' + ',q2s,'y^2'];
39
40 disp(' ')
41 disp('Quadratic form')
42 disp(' ')
43 disp(qf)
44
45 disp(' ')
46
47 disp('K = ')
48 disp(K)
49
50 [Q,D] = eig(K);
51
52 disp('Eigenvectors - principal axes')
53 disp(' ')
54 disp(Q)
55 disp('Eigenvalues')
56 ev = diag(D);
57 disp(' ')
```

```
58 disp(ev)
59 disp('Lengths of principal axes')
60 l = 1./sqrt(ev);
61 disp(' ')
62 disp(l)
63
64 W = Q * 1/sqrt(D);
65
66 t = linspace(0,2*pi,100);
67
68 plot(W(1,1).*cos(t)+W(1,2).* sin(t),W(2,1).* cos(t) + W(2,2) .* sin(t),'m')
69
70 W1 = W(1,:); W2 = W(2,:);
71 N = diag([1 + 1/norm(W(:,1)),1 + 1/norm(W(:,2))]);
72 WN = W * N;
73 WN1 = WN(1,:); WN2 = WN(2,:);
74
75 xa = [W1;WN1];
76 ya = [W2;WN2];
77 line(xa,ya);
78
79 xa = - xa;
80 ya = - ya;
81 line(xa,ya);
82
83 axis equal;
84
85 % title(['Ellipse:    ',qf,' = 1']);
```

A5: MATLAB code for Fuzzy TDS Stability Analysis—Theorem 5.1

```
1 clear all
2 clc
3 %%%%%%%%%%% Example-1 %%%%%%%%%%%%
4 A1=[-2 0;0 -0.9];
5 A2=[-1 0.5;0 -1];
6 Ad1=[-1 0;-1 -1];
7 Ad2=[-1 0;0.1 -1];
8 %%%%%%%%%%% Example-2 %%%%%%%%%%%%
9 % A1=[-2 0;0 -0.9];
10 % A2=[-1.5 1;0 -0.75];
11 % Ad1=[-1 0;-1 -1];
12 % Ad2=[-1 0;1 -0.85];
13 %%%%%%%%%%%%%%%%%%%%%%%%%%%%%%%%%
14 miu=0.00001;
15 d=2.5503;    % d=\tau_{0}
16 %%%%%%%%%%%%%%% LMI Variables %%%%%%
17 setlmis([]);
18 P11=lmivar(1,[2,1]);
19 P12=lmivar(2,[2,2]);
20 P13=lmivar(2,[2,2]);
21 P22=lmivar(1,[2,1]);
22 P23=lmivar(2,[2,2]);
23 P33=lmivar(1,[2,1]);
```

```
24 Q1=lmivar(1,[2,1]);
25 Q2=lmivar(1,[2,1]);
26 R=lmivar(1,[2,1]);
27 X11=lmivar(2,[2,2]);
28 X12=lmivar(2,[2,2]);
29 X22=lmivar(2,[2,2]);
30 M1=lmivar(2,[2,2]);
31 M2=lmivar(2,[2,2]);
32 %%%%%%%%%%%%%%%% LMI-1 %%%%%%%%%%%%%%%%%%%%%%%%%%%%%%
33 lmiterm([-1 1 1 R],1,1);
34 lmiterm([-1 1 3 X11],1,1);
35 lmiterm([-1 1 4 X12],1,1);
36 lmiterm([-1 2 2 R],3,1);
37 lmiterm([-1 2 3 -X12],1,1);
38 lmiterm([-1 2 4 X22],1,1);
39 lmiterm([-1 3 3 R],1,1);
40 lmiterm([-1 4 4 R],3,1);
41 %%%%%%%%%%%%%%%% LMI-2 %%%%%%%%%%%%%%%%%%%%
42 lmiterm([2 1 1 M1],1,A1,'s');
43 lmiterm([2 1 1 Q1],1,1);
44 lmiterm([2 1 1 Q2],1,1);
45 lmiterm([2 1 1 R],-4,1);
46 lmiterm([2 1 1 P12],1,1,'s');
47
48 lmiterm([2 1 2 M1],1,Ad1);
49 lmiterm([2 1 2 R],-2,1);
50 lmiterm([2 1 2 X11],-1,1);
51 lmiterm([2 1 2 X12],-1,1,'s');
52 lmiterm([2 1 2 X22],-1,1);
53 lmiterm([2 1 2 P12],-(1-miu),1);
54 lmiterm([2 1 2 P13],(1-miu),1);
55
56 lmiterm([2 1 3 P13],-1,1);
57 lmiterm([2 1 3 X11],1,1);
58 lmiterm([2 1 3 X12],-1,1);
59 lmiterm([2 1 3 -X12],1,1);
60 lmiterm([2 1 3 X22],-1,1);
61
62 lmiterm([2 1 4 R],6,1);
63
64 lmiterm([2 1 5 P23],d,1);
65 lmiterm([2 1 5 X12],2,1);
66 lmiterm([2 1 5 X22],2,1);
67
68 lmiterm([2 1 6 P11],1,1);
69 lmiterm([2 1 6 M1],-1,1);
70 lmiterm([2 1 6 -M2],A1',1);
71
72 lmiterm([2 2 2 Q1],-(1-miu),1);
73 lmiterm([2 2 2 R],-8,1);
74 lmiterm([2 2 2 X11],1,1,'s');
75 lmiterm([2 2 2 X22],-1,1,'s');
76
77 lmiterm([2 2 3 X11],-1,1);
78 lmiterm([2 2 3 X12],1,1,'s');
```

```
79  lmiterm([2 2 3 X22],-1,1);
80  lmiterm([2 2 3 R],-2,1);
81
82  lmiterm([2 2 4 R],6,1);
83  lmiterm([2 2 4 X12],2,1);
84  lmiterm([2 2 4 -X22],2,1);
85
86  lmiterm([2 2 5 R],6,1);
87  lmiterm([2 2 5 X12],-2,1);
88  lmiterm([2 2 5 X22],2,1);
89  lmiterm([2 2 5 P23],-d*(1-miu),1);
90  lmiterm([2 2 5 P33],d*(1-miu),1);
91
92  lmiterm([2 2 6 -M2],Ad1',1);
93
94  lmiterm([2 3 3 R],-4,1);
95  lmiterm([2 3 3 Q2],-1,1);
96
97  lmiterm([2 3 4 X12],-2,1);
98  lmiterm([2 3 4 -X22],2,1);
99
100 lmiterm([2 3 5 P33],-d,1);
101 lmiterm([2 3 5 R],6,1);
102
103 lmiterm([2 4 4 R],-12,1);
104
105 lmiterm([2 4 5 X22],-4,1);
106
107 lmiterm([2 5 5 R],-12,1);
108
109 lmiterm([2 5 6 -P13],d,1);
110
111 lmiterm([2 6 6 R],d^2,1);
112 lmiterm([2 6 6 M2],-1,1,'s');
113 %%%%%%%%%%%%%%%% LMI-3 %%%%%%%%%%%%%%%%%%%%
114 lmiterm([3 1 1 M1],1,A2,'s');
115 lmiterm([3 1 1 Q1],1,1);
116 lmiterm([3 1 1 Q2],1,1);
117 lmiterm([3 1 1 R],-4,1);
118 lmiterm([3 1 1 P12],1,1,'s');
119
120 lmiterm([3 1 2 M1],1,Ad2);
121 lmiterm([3 1 2 R],-2,1);
122 lmiterm([3 1 2 X11],-1,1);
123 lmiterm([3 1 2 X12],-1,1,'s');
124 lmiterm([3 1 2 X22],-1,1);
125 lmiterm([3 1 2 P12],-(1-miu),1);
126 lmiterm([3 1 2 P13],(1-miu),1);
127
128 lmiterm([3 1 3 P13],-1,1);
129 lmiterm([3 1 3 X11],1,1);
130 lmiterm([3 1 3 X12],-1,1);
131 lmiterm([3 1 3 -X12],1,1);
132 lmiterm([3 1 3 X22],-1,1);
133
```

```
134 lmiterm([3 1 4 R],6,1);
135
136 lmiterm([3 1 5 P23],d,1);
137 lmiterm([3 1 5 X12],2,1);
138 lmiterm([3 1 5 X22],2,1);
139
140 lmiterm([3 1 6 P11],1,1);
141 lmiterm([3 1 6 M1],-1,1);
142 lmiterm([3 1 6 -M2],A2',1);
143
144 lmiterm([3 2 2 Q1],-(1-miu),1);
145 lmiterm([3 2 2 R],-8,1);
146 lmiterm([3 2 2 X11],1,1,'s');
147 lmiterm([3 2 2 X22],-1,1,'s');
148
149 lmiterm([3 2 3 X11],-1,1);
150 lmiterm([3 2 3 X12],1,1,'s');
151 lmiterm([3 2 3 X22],-1,1);
152 lmiterm([3 2 3 R],-2,1);
153
154 lmiterm([3 2 4 R],6,1);
155 lmiterm([3 2 4 X12],2,1);
156 lmiterm([3 2 4 -X22],2,1);
157
158 lmiterm([3 2 5 R],6,1);
159 lmiterm([3 2 5 X12],-2,1);
160 lmiterm([3 2 5 X22],2,1);
161 lmiterm([3 2 5 P23],-d*(1-miu),1);
162 lmiterm([3 2 5 P33],d*(1-miu),1);
163
164 lmiterm([3 2 6 -M2],Ad2',1);
165
166 lmiterm([3 3 3 R],-4,1);
167 lmiterm([3 3 3 Q2],-1,1);
168
169 lmiterm([3 3 4 X12],-2,1);
170 lmiterm([3 3 4 -X22],2,1);
171
172 lmiterm([3 3 5 P33],-d,1);
173 lmiterm([3 3 5 R],6,1);
174
175 lmiterm([3 4 4 R],-12,1);
176
177 lmiterm([3 4 5 X22],-4,1);
178
179 lmiterm([3 5 5 R],-12,1);
180
181 lmiterm([3 5 6 -P13],d,1);
182
183 lmiterm([3 6 6 R],d^2,1);
184 lmiterm([3 6 6 M2],-1,1,'s');
185 %%%%%%%%%%%%%%%% LMI-4 %%%%%%%%%%%%%%%%%%%%
186 lmiterm([4 1 1 M1],1,A1,'s');
187 lmiterm([4 1 1 Q1],1,1);
188 lmiterm([4 1 1 Q2],1,1);
```

```
189 lmiterm([4 1 1 R],-4,1);
190 lmiterm([4 1 1 P12],1,1,'s');
191
192 lmiterm([4 1 2 M1],1,Ad1);
193 lmiterm([4 1 2 R],-2,1);
194 lmiterm([4 1 2 X11],-1,1);
195 lmiterm([4 1 2 X12],-1,1,'s');
196 lmiterm([4 1 2 X22],-1,1);
197 lmiterm([4 1 2 P12],-(1-miu),1);
198 lmiterm([4 1 2 P13],(1-miu),1);
199
200 lmiterm([4 1 3 P13],-1,1);
201 lmiterm([4 1 3 X11],1,1);
202 lmiterm([4 1 3 X12],-1,1);
203 lmiterm([4 1 3 -X12],1,1);
204 lmiterm([4 1 3 X22],-1,1);
205
206 lmiterm([4 1 4 R],6,1);
207 lmiterm([4 1 4 P22],d,1);
208
209 lmiterm([4 1 5 X12],2,1);
210 lmiterm([4 1 5 X22],2,1);
211
212 lmiterm([4 1 6 P11],1,1);
213 lmiterm([4 1 6 M1],-1,1);
214 lmiterm([4 1 6 -M2],A1',1);
215
216 lmiterm([4 2 2 Q1],-(1-miu),1);
217 lmiterm([4 2 2 R],-8,1);
218 lmiterm([4 2 2 X11],1,1,'s');
219 lmiterm([4 2 2 X22],-1,1,'s');
220
221 lmiterm([4 2 3 X11],-1,1);
222 lmiterm([4 2 3 X12],1,1,'s');
223 lmiterm([4 2 3 X22],-1,1);
224 lmiterm([4 2 3 R],-2,1);
225
226 lmiterm([4 2 4 R],6,1);
227 lmiterm([4 2 4 X12],2,1);
228 lmiterm([4 2 4 -X22],2,1);
229 lmiterm([4 2 4 P22],-d*(1-miu),1);
230 lmiterm([4 2 4 -P23],d*(1-miu),1);
231
232 lmiterm([4 2 5 R],6,1);
233 lmiterm([4 2 5 X12],-2,1);
234 lmiterm([4 2 5 X22],2,1);
235
236 lmiterm([4 2 6 -M2],Ad1',1);
237
238 lmiterm([4 3 3 R],-4,1);
239 lmiterm([4 3 3 Q2],-1,1);
240
241 lmiterm([4 3 4 X12],-2,1);
242 lmiterm([4 3 4 -X22],2,1);
243 lmiterm([4 3 4 -P23],-d,1);
```

```
244
245 lmiterm([4 3 5 R],6,1);
246
247 lmiterm([4 4 4 R],-12,1);
248
249 lmiterm([4 4 5 X22],-4,1);
250
251 lmiterm([4 4 6 -P12],d,1);
252
253 lmiterm([4 5 5 R],-12,1);
254
255 lmiterm([4 6 6 R],d^2,1);
256 lmiterm([4 6 6 M2],-1,1,'s');
257 %%%%%%%%%%%%%%%% LMI-5 %%%%%%%%%%%%%%%%%%%%
258 lmiterm([5 1 1 M1],1,A2','s');
259 lmiterm([5 1 1 Q1],1,1);
260 lmiterm([5 1 1 Q2],1,1);
261 lmiterm([5 1 1 R],-4,1);
262 lmiterm([5 1 1 P12],1,1,'s');
263
264 lmiterm([5 1 2 M1],1,Ad2);
265 lmiterm([5 1 2 R],-2,1);
266 lmiterm([5 1 2 X11],-1,1);
267 lmiterm([5 1 2 X12],-1,1,'s');
268 lmiterm([5 1 2 X22],-1,1);
269 lmiterm([5 1 2 P12],-(1-miu),1);
270 lmiterm([5 1 2 P13],(1-miu),1);
271
272 lmiterm([5 1 3 P13],-1,1);
273 lmiterm([5 1 3 X11],1,1);
274 lmiterm([5 1 3 X12],-1,1);
275 lmiterm([5 1 3 -X12],1,1);
276 lmiterm([5 1 3 X22],-1,1);
277
278 lmiterm([5 1 4 R],6,1);
279 lmiterm([5 1 4 P22],d,1);
280
281 lmiterm([5 1 5 X12],2,1);
282 lmiterm([5 1 5 X22],2,1);
283
284 lmiterm([5 1 6 P11],1,1);
285 lmiterm([5 1 6 M1],-1,1);
286 lmiterm([5 1 6 -M2],A2',1);
287
288 lmiterm([5 2 2 Q1],-(1-miu),1);
289 lmiterm([5 2 2 R],-8,1);
290 lmiterm([5 2 2 X11],1,1,'s');
291 lmiterm([5 2 2 X22],-1,1,'s');
292
293 lmiterm([5 2 3 X11],-1,1);
294 lmiterm([5 2 3 X12],1,1,'s');
295 lmiterm([5 2 3 X22],-1,1);
296 lmiterm([5 2 3 R],-2,1);
297
298 lmiterm([5 2 4 R],6,1);
```

```
299  lmiterm([5 2 4 X12],2,1);
300  lmiterm([5 2 4 -X22],2,1);
301  lmiterm([5 2 4 P22],-d*(1-miu),1);
302  lmiterm([5 2 4 -P23],d*(1-miu),1);
303
304  lmiterm([5 2 5 R],6,1);
305  lmiterm([5 2 5 X12],-2,1);
306  lmiterm([5 2 5 X22],2,1);
307
308  lmiterm([5 2 6 -M2],Ad2',1);
309
310  lmiterm([5 3 3 R],-4,1);
311  lmiterm([5 3 3 Q2],-1,1);
312
313  lmiterm([5 3 4 X12],-2,1);
314  lmiterm([5 3 4 -X22],2,1);
315  lmiterm([5 3 4 -P23],-d,1);
316
317  lmiterm([5 3 5 R],6,1);
318
319  lmiterm([5 4 4 R],-12,1);
320
321  lmiterm([5 4 5 X22],-4,1);
322
323  lmiterm([5 4 6 -P12],d,1);
324
325  lmiterm([5 5 5 R],-12,1);
326
327  lmiterm([5 6 6 R],d^2,1);
328  lmiterm([5 6 6 M2],-1,1,'s');
329  %%%%%%%%%%%% LMI Constraint %%%%%%%%%%%%%%%%%%
330  lmiterm([-6 1 1 P11],1,1);
331  lmiterm([-6 1 2 P12],1,1);
332  lmiterm([-6 1 3 P13],1,1);
333  lmiterm([-6 2 2 P22],1,1);
334  lmiterm([-6 2 3 P23],1,1);
335  lmiterm([-6 3 3 P33],1,1);
336  lmiterm([-7 1 1 Q1],1,1);
337  lmiterm([-8 1 1 Q2],1,1);
338  lmiterm([-9 1 1 R],1,1);
339
340  lmi_result=getlmis;
341  %%%%%%%%%%%%%% LMI Solver %%%%%%%%%%%%%%%%%%%%%%%%
342  [tmin,xfeas]=feasp(lmi_result);
```

A6: MATLAB code for Fuzzy TDS Stabilization—Theorem 5.2

```
1  clear all
2  clc
3  %%%%%%%%%%% Example-3 %%%%%%%%%%%%%
4  A1=[0 0.6;0 1];
5  A2=[1 0;1 0];
6  Ad1=[0.5 0.9;0 2];
7  Ad2=[0.9 0;1 1.6];
8  B1=[1;1];
9  B2=[1;1];
10 %%%%%%%%%%%%%%%%%%%%%%%%%%%%%%%%%%
11 miu=0.0001;
12 alpha=1;       % \alpha=\rho
13 d=0.8152;      % d=\tau_{0}
14 %%%%%%%%%%%%%%% LMI Variables %%%%%%%%%%%%%%%%%
15 setlmis([]);
16 P11=lmivar(1,[2,1]);
17 P12=lmivar(2,[2,2]);
18 P13=lmivar(2,[2,2]);
19 P22=lmivar(1,[2,1]);
20 P23=lmivar(2,[2,2]);
21 P33=lmivar(1,[2,1]);
22 Q1=lmivar(1,[2,1]);
23 Q2=lmivar(1,[2,1]);
24 R=lmivar(1,[2,1]);
25 X11=lmivar(2,[2,2]);
26 X12=lmivar(2,[2,2]);
27 X22=lmivar(2,[2,2]);
28 Y=lmivar(2,[2,2]);
29 N1=lmivar(2,[1,2]);
30 N2=lmivar(2,[1,2]);
31 %%%%%%%%%%%%%%%% LMI-1 %%%%%%%%%%%%%%%%%%%%%%%%%%%%%
32 lmiterm([-1 1 1 R],1,1);
33 lmiterm([-1 1 3 X11],1,1);
34 lmiterm([-1 1 4 X12],1,1);
35 lmiterm([-1 2 2 R],3,1);
36 lmiterm([-1 2 3 -X12],1,1);
37 lmiterm([-1 2 4 X22],1,1);
38 lmiterm([-1 3 3 R],1,1);
39 lmiterm([-1 4 4 R],3,1);
40 %%%%%%%%%%%%%%% LMI-2 %%%%%%%%%%%%%%%%%%%
41 lmiterm([2 1 1 Y],alpha*A1',1,'s');
42 lmiterm([2 1 1 Q1],1,1);
43 lmiterm([2 1 1 Q2],1,1);
44 lmiterm([2 1 1 R],-4,1);
45 lmiterm([2 1 1 P12],1,1,'s');
46 lmiterm([2 1 1 N1],alpha*B1,1,'s');
47
48 lmiterm([2 1 2 -Y],alpha*Ad1,1);
49 lmiterm([2 1 2 R],-2,1);
50 lmiterm([2 1 2 X11],-1,1);
51 lmiterm([2 1 2 X12],-1,1,'s');
52 lmiterm([2 1 2 X22],-1,1);
53 lmiterm([2 1 2 P12],-(1-miu),1);
54 lmiterm([2 1 2 P13],(1-miu),1);
55
56 lmiterm([2 1 3 P13],-1,1);
57 lmiterm([2 1 3 X11],1,1);
58 lmiterm([2 1 3 X12],-1,1);
59 lmiterm([2 1 3 -X12],1,1);
60 lmiterm([2 1 3 X22],-1,1);
61
62 lmiterm([2 1 4 R],6,1);
```

```
63
64 lmiterm([2 1 5 P23],d,1);
65 lmiterm([2 1 5 X12],2,1);
66 lmiterm([2 1 5 X22],2,1);
67
68 lmiterm([2 1 6 P11],1,1);
69 lmiterm([2 1 6 -Y],-alpha,1);
70 lmiterm([2 1 6 Y],1,A1');
71 lmiterm([2 1 6 -N1],1,B1');
72
73 lmiterm([2 2 2 R],-8,1);
74 lmiterm([2 2 2 Q1],-(1-miu),1);
75 lmiterm([2 2 2 X11],1,1,'s');
76 lmiterm([2 2 2 X22],-1,1,'s');
77
78 lmiterm([2 2 3 X11],-1,1);
79 lmiterm([2 2 3 X12],1,1,'s');
80 lmiterm([2 2 3 X22],-1,1);
81 lmiterm([2 2 3 R],-2,1);
82
83 lmiterm([2 2 4 R],6,1);
84 lmiterm([2 2 4 X12],2,1);
85 lmiterm([2 2 4 -X22],2,1);
86
87 lmiterm([2 2 5 R],6,1);
88 lmiterm([2 2 5 X12],-2,1);
89 lmiterm([2 2 5 X22],2,1);
90 lmiterm([2 2 5 P23],-d*(1-miu),1);
91 lmiterm([2 2 5 P33],d*(1-miu),1);
92
93 lmiterm([2 2 6 Y],1,Ad1');
94
95 lmiterm([2 3 3 R],-4,1);
96 lmiterm([2 3 3 Q2],-1,1);
97
98 lmiterm([2 3 4 X12],-2,1);
99 lmiterm([2 3 4 -X22],2,1);
100
101 lmiterm([2 3 5 P33],-d,1);
102 lmiterm([2 3 5 R],6,1);
103
104 lmiterm([2 4 4 R],-12,1);
105
106 lmiterm([2 4 5 X22],-4,1);
107
108 lmiterm([2 5 5 R],-12,1);
109
110 lmiterm([2 5 6 -P13],d,1);
111
112 lmiterm([2 6 6 R],d^2,1);
113 lmiterm([2 6 6 Y],-1,1,'s');
114 %%%%%%%%%%%%%%%% LMI-3 %%%%%%%%%%%%%%%%%%%
115 lmiterm([3 1 1 Y],alpha*A1',1,'s');
116 lmiterm([3 1 1 Q1],1,1);
117 lmiterm([3 1 1 Q2],1,1);
```

```
118 lmiterm([3 1 1 R],-4,1);
119 lmiterm([3 1 1 P12],1,1,'s');
120 lmiterm([3 1 1 N1],alpha*B1,1,'s');
121
122 lmiterm([3 1 2 -Y],alpha*Ad1,1);
123 lmiterm([3 1 2 R],-2,1);
124 lmiterm([3 1 2 X11],-1,1);
125 lmiterm([3 1 2 X12],-1,1,'s');
126 lmiterm([3 1 2 X22],-1,1);
127 lmiterm([3 1 2 P12],-(1-miu),1);
128 lmiterm([3 1 2 P13],(1-miu),1);
129
130 lmiterm([3 1 3 P13],-1,1);
131 lmiterm([3 1 3 X11],1,1);
132 lmiterm([3 1 3 X12],-1,1);
133 lmiterm([3 1 3 -X12],1,1);
134 lmiterm([3 1 3 X22],-1,1);
135
136 lmiterm([3 1 4 R],6,1);
137 lmiterm([3 1 4 P22],d,1);
138
139 lmiterm([3 1 5 X12],2,1);
140 lmiterm([3 1 5 X22],2,1);
141
142 lmiterm([3 1 6 P11],1,1);
143 lmiterm([3 1 6 -Y],-alpha,1);
144 lmiterm([3 1 6 Y],1,A1');
145 lmiterm([3 1 6 -N1],1,B1');
146
147 lmiterm([3 2 2 Q1],-(1-miu),1);
148 lmiterm([3 2 2 R],-8,1);
149 lmiterm([3 2 2 X11],1,1,'s');
150 lmiterm([3 2 2 X22],-1,1,'s');
151
152 lmiterm([3 2 3 X11],-1,1);
153 lmiterm([3 2 3 X12],1,1,'s');
154 lmiterm([3 2 3 X22],-1,1);
155 lmiterm([3 2 3 R],-2,1);
156
157 lmiterm([3 2 4 R],6,1);
158 lmiterm([3 2 4 X12],2,1);
159 lmiterm([3 2 4 -X22],2,1);
160 lmiterm([3 2 4 P22],-d*(1-miu),1);
161 lmiterm([3 2 4 -P23],d*(1-miu),1);
162
163 lmiterm([3 2 5 R],6,1);
164 lmiterm([3 2 5 X12],-2,1);
165 lmiterm([3 2 5 X22],2,1);
166
167 lmiterm([3 2 6 Y],1,Ad1');
168
169 lmiterm([3 3 3 R],-4,1);
170 lmiterm([3 3 3 Q2],-1,1);
171
172 lmiterm([3 3 4 X12],-2,1);
```

```
173 lmiterm([3 3 4 -X22],2,1);
174 lmiterm([3 3 4 -P23],-d,1);
175
176 lmiterm([3 3 5 R],6,1);
177
178 lmiterm([3 4 4 R],-12,1);
179
180 lmiterm([3 4 5 X22],-4,1);
181
182 lmiterm([3 4 6 -P12],d,1);
183
184 lmiterm([3 5 5 R],-12,1);
185
186 lmiterm([3 6 6 R],d^2,1);
187 lmiterm([3 6 6 Y],-1,1,'s');
188 %%%%%%%%%%%%%%% LMI-4 %%%%%%%%%%%%%%%%%%%%
189 lmiterm([4 1 1 Y],alpha*A2',1,'s');
190 lmiterm([4 1 1 Q1],1,1);
191 lmiterm([4 1 1 Q2],1,1);
192 lmiterm([4 1 1 R],-4,1);
193 lmiterm([4 1 1 P12],1,1,'s');
194 lmiterm([4 1 1 N2],alpha*B2,1,'s');
195
196 lmiterm([4 1 2 -Y],alpha*Ad2,1);
197 lmiterm([4 1 2 R],-2,1);
198 lmiterm([4 1 2 X11],-1,1);
199 lmiterm([4 1 2 X12],-1,1,'s');
200 lmiterm([4 1 2 X22],-1,1);
201 lmiterm([4 1 2 P12],-(1-miu),1);
202 lmiterm([4 1 2 P13],(1-miu),1);
203
204 lmiterm([4 1 3 P13],-1,1);
205 lmiterm([4 1 3 X11],1,1);
206 lmiterm([4 1 3 X12],-1,1);
207 lmiterm([4 1 3 -X12],1,1);
208 lmiterm([4 1 3 X22],-1,1);
209
210 lmiterm([4 1 4 R],6,1);
211
212 lmiterm([4 1 5 P23],d,1);
213 lmiterm([4 1 5 X12],2,1);
214 lmiterm([4 1 5 X22],2,1);
215
216 lmiterm([4 1 6 P11],1,1);
217 lmiterm([4 1 6 -Y],-alpha,1);
218 lmiterm([4 1 6 Y],1,A2');
219 lmiterm([4 1 6 -N2],1,B2');
220
221 lmiterm([4 2 2 Q1],-(1-miu),1);
222 lmiterm([4 2 2 R],-8,1);
223 lmiterm([4 2 2 X11],1,1,'s');
224 lmiterm([4 2 2 X22],-1,1,'s');
225
226 lmiterm([4 2 3 X11],-1,1);
227 lmiterm([4 2 3 X12],1,1,'s');
```

```
228 lmiterm([4 2 3 X22],-1,1);
229 lmiterm([4 2 3 R],-2,1);
230 
231 lmiterm([4 2 4 R],6,1);
232 lmiterm([4 2 4 X12],2,1);
233 lmiterm([4 2 4 -X22],2,1);
234 
235 lmiterm([4 2 5 R],6,1);
236 lmiterm([4 2 5 X12],-2,1);
237 lmiterm([4 2 5 X22],2,1);
238 lmiterm([4 2 5 P23],-d*(1-miu),1);
239 lmiterm([4 2 5 P33],d*(1-miu),1);
240 
241 lmiterm([4 2 6 Y],1,Ad2');
242 
243 lmiterm([4 3 3 R],-4,1);
244 lmiterm([4 3 3 Q2],-1,1);
245 
246 lmiterm([4 3 4 X12],-2,1);
247 lmiterm([4 3 4 -X22],2,1);
248 
249 lmiterm([4 3 5 P33],-d,1);
250 lmiterm([4 3 5 R],6,1);
251 
252 lmiterm([4 4 4 R],-12,1);
253 
254 lmiterm([4 4 5 X22],-4,1);
255 
256 lmiterm([4 5 5 R],-12,1);
257 
258 lmiterm([4 5 6 -P13],d,1);
259 
260 lmiterm([4 6 6 R],d^2,1);
261 lmiterm([4 6 6 Y],-1,1,'s');
262 %%%%%%%%%%%%%%%% LMI-5 %%%%%%%%%%%%%%%%%%%%
263 lmiterm([5 1 1 Y],alpha*A2',1,'s');
264 lmiterm([5 1 1 Q1],1,1);
265 lmiterm([5 1 1 Q2],1,1);
266 lmiterm([5 1 1 R],-4,1);
267 lmiterm([5 1 1 P12],1,1,'s');
268 lmiterm([5 1 1 N2],alpha*B2,1,'s');
269 
270 lmiterm([5 1 2 -Y],alpha*Ad2,1);
271 lmiterm([5 1 2 R],-2,1);
272 lmiterm([5 1 2 X11],-1,1);
273 lmiterm([5 1 2 X12],-1,1,'s');
274 lmiterm([5 1 2 X22],-1,1);
275 lmiterm([5 1 2 P12],-(1-miu),1);
276 lmiterm([5 1 2 P13],(1-miu),1);
277 
278 lmiterm([5 1 3 P13],-1,1);
279 lmiterm([5 1 3 X11],1,1);
280 lmiterm([5 1 3 X12],-1,1);
281 lmiterm([5 1 3 -X12],1,1);
282 lmiterm([5 1 3 X22],-1,1);
```

```
283
284 lmiterm([5 1 4 R],6,1);
285 lmiterm([5 1 4 P22],d,1);
286
287 lmiterm([5 1 5 X12],2,1);
288 lmiterm([5 1 5 X22],2,1);
289
290 lmiterm([5 1 6 P11],1,1);
291 lmiterm([5 1 6 -Y],-alpha,1);
292 lmiterm([5 1 6 Y],1,A2');
293 lmiterm([5 1 6 -N2],1,B2');
294
295 lmiterm([5 2 2 Q1],-(1-miu),1);
296 lmiterm([5 2 2 R],-8,1);
297 lmiterm([5 2 2 X11],1,1,'s');
298 lmiterm([5 2 2 X22],-1,1,'s');
299
300 lmiterm([5 2 3 X11],-1,1);
301 lmiterm([5 2 3 X12],1,1,'s');
302 lmiterm([5 2 3 X22],-1,1);
303 lmiterm([5 2 3 R],-2,1);
304
305 lmiterm([5 2 4 R],6,1);
306 lmiterm([5 2 4 X12],2,1);
307 lmiterm([5 2 4 -X22],2,1);
308 lmiterm([5 2 4 P22],-d*(1-miu),1);
309 lmiterm([5 2 4 -P23],d*(1-miu),1);
310
311 lmiterm([5 2 5 R],6,1);
312 lmiterm([5 2 5 X12],-2,1);
313 lmiterm([5 2 5 X22],2,1);
314
315 lmiterm([5 2 6 Y],1,Ad2');
316
317 lmiterm([5 3 3 R],-4,1);
318 lmiterm([5 3 3 Q2],-1,1);
319
320 lmiterm([5 3 4 X12],-2,1);
321 lmiterm([5 3 4 -X22],2,1);
322 lmiterm([5 3 4 -P23],-d,1);
323
324 lmiterm([5 3 5 R],6,1);
325
326 lmiterm([5 4 4 R],-12,1);
327
328 lmiterm([5 4 5 X22],-4,1);
329
330 lmiterm([5 4 6 -P12],d,1);
331
332 lmiterm([5 5 5 R],-12,1);
333
334 lmiterm([5 6 6 R],d^2,1);
335 lmiterm([5 6 6 Y],-1,1,'s');
336 %%%%%%%%%%%%%%%% LMI-6 %%%%%%%%%%%%%%%%%%%
337 lmiterm([6 1 1 Y],alpha*A1',1,'s');
```

```
338 lmiterm([6 1 1 Y],alpha*A2',1,'s');
339 lmiterm([6 1 1 Q1],2,1);
340 lmiterm([6 1 1 Q2],2,1);
341 lmiterm([6 1 1 R],-8,1);
342 lmiterm([6 1 1 P12],2,1,'s');
343 lmiterm([6 1 1 N1],alpha*B2,1,'s');
344 lmiterm([6 1 1 N2],alpha*B1,1,'s');
345
346 lmiterm([6 1 2 -Y],alpha*Ad1,1);
347 lmiterm([6 1 2 -Y],alpha*Ad2,1);
348 lmiterm([6 1 2 R],-4,1);
349 lmiterm([6 1 2 X11],-2,1);
350 lmiterm([6 1 2 X12],-2,1,'s');
351 lmiterm([6 1 2 X22],-2,1);
352 lmiterm([6 1 2 P12],-2*(1-miu),1);
353 lmiterm([6 1 2 P13],2*(1-miu),1);
354
355 lmiterm([6 1 3 P13],-2,1);
356 lmiterm([6 1 3 X11],2,1);
357 lmiterm([6 1 3 X12],-2,1);
358 lmiterm([6 1 3 -X12],2,1);
359 lmiterm([6 1 3 X22],-2,1);
360
361 lmiterm([6 1 4 R],12,1);
362
363 lmiterm([6 1 5 P23],2*d,1);
364 lmiterm([6 1 5 X12],4,1);
365 lmiterm([6 1 5 X22],4,1);
366
367 lmiterm([6 1 6 P11],2,1);
368 lmiterm([6 1 6 -Y],-2*alpha,1);
369 lmiterm([6 1 6 Y],1,A1');
370 lmiterm([6 1 6 Y],1,A2');
371 lmiterm([6 1 6 -N1],1,B2');
372 lmiterm([6 1 6 -N2],1,B1');
373
374 lmiterm([6 2 2 Q1],-2*(1-miu),1);
375 lmiterm([6 2 2 R],-16,1);
376 lmiterm([6 2 2 X11],2,1,'s');
377 lmiterm([6 2 2 X22],-2,1,'s');
378
379 lmiterm([6 2 3 X11],-2,1);
380 lmiterm([6 2 3 X12],2,1,'s');
381 lmiterm([6 2 3 X22],-2,1);
382 lmiterm([6 2 3 R],-4,1);
383
384 lmiterm([6 2 4 R],12,1);
385 lmiterm([6 2 4 X12],4,1);
386 lmiterm([6 2 4 -X22],4,1);
387
388 lmiterm([6 2 5 R],12,1);
389 lmiterm([6 2 5 X12],-4,1);
390 lmiterm([6 2 5 X22],4,1);
391 lmiterm([6 2 5 P23],-2*d*(1-miu),1);
392 lmiterm([6 2 5 P33],2*d*(1-miu),1);
```

```
393
394 lmiterm([6 2 6 Y],1,Ad1');
395 lmiterm([6 2 6 Y],1,Ad2');
396
397 lmiterm([6 3 3 R],-8,1);
398 lmiterm([6 3 3 Q2],-2,1);
399
400 lmiterm([6 3 4 X12],-4,1);
401 lmiterm([6 3 4 -X22],4,1);
402
403 lmiterm([6 3 5 P33],-2*d,1);
404 lmiterm([6 3 5 R],12,1);
405
406 lmiterm([6 4 4 R],-24,1);
407
408 lmiterm([6 4 5 X22],-8,1);
409
410 lmiterm([6 5 5 R],-24,1);
411
412 lmiterm([6 5 6 -P13],2*d,1);
413
414 lmiterm([6 6 6 R],2*d^2,1);
415 lmiterm([6 6 6 Y],-2,1,'s');
416 %%%%%%%%%%%%%%% LMI-7 %%%%%%%%%%%%%%%%%%%%
417 lmiterm([7 1 1 Y],alpha*A1',1,'s');
418 lmiterm([7 1 1 Y],alpha*A2',1,'s');
419 lmiterm([7 1 1 Q1],2,1);
420 lmiterm([7 1 1 Q2],2,1);
421 lmiterm([7 1 1 R],-8,1);
422 lmiterm([7 1 1 P12],2,1,'s');
423 lmiterm([7 1 1 N1],alpha*B2,1,'s');
424 lmiterm([7 1 1 N2],alpha*B1,1,'s');
425
426 lmiterm([7 1 2 -Y],alpha*Ad1,1);
427 lmiterm([7 1 2 -Y],alpha*Ad2,1);
428 lmiterm([7 1 2 R],-4,1);
429 lmiterm([7 1 2 X11],-2,1);
430 lmiterm([7 1 2 X12],-2,1,'s');
431 lmiterm([7 1 2 X22],-2,1);
432 lmiterm([7 1 2 P12],-2*(1-miu),1);
433 lmiterm([7 1 2 P13],2*(1-miu),1);
434
435 lmiterm([7 1 3 P13],-2,1);
436 lmiterm([7 1 3 X11],2,1);
437 lmiterm([7 1 3 X12],-2,1);
438 lmiterm([7 1 3 -X12],2,1);
439 lmiterm([7 1 3 X22],-2,1);
440
441 lmiterm([7 1 4 R],12,1);
442 lmiterm([7 1 4 P22],2*d,1);
443
444 lmiterm([7 1 5 X12],4,1);
445 lmiterm([7 1 5 X22],4,1);
446
447 lmiterm([7 1 6 P11],2,1);
```

```
448 lmiterm([7 1 6 -Y],-2*alpha,1);
449 lmiterm([7 1 6 Y],1,A1');
450 lmiterm([7 1 6 Y],1,A2');
451 lmiterm([7 1 6 -N1],1,B2');
452 lmiterm([7 1 6 -N2],1,B1');
453
454 lmiterm([7 2 2 Q1],-2*(1-miu),1);
455 lmiterm([7 2 2 R],-16,1);
456 lmiterm([7 2 2 X11],2,1,'s');
457 lmiterm([7 2 2 X22],-2,1,'s');
458
459 lmiterm([7 2 3 X11],-2,1);
460 lmiterm([7 2 3 X12],2,1,'s');
461 lmiterm([7 2 3 X22],-2,1);
462 lmiterm([7 2 3 R],-4,1);
463
464 lmiterm([7 2 4 R],12,1);
465 lmiterm([7 2 4 X12],4,1);
466 lmiterm([7 2 4 -X22],4,1);
467 lmiterm([7 2 4 P22],-2*d*(1-miu),1);
468 lmiterm([7 2 4 -P23],2*d*(1-miu),1);
469
470 lmiterm([7 2 5 R],12,1);
471 lmiterm([7 2 5 X12],-4,1);
472 lmiterm([7 2 5 X22],4,1);
473
474 lmiterm([7 2 6 Y],1,Ad1');
475 lmiterm([7 2 6 Y],1,Ad2');
476
477 lmiterm([7 3 3 R],-8,1);
478 lmiterm([7 3 3 Q2],-2,1);
479
480 lmiterm([7 3 4 X12],-4,1);
481 lmiterm([7 3 4 -X22],4,1);
482 lmiterm([7 3 4 -P23],-2*d,1);
483
484 lmiterm([7 3 5 R],12,1);
485
486 lmiterm([7 4 4 R],-24,1);
487
488 lmiterm([7 4 5 X22],-8,1);
489
490 lmiterm([7 4 6 -P12],2*d,1);
491
492 lmiterm([7 5 5 R],-24,1);
493
494 lmiterm([7 6 6 R],2*d^2,1);
495 lmiterm([7 6 6 Y],-2,1,'s');
496 %%%%%%%%%%% LMI Constraint %%%%%%%%%%%%%%%%%%
497 lmiterm([-8 1 1 P11],1,1);
498 lmiterm([-8 1 2 P12],1,1);
499 lmiterm([-8 1 3 P13],1,1);
500 lmiterm([-8 1 1 P22],1,1);
501 lmiterm([-8 2 3 P23],1,1);
502 lmiterm([-8 1 1 P33],1,1);
```

```
503 lmiterm([-9 1 1 Q1],1,1);
504 lmiterm([-10 1 1 Q2],1,1);
505 lmiterm([-11 1 1 R],1,1);
506
507 lmi_resultr=getlmis;
508 %%%%%%%%%%%%%% LMI Solver %%%%%%%%%%%%%%%%%%%%%%%%%
509 [tmin,xfeas]=feasp(lmi_resultr);
510 x1=dec2mat(lmi_resultrd67,xfeas,Y)
511 x2=dec2mat(lmi_resultrd67,xfeas,N1)
512 x3=dec2mat(lmi_resultrd67,xfeas,N2)
513 p=inv(x1)
514 K1=x2*p'
515 K2=x3*p'
516 a1=A1+B1*K1
517 a2=A2+B2*K2
```

References

1. P. Gahinet, A. Nemirovski, A.J. Laub, M. Chilali, *LMI Control Toolbox Users Guide* (Mathworks, Cambridge, 1995)
2. M.N.A. Parlakci, Improved robust stability criteria and design of robust stabilizing controller for uncertain linear time-delay system. Int. J. Robust Nonlinear Control **16**, 599–636 (2006)

Printed in the United States
By Bookmasters